ADAPTIVE FILTERS

PRENTICE-HALL SIGNAL PROCESSING SERIES

Alan V. Oppenheim, Editor

ANDREWS and HUNT *Digital Image Restoration*
BRIGHAM *The Fast Fourier Transform*
BURDIC *Underwater Acoustic System Analysis*
CASTLEMAN *Digital Image Processing*
CROCHIERE and RABINER *Multirate Digital Signal Processing*
DUDGEON and MERSEREAU *Multidimensional Digital Signal Processing*
HAMMING *Digital Filters, 2e*
HAYKIN, ED. *Array Signal Processing*
JAYANT and NOLL *Digital Coding of Waveforms*
LEA, ED. *Trends in Speech Recognition*
LIM, ED. *Speech Enhancement*
MCCLELLAN and RADER *Number Theory in Digital Signal Processing*
OPPENHEIM, ED. *Applications of Digital Signal Processing*
OPPENHEIM, WILLSKY, with YOUNG *Signals and Systems*
OPPENHEIM and SCHAFER *Digital Signal Processing*
PHILLIPS and NAGLE *Digital Control System Analysis*
RABINER and GOLD *Theory and Applications of Digital Signal Processing*
RABINER and SCHAFER *Digital Processing of Speech Signals*
ROBINSON and TREITEL *Geophysical Signal Analysis*
TRIBOLET *Seismic Applications of Homomorphic Signal Processing*
WIDROW and STEARNS *Adaptive Signal Processing*

ADAPTIVE FILTERS

Edited by
C. F. N. Cowan and P. M. Grant

With contributions from:
P. F. Adams
C. F. N. Cowan
E. R. Ferrara, Jr.
B. Friedlander
P. M. Grant
J. R. Treichler
J. M. Turner

Prentice-Hall, Inc., Englewood Cliffs, New Jersey 07632

Library of Congress Cataloging in Publication Data

Adaptive filters.

(Prentice-Hall series in signal processing)
Bibliography: p. 283
Includes index.
1. Adaptive filters. I. Cowan, C. F. N. (Colin F. N.)
II. Grant, Peter M. III. Adams, P. F. (Peter F.)
IV. Series.
TK7872.F5A33 1985 621.3815'324 84-18039

ISBN 0-13-004037-1

Editorial/production supervision: *Fred Dahl*
Manufacturing buyer: *Anthony Caruso*

PRENTICE-HALL SERIES IN SIGNAL PROCESSING
Alan V. Oppenheim, Editor

Printed in the United States of America

10 9 8 7 6 5 4 3 2 1

ISBN 0-13-004037-1 01

Prentice-Hall International, Inc., *London*
Prentice-Hall of Australia Pty. Limited, *Sydney*
Editora Prentice-Hall do Brasil, Ltda., *Rio de Janeiro*
Prentice-Hall Canada Inc., *Toronto*
Prentice-Hall Hispanoamericana, S.A., *Mexico*
Prentice-Hall of India Private Limited, *New Delhi*
Prentice-Hall of Japan, Inc., *Tokyo*
Prentice-Hall of Southeast Asia Pte. Ltd., *Singapore*
Whitehall Books Limited, *Wellington, New Zealand*

CONTRIBUTORS

Peter F. Adams
British Telecom Research Laboratories
Martlesham Heath
Ipswich IP5 7RE, England

Colin F. N. Cowan
Department of Electrical Engineering
King's Buildings
University of Edinburgh
Edinburgh EH9 3JL, Scotland

Earl R. Ferrara, Jr.
ESL Inc.
495 Java Drive
Sunnyvale, CA 94086

Benjamin Friedlander
Systems Control Technology Inc.
1801 Page Mill Road
Palo Alto, CA 94304

Peter M. Grant
Department of Electrical Engineering
King's Buildings
University of Edinburgh
Edinburgh EH9 3JL, Scotland

*John R. Treichler**
ARGOSystems, Inc.
884 Hermosa Court
Sunnyvale, CA 94086

John M. Turner†
Information Systems Laboratory
Department of Electrical Engineering
Stanford University
Stanford, CA 94305

*Now at Applied Signal Technology, Inc., Sunnyvale, California.
†Now at Allophonix, Inc., Palo Alto, California.

CONTENTS

2

OPTIMUM ESTIMATION TECHNIQUES, 15

Colin F. N. Cowan

3

ADAPTIVE ALGORITHMS FOR FINITE IMPULSE RESPONSE FILTERS, 29

Benjamin Friedlander

4

ADAPTIVE ALGORITHMS FOR INFINITE IMPULSE RESPONSE FILTERS, 60

John R. Treichler

5

RECURSIVE LEAST-SQUARES ESTIMATION AND LATTICE FILTERS, 91

John M. Turner

8

ADAPTIVE FILTERS IN TELECOMMUNICATIONS, 216

Peter F. Adams

9

OTHER ADAPTIVE FILTER APPLICATIONS, 257

Peter M. Grant

REFERENCES, 283

INDEX, 303

PREFACE

The subject of adaptive processors has been a research topic since the 1960s and these have subsequently been applied in many practical systems mainly as adaptive filters or adaptive antennas. This text is concerned primarily with adaptive filters, which have been widely used, for example, as adaptive equalizers for telecommunications data transmission systems. However, even the basic background information regarding the various designs of adaptive processors may only be found in diverse locations in the research literature, such as signal processing, antenna, and control theory publications. The result is that newcomers to this field find great difficulty in compiling an overview of the subject area on which to base design decisions. The major objective of this text is, therefore, to provide a coherent and comprehensive introduction to the subject of adaptive filtering covering the basic theory, practical realizations, and applications.

One of the distinguishing features that we have attempted to incorporate into this text is its broad scope combined with a good balance between theoretical and practical chapters. We hope that this should ensure that it is of interest both in the academic community for graduate-level teaching and research as well as for practicing engineers, who quickly need to become aware of the practical possibilities of these processors.

The text has been structured to include a general introduction to the subject followed by five in-depth chapters on the theoretical development of several alternative adaptive algorithm formulations. Chapter 2 on estimation theory forms a basis for this theoretical treatment and it is followed by three chapters dedicated to adaptive finite impulse response, infinite impulse response, and lattice filters. In addition, the recent growing interest in applying frequency-domain processing to realize adaptive filters with reduced computational complexity has encouraged us to include Chapter 6, which deals with signal transformation techniques for adaptive filtering. We believe that this approach not only ensures that the book represents state-of-the-art developments in adaptive filtering but that it provides a very broad coverage of different techniques for realizing these filters.

These more theoretical chapters are balanced by a set of three practical chapters which present some of the possible applications of adaptive filters. These cover in-depth adaptive filter implementations, Chapter 7, and their main applications in communications equalization and echo cancellation, Chapter 8. Other application areas, such as fast tracking filters for HF and microwave digital radio, linear predictive coding, and maximum-entropy and maximum-likelihood analysis techniques, are also reported briefly in Chapter 9.

In order to prepare a current text on these diverse topics, we have approached experts who are familiar with these areas and invited them to prepare the individual chapters as reviews of their specialist area. This ensures that the included material is considerably more comprehensive and understandable than the individual research papers on which it is based. The authors have been encouraged to contrast the different filter design approaches in terms of achievable performance and implementation complexity. By defining a common notation in advance we hope that we have ensured that it is possible easily to cross-reference between chapters to comprehend the practical significance of the different filter approaches, without encountering too much confusing terminology.

All the references have been combined into a single listing at the end of the text to enable the reader to refer back easily to the key papers in the literature.

ACKNOWLEDGMENTS

The editors wish to acknowledge their thanks to colleagues in the University of Edinburgh for their support and encouragement of our research in this subject area. Professors J. H. Collins and J. Mavor provided the initial stimulus to us to enter this field and encouraged us in the undertaking of this text. Many other colleagues have worked with or alongside us, and in particular we wish to acknowledge the contributions of M. J. Rutter, A. Morgul, W. K. Wong, A. Alvarez, B. Mulgrew, S. G. Smith, and J. H. Dripps. The authors would also like to thank Tom Alexander of North Carolina State University for his many valuable comments in reviewing this text for Prentice-Hall. Acknowledgment is made for the encouragement, assistance, and advice of B. Dickinson and M. Morf on portions of Chapter 5. Finally, we express our thanks to Gillian Erskine for her considerable work in the preparation of the various versions of the manuscript for this text.

The authors wish to acknowledge the permission of the following organizations to reproduce figures on which they hold the copyrights: Institution of Electrical Engineers, London, Figures 7.4, 7.6, 7.10, 7.16, 7.18, 7.21, 7.24, 7.25, 8.20, 8.27, and 9.4; Institute of Electrical and Electronics Engineers, New York, Figures 4.3 through 4.10, 5.9, 5.16, 6.1, 7.8, 7.19, 7.20, 7.23, 7.26, 7.27, 8.16, 9.7, and 9.12; EW Communications Inc., Palo Alto, Figures 7.7 and 9.11; British Telecom Figures 8.10, 8.17, 8.22, and 8.25.

ABBREVIATIONS

ADC	Analog-to-digital converter
AGC	Automatic gain control
ALCE	Adaptive lattice channel equaliser
ALE	Adaptive line enhancement
AR	Autoregressive
ARC	Automatic reference control
ARMA	Autoregressive moving average
BBD	Bucket-brigade device
BIBO	Bounded-input bounded-output
CAD	Computer-aided design
CCD	Charge-coupled device
CCITT	International Telegraph and Telephone Cousultative Committee
CMOS	Complementary metal-oxide silicon (transistor)
CORDIC	Coordinate rotation digital computer
CSLC	Coherent sidelobe canceler
CW	Continuous wave (sinusoid)
CZT	Chirp *z*-transform
DAC	Digital-to-analog converter
DC	Direct current (i.e., at zero frequency)
DFB	Decision feedback
DFT	Discrete Fourier transform
DSBAM	Double-sideband amplitude modulation
DSP	Digital signal processing
EEG	Electroencephalograph
FDM	Frequency-division multiplex
FIR	Finite impulse response
FFT	Fast Fourier transform
FLMS	Fast least mean squares
FT	Fractional tap
GaAs	Gallium arsenide
HARF	Hyperstable adaptive recursive filter
HF	High frequency
IC	Integrated circuit
IF	Intermediate frequency
IIR	Infinite impulse response
ISI	Intersymbol interference
LC	Inductor capacitor
LMS	Least-mean-squares
LOS	Line of sight
LPC	Linear predictive coder
LSALE	Least-squares adaptive lattice equaliser

LSI	Large-scale integration
LSL	Least-squares lattice
LST	Loud-speaking telephone
MA	Moving average
MDAC	Multiplying digital-to-analog converter
MEM	Maximum-entropy method
MLM	Maximum-likelihood method
MMSE	Minimum mean-square error
MR	Multiple response
MSE	Mean-square error
MSI	Medium-scale integration
MSTMP	Mean-square tap misadjustment power
NEC	Nippon Electric Company
NMOS	Negative (doped) metal-oxide silicon (transistor)
ODE	Ordinary differential equation
OEM	Own equipment manufacture
PARCOR	Partial correlation
PCM	Pulse-code modulation
PN	Pseudonoise
PRBS	Pseudorandom bit sequence
PSK	Phase-shift keying
PTF	Programmable transversal filter
QAM	Quadrature amplitude modulation
RAM	Random access memory
RF	Radio frequency
RLS	Recursive least squares
RNS	Residue number system
ROM	Read-only memory
S	Radar frequency band
SAW	Surface acoustic wave
SHARF	Simple hyperstable adaptive recursive filter
SNR	Signal-to-noise ratio
SOS	Silicon on saphire
SPR	Strictly positive real
SQNLSL	Square-root normalized least-squares lattice
TDL	Tapped delay line
TDM	Time-division multiplex
TDOA	Time difference of arrival
TM	Transmultiplexer
TTL	Transistor-transistor logic
UFLMS	Unconstrained frequency-domain least mean squares
UHF	Ultra high frequency
ULA	Uncommited logic array
VHF	Very high frequency
VLSI	Very large scale integration
VSBAM	Vestigial sideband amplitude modulated
WAL2	Walsh 2–signal format for telecommunications transmission
X	Radar frequency band

SYMBOLS

The notation has, as far as possible, been unified throughout the text. This list of symbols presents a common set of mathematical symbols and operators which are found in a number of the chapters.

The notation used in Chapter 6, describing frequency-domain adaptive algorithms, is notably different in many respects from that adopted in the rest of the text. The notation in Chapter 6 is therefore individually explained in the introduction to that chapter. The specific time index, t, has been omitted in the text, with the exception of Chapter 5. An explanation of notation is included in the introduction to this chapter.

Scalar Variables

a_i	ith feedforward coefficient in an IIR filter
b_i	ith feedback coefficient in an IIR filter
$b_j(t)$	Backward lattice filter signal at stage j and time t
$d(n)$	Desired output from an adaptive filter
$e(n)$	Error signal at adaptive filter output
$f_j(t)$	Forward lattice filter signal at stage j and time t
h_i	ith value of the impulse response of a filter
$J_{\mathbf{H}}$	Cost function specified in terms of the parameter vector $\mathbf{H}$
k_i	Lattice reflection (PARCOR) coefficient at stage i
$K(n)$	Kalman gain
M	Misadjustment factor for an adaptive process
$r(\tau)$	Autocorrelation value at Lag τ
$s(n)$, $x(n)$	Filter input signal
$y(n)$	Desired output from an adaptive filter
$\hat{y}(n)$	Estimation process output

$\gamma_{i,T}$	Lattice filter likelihood variable
$\eta(n)$, $\nu(n)$	White, Gaussian noise signals
λ	Memory factor in exponentially weighted RLS algorithm
λ_i	Eigenvalues of the autocorrelation matrix **R**
μ, ρ	Convergence coefficients of an adaptive algorithm
σ_ν^2	Mean-square value (power) of signal $\nu(n)$
τ_i	Time constant describing the convergence rate of the adaption mode corresponding to the ith eigenvalue

Operators

$E\{\cdot\}$	Statistical expection of $\{\cdot\}$
$\mathrm{Im}\,[\cdot]$	Imaginary part of $[\cdot]$
$\mathrm{Re}\,[\cdot]$	Real part of $[\cdot]$
$\mathrm{sgn}\,[\cdot]$	Sign of $[\cdot]$
$\mathrm{tr}\,[\cdot]$	Trace of the matrix $[\cdot]$
$\mathbf{z}$	Downshift operator
z^{-1}	Unit sample delay
$\nabla_{\mathbf{H}}$	Gradient operator specified in terms of the parameter vector **H**
$\sum$	Summation operator
$\prod$	Product operator

Matrices

All vectors are specified as column vectors unless otherwise defined in the text. The matrix transpose operation is defined by the superscript T; hence $\mathbf{X}^{\mathrm{T}} = [x_1 \quad x_2 \quad x_3 \quad \cdots \quad x_{\mathrm{N}}]$

$\mathbf{H}(n)$	Vector of weight values of an FIR filter
$\mathbf{H}_{\mathrm{opt}}$	Vector of filter weights for the optimum (Wiener) estimator
$\tilde{\mathbf{H}}(n)$	Error in vector $\mathbf{H}(n)$ from $\mathbf{H}_{\mathrm{opt}}$
$\mathbf{K}(n)$	Kalman gain vector
$\mathbf{P}$	Cross-correlation matrix
$\mathbf{R}$	Autocorrelation matrix
$\mathbf{S}(n)$, $\mathbf{X}(n)$	$N \times 1$ vectors of filter input signals
$\boldsymbol{\theta}(n)$	Vector of weight values for an IIR filter
L	Diagonal matrix of eigenvalues
$\boldsymbol{\phi}$	Extended matrix of observed parameters used in the RLS algorithm

ADAPTIVE FILTERS

1

INTRODUCTION TO ADAPTIVE FILTERS

Peter M. Grant and Colin F. N. Cowan

1.1 ADAPTIVE PROCESSING

Conventional signal processing systems for the extraction of information from an incoming signal such as a matched filter operate in an open-loop fashion. That is, the same processing function is carried out in the present time interval regardless of whether that function produced the correct result in the preceding time interval. In other words, conventional signal processing techniques make the basic assumption that the signal degradation is a known and time-invariant quantity.

Adaptive processors [Widrow and Stearns 1984, Masenten], on the other hand, operate with a closed-loop (feedback) arrangement. The incoming signal $s(n)$ is filtered or weighted in a programmable filter to yield an output $\hat{y}(n)$ which is then compared against a desired, conditioning or training signal, $y(n)$, to yield an error signal, $e(n)$. This error is then used to update the processor weighting parameters (usually in an iterative way) such that the error is progressively minimized (i.e., the processor output more closely approximates to the training signal). Such processors fall into the two broad classes of adaptive filters (Figure 1.1) and adaptive antennas [Monzingo and Miller, Hudson, Gabriel 1976(2), and Taylor].

Adaptive filters, which are the subject of this text, are concerned with the use of a programmable filter whose frequency response or transfer function is altered, or adapted, to pass without degradation the desired components of the signal and to attenuate the undesired or interfering signals, or to reduce any distortion on the input signal. Adaptive antennas employ the spatial processing in an antenna array to steer the mainlobe toward the signal and generate nulls in the

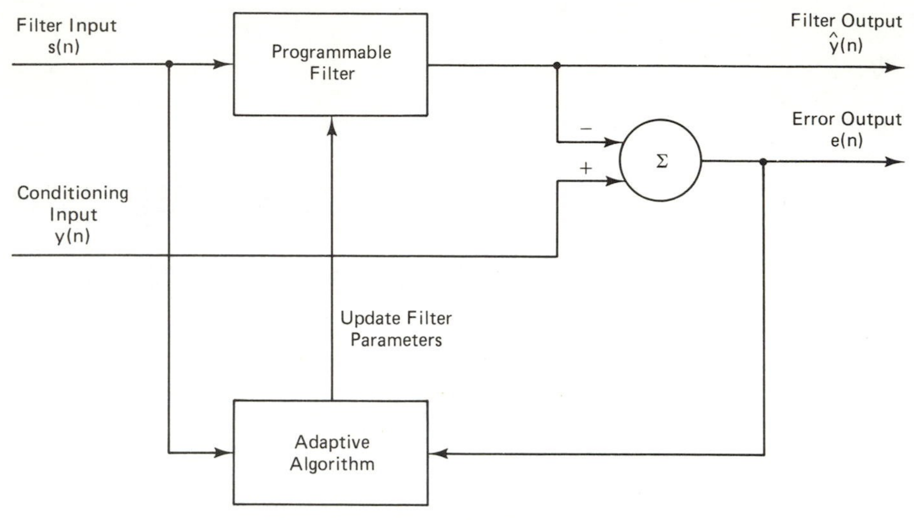

Figure 1.1 Schematic of adaptive filter.

beam pattern in the direction of the interfering sources. Thus they use spatial processing techniques for interference reduction.

In an adaptive system an absolute minimum of a priori information is necessary about the incoming signal. The adaptive filter operates by estimating the statistics of the incoming signal and adjusting its own response in such a way as to minimize some cost function. This cost function may be derived in a number of ways depending on the intended application, but normally it is derived by the use of the secondary signal source or conditioning input shown in Figure 1.1. This secondary signal input $y(n)$ may be defined as the desired output of the filter, in which case the task of the adaptive algorithm is to adjust the weights in the programmable filter device in such a way as to minimize the difference or error $e(n)$ between the filter output $\hat{y}(n)$ and the $y(n)$ input. These adaptive filters are frequently used to recover signals from channels whose characteristic is time varying.

All the systems considered in this text are sampled-data (discrete-time) systems. For convenience the explicit time index has thus been omitted from all mathematical expressions.

1.1.1 Adaptive Filters

This text examines the theory, design, and application of adaptive filters. The first adaptive or self-training filter is often credited to Lucky for his design in 1966 of a zero-forcing equalizer which compensated for distortion in data transmission systems. However, prior relevant work was reported in 1960 by Jakowatz et al. on adaptive waveform recognition. Theoretical work on adaptive filters was reported in 1961 in the United States by Glaser and in the same year in the United Kingdom, Gabor et al. used an analog tape transport mechanism to adjust the weights of a nonlinear "learning" filter. We may interpret the title "learning" as a reference to an adaptive processor.

Much of this early work on adaptive filters was arrived at by independent study in different research organizations. Other notable early developments occurred at the Technische Hochschule Karlsruhe in Germany and at Stanford University, where adaptive pattern recognition systems were initiated in 1959. Collaboration in 1964 between these institutions produced a comparative evaluation of their respective techniques [Steinbuch and Widrow] which subsequently led to the development of the most widely used algorithm for processor weight adjustment. Further relevant work was being conducted simultaneously at the Institute of Automatics and Telemechanics in Moscow. An excellent summary of the status, in the middle 1960s, of adaptive filters and early references for their use in adaptive or automatic equalization is provided by [Rudin]. More recently, simple review articles have been prepared on echo cancellation in telephony [Weinstein 1977(1)] and adaptive equalization [Qureshi 1982].

Many methods exist to adjust the filter weight values to obtain the optimum solution. Random perturbation techniques have been applied [Widrow et al. 1976(2)], where the weights are altered and the output is examined to ascertain whether the random perturbation moves it toward or away from the desired solution. Chapter 3 details the development of the least-mean-squares (LMS) adaptive algorithm, which came from the Stanford University pattern recognition work and was first formally reported by Widrow in 1967 in the context of adaptive arrays and in 1971 in the adaptive filter situation. It is now widely applied to the calculation of the adaptive filter weights, as it uses gradient search techniques, which converge toward the optimum solution much more efficiently than do other algorithms. It can be shown that this technique is very similar to the technique of maximizing signal-to-noise ratio which was developed concurrently by Applebaum for use when obtaining the optimum weights for an adaptive array. It has also been shown that Lucky's zero-forcing equalizer employs a simplification of the more general LMS gradient search technique.

1.1.2 Adaptive Filter Operation

The primary function that is performed by the adaptive filter is system modeling. This is illustrated in Figure 1.2, where a spectrally white primary signal is supplied direct to either the s or y input of the adaptive filter. The primary signal is also input to a system with impulse response $\mathbf{H}(n)$, and its output is connected to the other adaptive filter input. Two different options may then be implemented which result in quite distinct optimum weight vectors, $\mathbf{H}_{\text{opt}}$, in the adaptive filter. These occur when:

1. The unknown system, $\mathbf{H}(n)$, is in the y input path to the adaptive filter [Figure 1.2(a)]. Here the optimum impulse response of the adaptive filter is a direct model of the system response $\mathbf{H}(n)$.

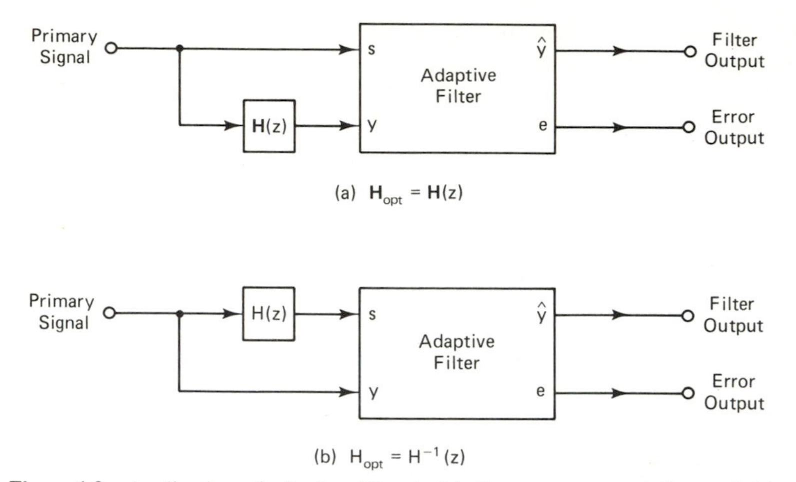

Figure 1.2 Application of adaptive filter to (a) direct system modeling and (b) inverse system modeling.

2. Alternatively, the unknown system, $\mathbf{H}(n)$, can be inserted in the s input path to the adaptive filter [Figure 1.2(b)]. In this case the optimum response of the adaptive filter is the inverse of the unknown system response.

A practical example that illustrates adaptive filter operation in the first mode (i.e., direct system modeling) is in echo cancellation across a telephone line hybrid, which is discussed further in Chapter 8.

An example that may be used to illustrate operation of the adaptive filter in the inverse response modeling mode of operation is that of equalization, for distortion in data transmission over telephone lines. Here, the input to the telephone line is excited by a known signal and the distorted line output is supplied to the $s(n)$ input of the adaptive filter. The filter is then trained by supplying a replica of the known (perfect) primary signal at the $y(n)$ input. The adaptive filter models the inverse of the line response to output equalized (distortion-free) data, which is described further in Chapter 8.

A further area where adaptive filters are applied is for noise cancellation [Sondhi and Berkley]. In this configuration the primary signal, which contains the desired information contaminated by a spurious signal, is applied to the $y(n)$ input. A separate, correlated, sample of this spurious signal is then obtained from another source which does not contain any of the required signal components. If this correlated signal is applied directly to the $s(n)$ adaptive filter input, the filter forms an impulse response that yields an output $\hat{y}(n)$ which subtracts coherently the unwanted component from $y(n)$, leaving only the desired signal on the $e(n)$ output.

One example of the use of this technique is in the monitoring of fetal heartbeat [Widrow et al. 1975(2)]. The primary signal is obtained from a transducer on the mother's abdomen. This transducer yields a signal containing the fetal heart signal, but it is heavily masked by the mother's heartbeat. The secondary signal, registering only the mother's heartbeat, is then obtained from a second transducer on the mother's chest. The adaptive filter then models the distorting path from the chest to the abdominal transducer to produce the signal that coherently subtracts from the abdominal signal. Further examples are in the removal of engine noise from a pilot's microphone in an aircraft cockpit [Arndt et al.] or analogous applications in other acoustically noisy environments such as in large power plants.

Another application of adaptive filters is in realizing a self-tuning filter which can be used to detect a sinusoid that is obscured by wideband noise. This application to adaptive line enhancement (ALE) [Zeidler et al.] is realized by feeding the signal directly into the $y(n)$ input of the filter and a delayed version of the signal to the filter $s(n)$ input. Provided that the delay is longer than the reciprocal of the filter bandwidth, the noise components will no longer be correlated at the two filter inputs but the sinusoidal components will still be correlated. The adaptive filter provides at its output a sinusoid with an enhanced signal-to-noise ratio, while the sinusoidal components are reduced at the error output. These and other examples of adaptive filter applications are discussed in Chapters 8 and 9.

1.2 PROGRAMMABLE FILTER DESIGNS

1.2.1 Recursive Filters

There are several types of programmable filters [Hamming] that can be used in the design of the adaptive filters which are described in Chapters 3 to 6. Here we summarize initially the two basic filter designs which are covered in Chapters 3 and 4 and then expand them into the processors described in the other two chapters.

The most generalized digital filter structure is the recursive filter design (Figure 1.3) [Oppenheim and Schafer, Gold and Rader, Peled and Liu 1976]. This comprises both feedforward multipliers, whose weights are controlled by the a coefficients, and feedback multipliers, which are controlled by the b coefficients. The response of this n-stage filter is governed by the nth-order difference equation which shows that the value of the present filter output sample is given by a linear combination of the weighted present and past input samples as well as the previous output samples. This structure results in a pole–zero filter design where the pole locations are controlled by the b coefficients and the zero locations by the a coefficients. The number of poles and zeros, or order of the filter, is given by the number of delay stages. Second-order integrated filters are commercially available for input sample rates (64 kilobaud) which are compatible with digital telephony systems [Adams et al. 1981(2)].

This recursive structure has theoretically an infinite memory and hence it is referred to as an infinite impulse response (IIR) filter design. It is not unconditionally stable unless restrictions are placed on the values of the b coefficients.

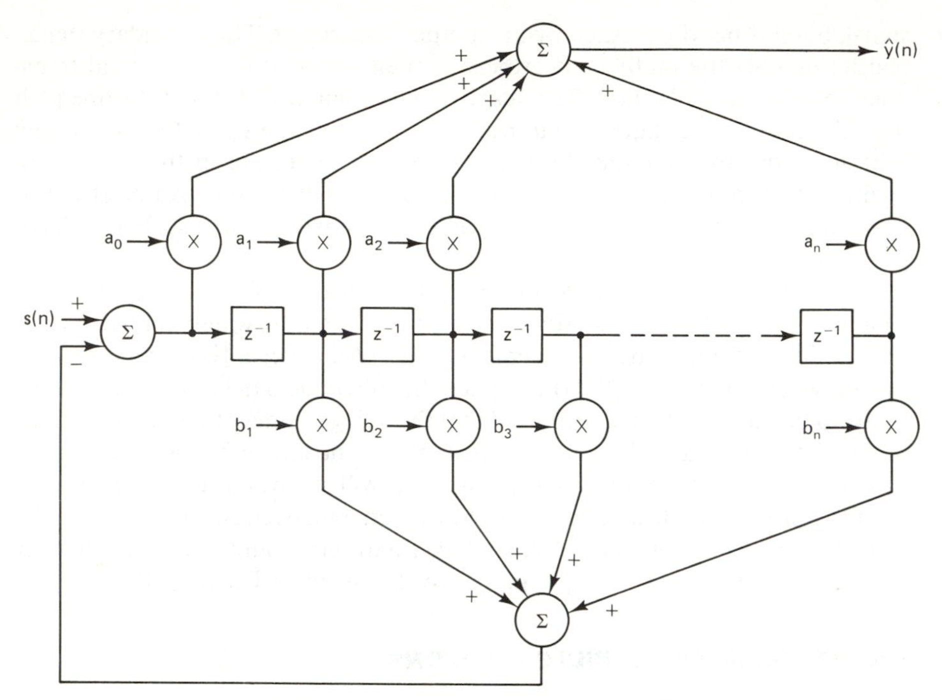

Figure 1.3 Infinite impulse response (IIR) recursive filter structure.

However, the inclusion of poles as well as zeros makes it possible to realize sharp cutoff filter characteristics incorporating a low transition bandwidth with only a modest number of delay stages (i.e., low filter complexity). One drawback of the IIR design is that no control is offered on the phase (group delay) response of the filter. However, the major problem with adaptive IIR filter design is the possible instability of the filter due to poles straying outside the stable region.

1.2.2 Nonrecursive Filters

One way to overcome the drawback of potential instability in the filter is to design an all-zero filter which uses only feedforward multipliers and is unconditionally stable (Figure 1.4). This has only a limited memory, which is controlled by the number of delay stages and it results in the finite impulse response (FIR) or transversal filter design [Kallman]. The input signal is delayed by a number of delay elements, which may be continuous but in the present text will be restricted to discrete values. The outputs of these time-delay elements are subsequently multiplied by a set of stored weights and the products summed to form the output signal. This implies that the output is given by the convolution of the input signal with the stored weights or impulse response values. This filter incorporates only zeros (as there are no recursive feedback elements) and hence a large number of delay elements are required to obtain a sharp cutoff frequency response.

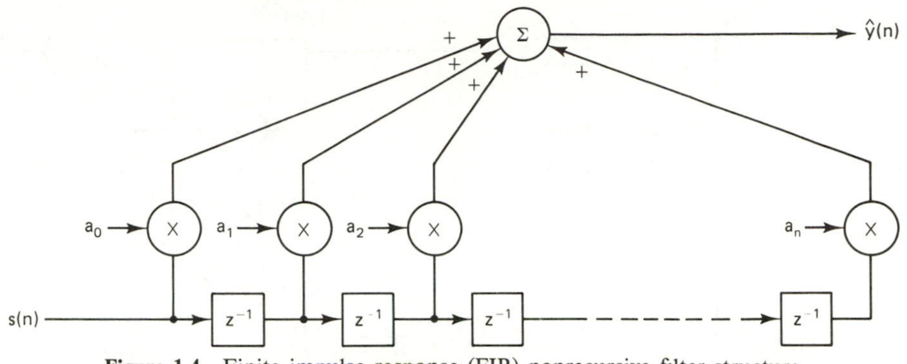

Figure 1.4 Finite impulse response (FIR) nonrecursive filter structure.

However, the filter is always stable and it can provide a linear-phase response.

An important contribution to the classification of filter designs was made by Chang in 1971 when he attempted to unify all approaches into a generalized equalizer or filter structure (Figure 1.5). This comprises a set of undefined filters feeding a linear weighting and combining network. The FIR filter can be configured from this generalized structure by replacing the undefined filter structure by a tapped delay line which outputs a series of time-delayed signal samples. The IIR filter structure incorporates further processing, due to the recursive feedback connections, into the time-delayed signal samples which subsequently feed to the weighted combiner.

An alternative FIR filter realization is the lattice structure [Griffiths 1977 and 1978, Makhoul 1977 and 1978(2), Friedlander and Morf 1982(2)], which can be considered as a cascade of single tap prediction error filters (Figure 1.6). This

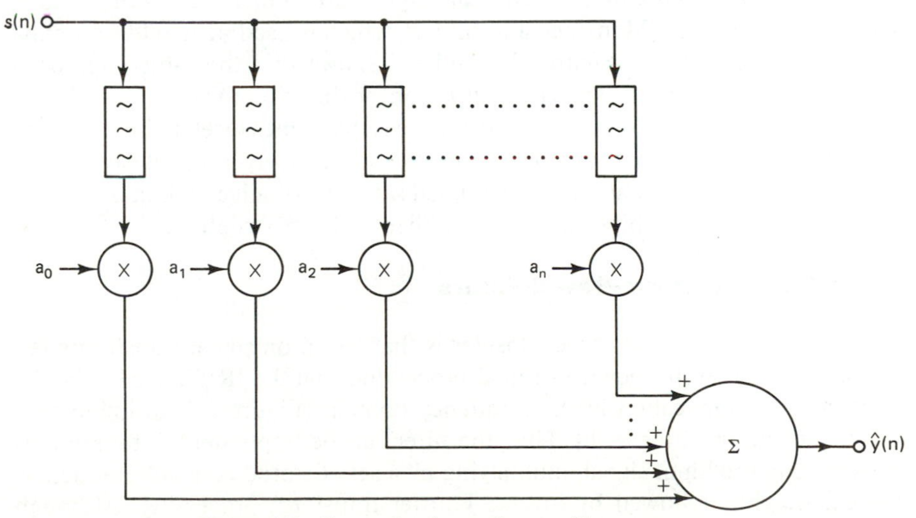

Figure 1.5 Generalized equalizer or filter structure.

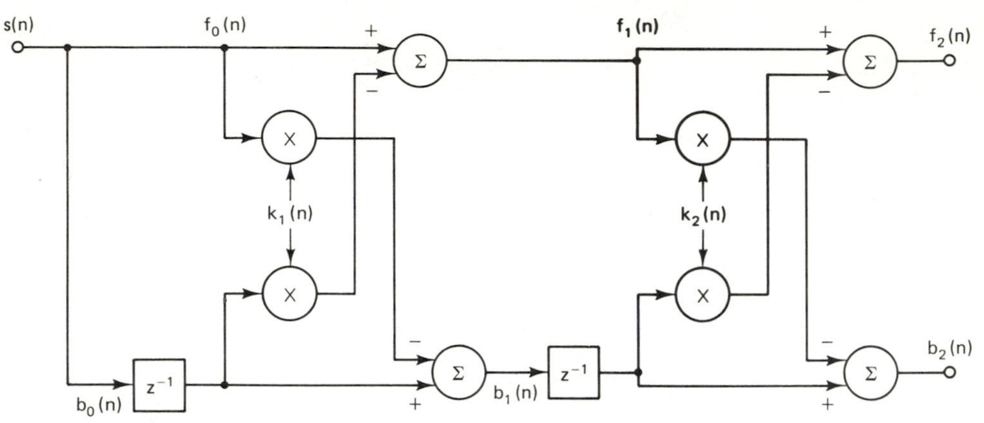

Figure 1.6 Lattice finite impulse response filter structure.

structure, which is used extensively in linear predictive coders for speech processing [Blankenship, Makhoul 1975], splits the signal into sets of forward (f) and backward (b) residual signal samples, with delays added into the backward channel. These signals are multiplied by Parcor coefficients, $k(n)$, which are so called because of their correspondence with the reflection coefficients in a discrete lattice. The forward residual Parcor coefficient for any stage is normally equal to the complex conjugate of the backward coefficient, except in sampled data (baseband) processors, where they are equal.

The calculation of these Parcor coefficient values by recursive techniques has been reported by Itakura and Saito in 1970 and refined by both Makhoul and Mead, who have further suggested methods to simplify the algorithm computation. It can be shown that the lattice structure of Figure 1.6 has an equivalent FIR filter realization (Figure 1.7) which bears significant similarities to the Gram-Schmidt preprocessor [Monzingo and Miller], which is used in adaptive antenna arrays. A brief study of Figures 1.6 and 1.7 shows that the lattice approach provides a very compact hardware realization of this structure.

The key attraction of the lattice structure is that it measures at the backward residual outputs the signal autocorrelation over successively longer delays to output a set of data-dependent orthogonal signal samples, which can then be used to feed the weighted combiner of the generalized Chang structure.

1.2.3 Transform-Based Filters

The final type of filter we consider is that based on signal transformation techniques. It is well known in signal processing that a FIR filter can also be realized by multiplication in the frequency domain (Figure 1.8) to implement "fast" convolution [Bracewell]. Thus the filter can be implemented by Fourier-transforming the input signal, multiplying by a set of stored complex frequency-domain weights, followed by inverse Fourier transform processing. Although this appears at first sight to be more complicated than time-domain convolution,

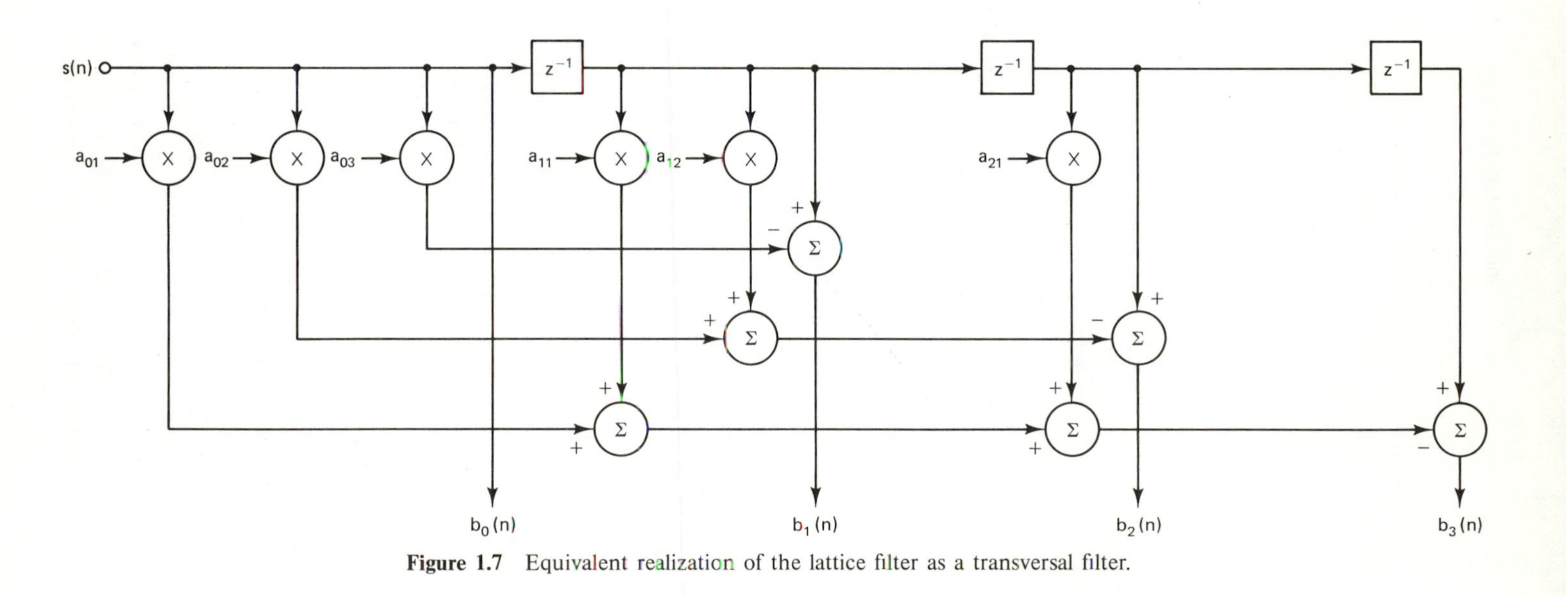

Figure 1.7 Equivalent realization of the lattice filter as a transversal filter.

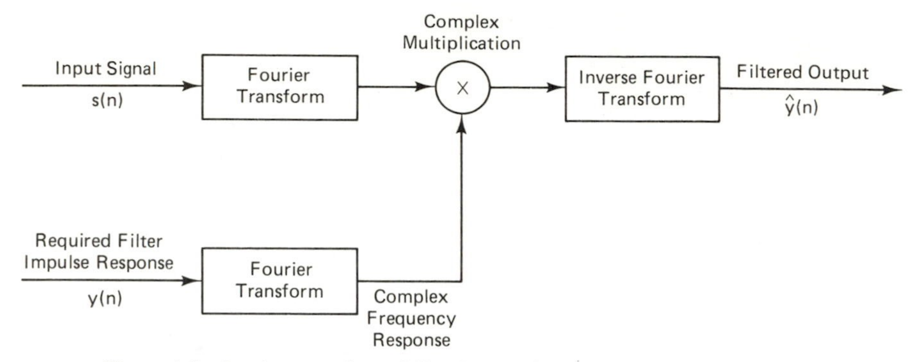

Figure 1.8 Implementation of filtering by frequency-domain processing.

the reduced computational load of a fast Fourier transform (FFT) processor [Brigham] produces an overall simplification in the processor. It can be shown that for intermediate or large transform sizes (e.g., >128-point processors) the frequency-domain approach involves many fewer multiplications than the N parallel multiplications per sample period which are required to implement the N-point FIR-based convolution. Thus the frequency-domain technique offers wider filter bandwidth for a given speed of logic, but it does implement circular convolution and care is required to configure it for linear convolution, which is equivalent to FIR filtering. However, it has been extensively applied in radar systems [Martinson] in the design of matched filter-based receivers for the correlation of wideband coded waveforms.

1.3 OPTIMUM LINEAR ESTIMATION

The concept of optimum linear estimation [Assefi] is fundamental to any consideration of adaptive filters. The actual estimation processes that take place may be divided into two classes. The first of these is the overall process carried out by the filter in estimating the secondary input signal, $y(n)$ (see Figure 1.2). The second process, which is interlinked with the primary objective, is the estimation of the filter tap weights. These estimations occur in a continuous time-varying manner in adaptive filters.

Chapter 2 deals in some detail with the theory involved in optimum linear estimation. The first part of this chapter deals with the Wiener estimator [Wiener, Bode and Shannon] which is an estimation process based on a finite set of past inputs (i.e., a block process). Consideration of the Wiener process is essential to an understanding of the adaption process in FIR adaptive filters.

The second part of Chapter 2 deals with the formulation of the optimum recursive (Kalman) estimator [Kalman, Kailath 1981]. This is particularly important in terms of the tap weight estimation used in the fast-tracking (high-convergence-rate) adaptive filters referred to in Chapters 3 and 5. The adaptive form of the Kalman filter is also important in its own right and this section includes an example of a Kalman equalizer [Lawrence and Kaufman].

1.4 ADAPTIVE FILTERS

Any of the four programmable filter designs described earlier may now be incorporated as the programmable filter in Figure 1.1 to realize an adaptive filter. The theory of these adaptive filter designs is detailed in Chapters 3 to 6, but some of the advantages and disadvantages of the different filter designs are discussed briefly here to illustrate their areas of application.

1.4.1 Adaptive Infinite Impulse Response Filters

Adaptive IIR filters have been principally applied to solving problems such as the reduction of multipath interference in radar and radio communication systems. Here the received signal comprises the original transmitted signal convolved with a channel response, which for multipath is an all-zero model. The adaptive receiver now models the inverse of the channel [Figure 1.2(b)] to remove the interference, and this is most efficiently accomplished using an all-pole adaptive filter model, where the pole positions are adjusted until they coincide with the zeros in the channel.

This can also be accomplished with an adaptive FIR filter design, but it is more economical to use the recursive structure, as it implements an inverse filter structure with a lower order of filter with fewer weights and it may reasonably be said that it will converge more rapidly than its transversal counterpart. However, the adaptive recursive filter requires a high degree of precision or accuracy in the digital circuit design to ensure stability of the filter. The adaptive processing technique based on IIR filters, which is detailed here in Chapter 4, is used in radar electronic support measure receivers for pulse sorting while adaptive Kalman filters are attractive for identifying the scan mode employed by specific emitters. They also find application in equalization and multipath reduction in high-frequency (3 to 30 MHz) digital communication channels, where their high speed of convergence is of primary importance.

1.4.2 Adaptive Finite Impulse Response Filters

Much of the reported literature on adaptive filters has been based on the FIR filter-based approaches [Widrow et al. 1975(2), Lucky 1966 and 1968] which are discussed here in Chapter 3. They are relatively simple to design and construct, and well-understood adaption algorithms (e.g., LMS) exist whose performance, with regard to rate of convergence, converged error, and so on, is well documented. Thus this is the approach most widely applied in the telecommunications applications of adaptive filters such as equalization and echo cancellation, which are discussed in Chapter 8.

One of the major deficiencies of this approach is that the filter weights are altered according to a single global error. The weights are therefore interrelated and this is one of the reasons for the relatively slow convergence response of LMS adaptive filters. One technique for overcoming this is to employ the lattice filter (Figure 1.6). The adaptive lattice structure, which incorporates recursive calculation of the internal Parcor coefficients, is an adaptive prediction error filter which performs spectral whitening. This property permits it to model the input signal and act as a parametric spectral estimator (see Section 9.3).

The structure of Figure 1.6 may be further modified to use it as a Chang equalizer by adding a linear weighting and combining network to the backward samples (Figure 1.9). This shows a distributed combining network which provides separate error outputs to update each of the weight values. This permits the convergence coefficient to be adjusted independently in each of the individual adaptive loops, under the control of the amplitude or power of the backward residual signals to maintain the same convergence rate for all the loops [Satorius and Alexander 1979(2)]. The updating of the k values and the g weights can be accomplished with different convergence coefficients to provide controlled convergence for the adaptive filter even when there is a large spread in the eigenvalues of the input signal autocorrelation matrix (i.e., the input signal is spectrally colored).

Chapter 5 describes the design and operation of these processors and in addition it covers the exact least-squares lattice [Makhoul 1977, Morf et al. 1977, Satorius and Pack 1981], which provides even faster convergence properties at the expense of the further computational complexity of this open-loop matrix-inversion-type processor.

1.4.3 Transform-Based Adaptive Filters

Another approach that provides approximately orthogonal signal samples is the implementation of the filter by frequency-domain multiplication (Figure 1.8). Chapter 6 describes the use of these processors for adaptive filters where separate updating loops are used for each of the transformed signal components. This again provides more uniform convergence rate across the filter. Frequency-domain filtering is shown to introduce problems due to the introduction of circular convolution and block processing in the discrete Fourier transform (DFT) [Bracewell], combined with the need to specify whether overlap-and-add or overlap-and-save arithmetic is employed. Chapter 6 reviews the design of these filters when implementing both circular and linear convolution. In addition, transform-based adaptive filters have been reported [Wong and Jan] which use the simpler computation of a Walsh [Beauchamp] transform processor.

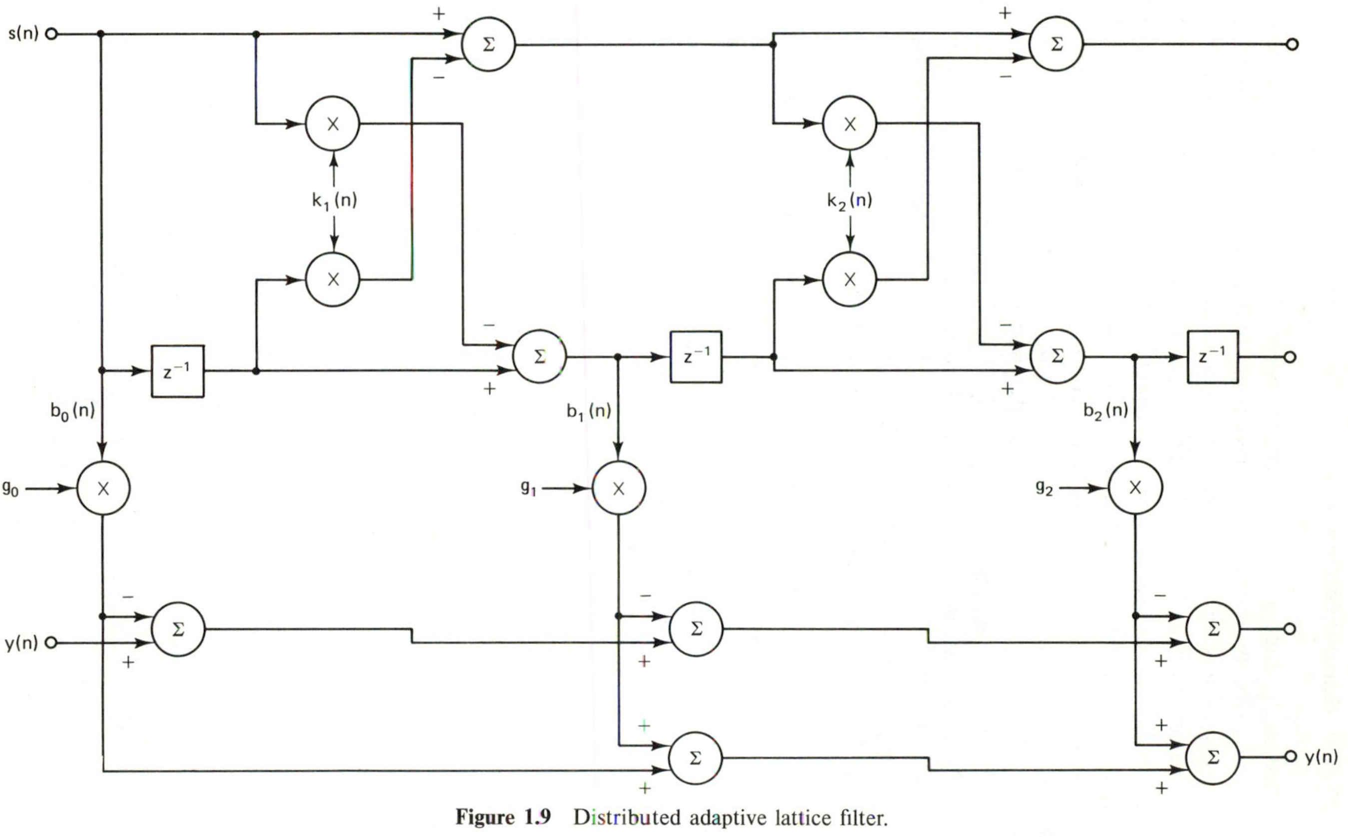

Figure 1.9 Distributed adaptive lattice filter.

1.4.4 Hardware Designs

Although adaptive filtering techniques have been reported in the technical literature for over two decades and have been implemented for some time as off-line processors, it is only the recent advances in large-scale integrated (LSI) and very large scale integrated (VLSI) circuit design techniques that have spurred interest in their hardware realisation. Chapter 7 discusses both analog and digital realizations of adaptive filters based on all four programmable filter design techniques covered in the earlier chapters. The interest in analog techniques was aided by the development in the early 1970s of analog FIR programmable filters based on charge-coupled device (CCD), [Beynon and Lamb] and bucket-brigade device (BBD) techniques [Weckler and Walby]. Several authors have subsequently shown that it was relatively simple to extend these designs into adaptive filters [White et al. 1979, Cowan et al. 1978–1981(1)]. However, the performance of these analog filters is ultimately limited by restrictions in dynamic range such as nonlinearities and noise effects, and this has fostered the development of hybrid or fully digital filter approaches.

At present all digital designs are generally larger and more complex than analog processors, but ultimately the progress of VLSI fabrication techniques to less than 1-μm feature sizes, and the development of computationally efficient circuit architectures such as bit serial systolic processors will result in the digital adaptive filters [Duttweiler, 1980] discussed in Chapter 7 being the most significant design approaches for application as real-time adaptive processors.

2

OPTIMUM ESTIMATION TECHNIQUES

Colin F. N. Cowan

2.1 INTRODUCTION

The concepts of optimum linear estimation are fundamental to any treatment of adaptive filters. In this chapter the theoretical basis of estimation theory is presented in terms of both optimum nonrecursive and recursive estimators. The fundamental definitions of the use of random variables is not covered here but may be found in [Assefi].

The adaptive filtering problem involves two estimation procedures: (1) the estimation of the required filter output, and (2) estimation of the filter weights required to achieve the former objective. The second of these two objectives is required because the input signal characteristics are not known a priori in the adaptive filtering situation.

The most commonly used type of adaptive filter structure is that using finite impulse response (FIR) architectures (considered in Chapter 3). These filters should converge to the optimum nonrecursive estimator solution given by the Wiener-Hopf equation [Wiener], which is derived in Section 2.2. Infinite impulse response (IIR) adaptive filters are considered in Chapter 4, and in Section 2.3 the optimum recursive estimator (Kalman filter) is discussed.

The definition of these estimators is critically dependent on the definition of a cost function which defines the quality of the estimate in terms of the difference between the estimator output and the actual parameter to be estimated:

$$e(n) = x(n) - \hat{x}(n) \tag{2.1}$$

Here $e(n)$ is the estimation error; $x(n)$ is the random variable to be estimated, which may be deterministic; and $\hat{x}(n)$ is the estimate of $x(n)$ provided by our estimation system, where

$$\hat{x}(n) = f\{y(n), h(n)\} \tag{2.2}$$

That is, $\hat{x}(n)$ is a linear function of the input signal sequence, $y(n)$, and a set of filter weights, $h(n)$. The observed signal sequence, $y(n)$, may generally be modeled as an originating sequence, $x(n)$, which is corrupted by additive white noise, $v(n)$, with variance σ_v^2.

$$y(n) = x(n) + v(n) \tag{2.3}$$

The most commonly used cost function to produce the optimum estimate $\hat{x}(n)$ is the least-mean-squares (LMS) function. The mean-square error is given by

$$E\{e^2(n)\} = \overline{[x(n) - \hat{x}(n)]^2} \tag{2.4}$$

This is minimized, with respect to the estimator weighting variables, to yield the optimum LMS estimation [Wiener, Kailath 1981, Bode and Shannon]. It should be noted that this is not the only cost function that may be used. Alternatives are functions such as modulus of error and nonlinear threshold functions. These error surfaces are shown in Figure 2.1. The nonlinear threshold type is used in the

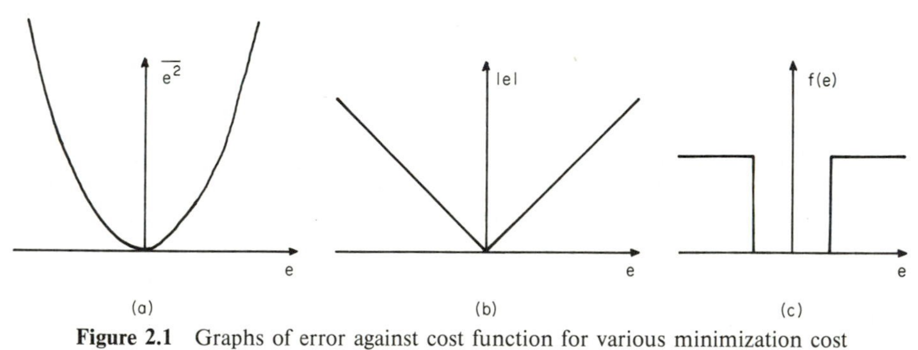

Figure 2.1 Graphs of error against cost function for various minimization cost functions.

instance, where a band of error may be acceptable (i.e., a defined error tolerance exists). In the least-squares approach small errors have less emphasis than do large errors (in contrast to modulus error, which gives equal weight to all errors).

2.2 OPTIMUM NONRECURSIVE (WIENER) ESTIMATION

In a nonrecursive estimator the estimate, $\hat{x}(n)$, is defined in terms of a finite linear polynomial in $y(n)$:

$$\hat{x}(n) = \sum_{k=0}^{N-1} y(n-k)h_k \tag{2.5}$$

where h_k are individual weights in a nonrecursive (FIR) filter structure illustrated in Figure 2.2. Equation (2.5) may be rewritten in matrix-vector notation as

$$\hat{x}(n) = \mathbf{Y}^{\mathrm{T}}(n)\mathbf{H} = \mathbf{H}^{\mathrm{T}}\mathbf{Y}(n) \tag{2.6}$$

where

$\mathbf{Y}^{\mathrm{T}}(n) = [y(n) \quad y(n-1) \quad \cdots \quad y(n-N-1)]$ and $\mathbf{H}^{\mathrm{T}} = [h_0 \quad h_1 \quad \cdots \quad h_{N-1}]$

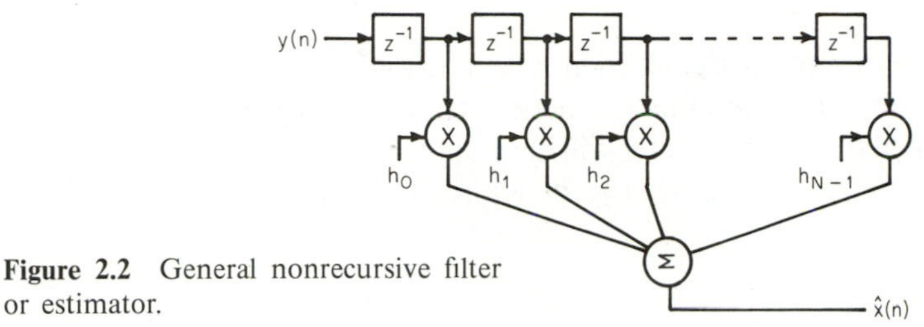

Figure 2.2 General nonrecursive filter or estimator.

and the superscript T denotes matrix transposition. The mean-square error function (2.4) then becomes

$$E\{e^2(n)\} = E\{x(n) - \mathbf{H}^{\mathrm{T}}\mathbf{Y}(n)\}^2 \tag{2.7}$$

Note: This represents a well-behaved quadratic error surface with a single unique minimum. Differentiating (2.7) with respect to $\mathbf{H}^{\mathrm{T}}$ yields

$$\frac{\partial \overline{e^2}(n)}{\partial \mathbf{H}^{\mathrm{T}}} = -2E[\{x(n) - \mathbf{H}^{\mathrm{T}}\mathbf{Y}(n)\}\mathbf{Y}^{\mathrm{T}}(n)] \tag{2.8}$$

and setting (2.8) equal to zero gives us

$$\begin{aligned} E\{[x(n) - \mathbf{H}^{\mathrm{T}}\mathbf{Y}(n)]\mathbf{Y}^{\mathrm{T}}(n)\} &= 0 \\ E\{x(n)\mathbf{Y}^{\mathrm{T}}(n)\} &= E\{\mathbf{H}^{\mathrm{T}}\mathbf{Y}(n)\mathbf{Y}^{\mathrm{T}}(n)\} \end{aligned} \tag{2.9}$$

Assuming that the weight vector $\mathbf{H}^{\mathrm{T}}$ and the signal vector $\mathbf{Y}(n)$ are uncorrelated, then

$$E\{x(n)\mathbf{Y}^{\mathrm{T}}(n)\} = \mathbf{H}_{\mathrm{opt}}^{\mathrm{T}}E\{\mathbf{Y}(n)\mathbf{Y}^{\mathrm{T}}(n)\} \tag{2.10}$$

The expectation terms in (2.10) may be defined as follows:

$\mathbf{P} = E\{x(n)\mathbf{Y}(n)\}$: the cross-correlation between the input signal and the parameter estimated

$\mathbf{R} = E\{\mathbf{Y}(n)\mathbf{Y}^{\mathrm{T}}(n)\}$: the autocorrelation matrix of the input signal sequence

Then (2.10) may be rewritten as

$$\mathbf{P}^{\mathrm{T}} = \mathbf{H}_{\mathrm{opt}}^{\mathrm{T}}\mathbf{R} \tag{2.11}$$

Equation (2.11) is the commonly recognized Wiener–Hopf equation [Wiener], which yields the optimum (least-mean-squares) Wiener solution for $\mathbf{H}$:

$$\mathbf{H}_{\mathrm{opt}} = \mathbf{R}^{-1}\mathbf{P} \tag{2.12}$$

2.2.1 Practical Example of a Wiener Estimator

In this section a practical example of a Wiener estimator is considered where the observed signal is a sinusoid with additive white noise. Thus

$$y(n) = \sin\frac{n\pi}{4} + \eta(n) \tag{2.13}$$

Therefore, the input signal, $x(n)$, is a sinusoid with a frequency of exactly one-eighth of the filter sampling frequency. To simplify the following derivation it will be assumed that the final filter will have only four weight values. The Wiener–Hopf equation (2.12) demands that we generate the autocorrelation matrix $\mathbf{R}$ defined by

$$\mathbf{R} = E\left\{\begin{bmatrix} y(n) \\ y(n-1) \\ y(n-2) \\ y(n-3) \end{bmatrix} [y(n) \quad y(n-1) \quad y(n-2) \quad y(n-3)]\right\} \tag{2.14}$$

$$= E\begin{bmatrix} y^2(n) & y(n)y(n-1) & y(n)y(n-2) & y(n)y(n-3) \\ y(n)y(n-1) & y^2(n-1) & y(n-1)y(n-2) & y(n-1)y(n-3) \\ y(n)y(n-2) & \cdots\cdots\cdots & y^2(n-2) & \cdots\cdots\cdots \\ \cdots\cdots\cdots & \cdots\cdots\cdots & \cdots\cdots\cdots & y^2(n-3) \end{bmatrix} \tag{2.15}$$

It should be noted that all the terms in (2.15) may be generated from the first row of the matrix since

$$E\{y^2(n)\} = E\{y^2(n-1)\} = E\{y^2(n-2)\} = E\{y^2(n-3)\}$$
$$E\{y(n)y(n-1)\} = E\{y(n-1)y(n-2)\} = E\{y(n-2)y(n-3)\}$$
$$E\{y(n)y(n-2)\} = E\{y(n-1)y(n-3)\}$$

This is simply because of the theoretically infinite average applied by the expectation operator, which means that only the time difference between the two operands is significant, provided it is assumed that the time series, $y(n)$, is stationary. Thus (2.15) is reduced to

$$\mathbf{R} = E\begin{bmatrix} y^2(n) & y(n)y(n-1) & y(n)y(n-2) & y(n)y(n-3) \\ y(n)y(n-1) & y^2(n) & y(n)y(n-1) & y(n)y(n-2) \\ y(n)y(n-2) & y(n)y(n-1) & y^2(n) & y(n)y(n-1) \\ y(n)y(n-3) & y(n)y(n-2) & y(n)y(n-1) & y^2(n) \end{bmatrix} \tag{2.16}$$

Such a symmetric matrix is said to be Toeplitz in nature. The generation of the elemental values in this matrix is particularly simple in the case of our example since only the diagonal terms are affected by the noise signal and all other terms may be generated from the deterministic signal component, $x(n)$, which has only eight discrete states, with the specified sampling interval. The actual values are given by

$$\mathbf{R} = \begin{bmatrix} \frac{1}{2} + \sigma_n^2 & \frac{1}{2\sqrt{2}} & 0 & \frac{-1}{2\sqrt{2}} \\ \frac{1}{2\sqrt{2}} & \frac{1}{2} + \sigma_n^2 & \frac{1}{2\sqrt{2}} & 0 \\ 0 & \frac{1}{2\sqrt{2}} & \frac{1}{2} + \sigma_n^2 & \frac{1}{2\sqrt{2}} \\ -\frac{1}{2\sqrt{2}} & 0 & \frac{1}{2\sqrt{2}} & \frac{1}{2} + \sigma_n^2 \end{bmatrix} \tag{2.17}$$

where $\sigma_n^2 = E\{\eta^2(n)\}$. Further manipulation is simplified by applying the substitution

$$\mathbf{R} = \begin{bmatrix} \nu & \rho & 0 & -\rho \\ \rho & \nu & \rho & 0 \\ 0 & \rho & \nu & \rho \\ -\rho & 0 & \rho & \nu \end{bmatrix} \tag{2.18}$$

The inverse of $\mathbf{R}$ is then given by

$$\mathbf{R}^{-1} = \begin{bmatrix} \frac{-\nu}{2\rho^2 - \nu^2} & \frac{\rho}{2\rho^2 - \nu^2} & 0 & \frac{-\rho}{2\rho^2 - \nu^2} \\ \frac{\rho}{2\rho^2 - \nu^2} & \frac{-\nu}{2\rho^2 - \nu^2} & \frac{\rho}{2\rho^2 - \nu^2} & 0 \\ 0 & \frac{\rho}{2\rho^2 - \nu^2} & \frac{-\nu}{2\rho^2 - \nu^2} & \frac{\rho}{2\rho^2 - \nu^2} \\ \frac{-\rho}{2\rho^2 - \nu^2} & 0 & \frac{\rho}{2\rho^2 - \nu^2} & \frac{-\nu}{2\rho^2 - \nu^2} \end{bmatrix} \tag{2.19}$$

In order to generate the Wiener weight vector the cross-correlation matrix, $\mathbf{P}$, must also be generated, where $\mathbf{P}$ is defined by

$$\mathbf{P} = E\left\{ x(n) \begin{bmatrix} y(n) \\ y(n-1) \\ y(n-2) \\ y(n-3) \end{bmatrix} \right\} \tag{2.20}$$

Again substituting actual values from (2.13) gives us

$$\mathbf{P} = \begin{bmatrix} \dfrac{1}{2} \\ \dfrac{1}{2\sqrt{2}} \\ 0 \\ -\dfrac{1}{2\sqrt{2}} \end{bmatrix} = \begin{bmatrix} \nu - \sigma_n^2 \\ \rho \\ 0 \\ -\rho \end{bmatrix} \tag{2.21}$$

Premultiplying $\mathbf{P}$ by $\mathbf{R}^{-1}$ from (2.19) yields

$$\mathbf{H}_{\text{opt}} = \mathbf{R}^{-1}\mathbf{P} = \frac{1}{2(1 + \sigma_n^2)} \begin{bmatrix} 1 \\ \dfrac{1}{\sqrt{2}} \\ 0 \\ -\dfrac{1}{\sqrt{2}} \end{bmatrix} \tag{2.22}$$

The result for $\mathbf{H}_{\text{opt}}$ in (2.22) is the expected matched filter in the sinusoidal input case. The residual mean-square error may easily be evaluated from (2.22) and (2.7) by substituting $\mathbf{H} = \mathbf{H}_{\text{opt}}$ in (2.7).

From (2.9) the following expression may be written:

$$E\{e(n)\mathbf{Y}(n)\} = 0 \tag{2.23}$$

and the mean-square error may be written as

$$E\{e^2(n)\} = E\{e(n)[x(n) - \mathbf{H}_{\text{opt}}\mathbf{Y}(n)]\} \tag{2.24}$$

Using this expression in (2.24), the mean-square error becomes

$$\begin{aligned} E\{e^2(n)\} &= E\{e(n)x(n)\} \\ &= E\{x^2(n)\} - \mathbf{H}_{\text{opt}}^{\mathrm{T}} E\{x(n)\mathbf{Y}(n)\} \\ &= E\{x^2(n)\} - \mathbf{H}_{\text{opt}}^{\mathrm{T}}\mathbf{P} \end{aligned} \tag{2.25}$$

Substituting values for $\mathbf{P}$ and $\mathbf{H}_{\text{opt}}$ from (2.21) and (2.22), the residual mean-square error is given by

$$\begin{aligned} E\{e^2(n)\} &= \frac{1}{2} - \frac{1}{2(1 + \sigma_n^2)} \\ &= \frac{\sigma_n^2}{2(1 + \sigma_n^2)} \end{aligned} \tag{2.26}$$

Therefore, with $\sigma_n^2 = 0$ the mean-square error (MSE) is zero, and with a 0-dB signal-to-noise ratio (i.e., $\sigma_n^2 = \frac{1}{2}$) the final (MSE) is $\frac{1}{6}$ (with the MSE before the estimation process being $\frac{1}{2}$).

Increasing the order of $\mathbf{H}_{opt}$ (i.e., using a longer estimation filter) results in a corresponding reduction in the optimum residual mean-square error. The example quoted here has been used only to demonstrate practically the necessary data manipulation to realize the Wiener estimator. Although better estimates could be obtained by using higher orders of $\mathbf{H}_{opt}$, this would require vastly more computation.

2.3 OPTIMUM RECURSIVE (KALMAN) ESTIMATION

The Wiener estimate considered in Section 2.2 is essentially a block process estimate which is best suited to the situation where only a finite block of data is available and may be processed in an "off-line" computer. In dealing with an infinite time series, however, the Wiener estimate would require a complete recalculation of all auto- and cross-correlation terms for each new input sample. The optimum recursive (or Kalman) estimator uses this new knowledge to update a recursive estimate [Kalman, Bozic, Kailath 1981].

In this section the optimum scalar Kalman filter will be derived and this will be extended in Section 2.4 to cover the vector Kalman formulation. Section 2.5 presents an example using a vector Kalman filter for communications channel equalization.

2.3.1 Scalar Kalman Filter

The Kalman estimator is essentially a parametric estimation process which relies on an autoregressive (AR) model of the signal generation process. The AR model of a first-order process of this type is shown in Figure 2.3(a) with the corresponding measurement model in Figure 2.3(b). The measurement model is simply a gain term, c, with additive white noise $v(n)$. Given this signal generation

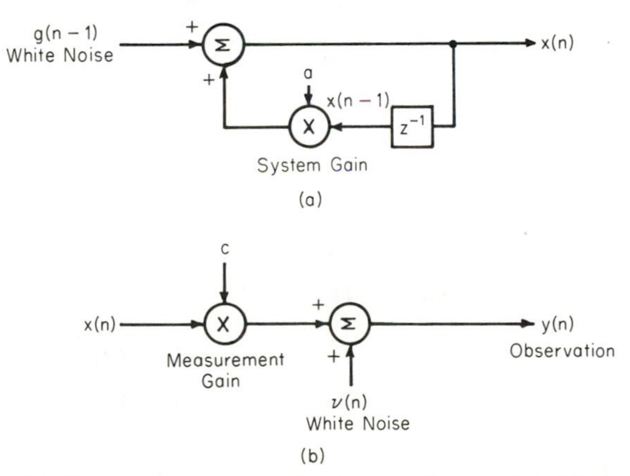

Figure 2.3 (a) First-order recursive signal generation model; (b) data measurement model.

model the observed data at sample instant n are given by

$$y(n) = cx(n) + v(n) \tag{2.27}$$

The first-order recursive estimator has the form

$$\hat{x}(n) = b(n)\hat{x}(n-1) + k(n)y(n) \tag{2.28}$$

Note that in (2.28) both filter gain terms are time varying (a schematic diagram of this general estimator is shown in Figure 2.4). To obtain the optimum (in the least-

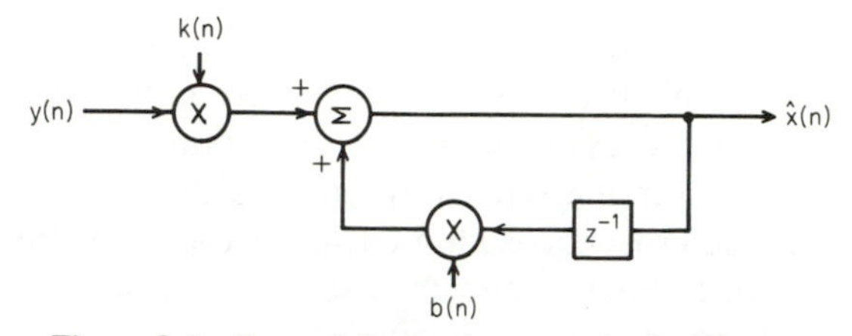

Figure 2.4 General first-order recursive estimator.

mean-squares sense) estimator the mean-square error, $p(n)$, is differentiated with respect to $b(n)$ and $k(n)$ and the results are equated to zero.

$$\begin{aligned} p(n) &= E[\hat{x}(n) - x(n)]^2 \\ &= E[b(n)\hat{x}(n-1) + k(n)y(n) - x(n)]^2 \end{aligned} \tag{2.29}$$

$$\frac{\partial p(n)}{\partial b(n)} = 2E\{[b(n)\hat{x}(n-1) + k(n)y(n) - x(n)]\hat{x}(n-1)\} = 0 \tag{2.30}$$

$$\frac{\partial p(n)}{\partial k(n)} = 2E\{[b(n)\hat{x}(n-1) + k(n)y(n) - x(n)]y(n)\} = 0 \tag{2.31}$$

A relationship between $b(n)$ and $k(n)$ may be derived using (2.30):

$$2E\{[b(n)\hat{x}(n-1) + k(n)y(n) - x(n)]\hat{x}(n-1)\} = 0$$

$$\Rightarrow E\{[b(n)\hat{x}(n-1)]\hat{x}(n-1)\} = E\{[k(n)y(n) - x(n)]\hat{x}(n-1)\} \tag{2.32}$$

$$\begin{aligned} &\Rightarrow E\{[b(n)[\hat{x}(n-1) - x(n-1)] + b(n)x(n-1)]\hat{x}(n-1)\} \\ &= E\{[x(n) - k(n)y(n)]\hat{x}(n-1)\} \end{aligned} \tag{2.33}$$

Substituting the value for $y(n)$ from (2.27) gives us

$$\begin{aligned} &b(n)E\{e(n-1)\hat{x}(n-1) + x(n-1)\hat{x}(n-1)\} \\ &\qquad = E\{[x(n)[1 - ck(n)] - k(n)v(n)]\hat{x}(n-1)\} \end{aligned} \tag{2.34}$$

For the optimum estimator the orthogonality principle must hold, yielding the following relationships:

$$E[e(n)\hat{x}(n-1)] = 0 \quad \text{and} \quad E[v(n)\hat{x}(n-1)] = 0$$

Equation (2.34) then becomes

$$b(n)E[x(n-1)\hat{x}(n-1)] = [1 - ck(n)]E[x(n)\hat{x}(n-1)] \tag{2.35}$$

From our signal generation model,

$$x(n) = ax(n-1) + g(n-1) \tag{2.36}$$

Substituting (2.36) in (2.35), we have

$$\begin{aligned} b(n)&E[x(n-1)\hat{x}(n-1)] \\ &= [1 - ck(n)]E[ax(n-1)\hat{x}(n-1) + g(n-1)\hat{x}(n-1)] \end{aligned} \tag{2.37}$$

From equations (2.27) and (2.28)

$$\hat{x}(n) = b(n)\hat{x}(n-1) + k(n)cx(n) + k(n)v(n) \tag{2.38}$$

and substituting for $x(n)$ from (2.36) gives us

$$\begin{aligned} \hat{x}(n) &= b(n)\hat{x}(n-1) + k(n)acx(n-1) + k(n)cg(n-1) + k(n)v(n) \\ \hat{x}(n-1) &= b(n-1)\hat{x}(n-2) + ack(n-1)x(n-2) + ck(n-1)g(n-2) \\ &\quad + k(n-1)v(n-1) \end{aligned} \tag{2.39}$$

and since the average of all products of terms in (2.39) with $g(n-1)$ are zero,

$$E\{\hat{x}(n-1)g(n-1)\} = 0$$

Using this relationship, (2.37) becomes

$$b(n)E[x(n-1)\hat{x}(n-1)] = a[1 - ck(n)]E[x(n-1)\hat{x}(n-1)] \tag{2.40}$$

This leads to the final relationship between $b(n)$ and $k(n)$:

$$b(n) = a[1 - ck(n)] \tag{2.41}$$

Substituting (2.41) into (2.28) yields

$$\boxed{\hat{x}(n) = a\hat{x}(n-1) + k(n)[y(n) - ac\hat{x}(n-1)]} \tag{2.42}$$

Equation (2.42) is the definition of the optimum first-order recursive estimator, or scalar Kalman filter. The first term, $a\hat{x}(n-1)$, is a prediction of the current sample and the second term is an adjustment based on an error estimate modified by the Kalman gain, $k(n)$. The form of this filter is illustrated in Figure 2.5.

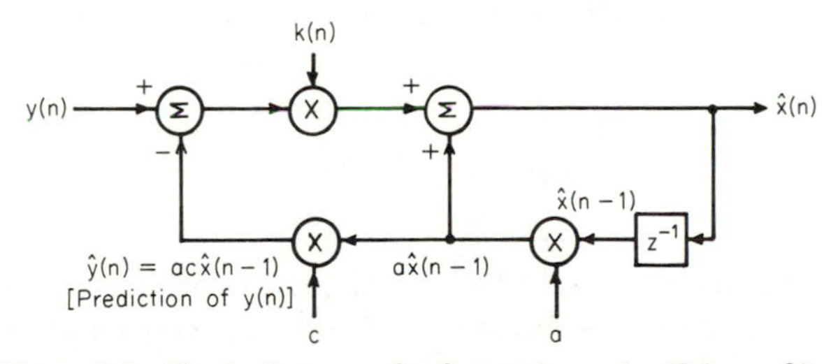

Figure 2.5 Block diagram of a first-order scalar Kalman filter.

2.3.2 Derivation of the Kalman Gain

Having defined the form of the Kalman filter the factor that remains to be derived is the time-varying Kalman gain term, $k(n)$. First, substituting (2.28) in (2.29), the mean-square error is given by

$$p(n) = E\{e(n)[b(n)\hat{x}(n-1) + k(n)y(n) - x(n)]\} \tag{2.43}$$

$$= -E\{e(n)x(n)\} \tag{2.44}$$

Using (2.27) to substitute for $x(n)$ in (2.44) yields

$$cE[e(n)x(n)] = -E[e(n)v(n)] \tag{2.45}$$

Substituting (2.45) in (2.44) gives us

$$p(n) = \frac{1}{c}E[e(n)v(n)] \tag{2.46}$$

and expanding $e(n)$ and substituting for $\hat{x}(n)$ from (2.28), we obtain

$$p(n) = \frac{1}{c}k(n)E[y(n)v(n)] = \frac{1}{c}k(n)\sigma_v^2 \tag{2.47}$$

$$k(n) = \frac{cp(n)}{\sigma_v^2} \tag{2.48}$$

where $\sigma_v^2 = E[v(n)]^2$. Now, substituting (2.42) into the mean-square error, (2.29), we have

$$p(n) = E\{a\hat{x}(n-1) + k(n)[y(n) - ac\hat{x}(n-1)] - x(n)\}^2 \tag{2.49}$$

and using (2.27) and (2.36) yields

$$p(n) = E\{a[1 - ck(n)]e(n-1) - [1 - ck(n)]g(n-1) + k(n)v(n)\}^2 \tag{2.50}$$

$$= a^2[1 - ck(n)]^2p(n-1) + [1 - ck(n)]^2\sigma_g^2 + k^2(n)\sigma_v^2 \tag{2.51}$$

Substituting (2.47) in (2.51) gives us

$$\boxed{k(n) = \frac{c[a^2p(n-1) + \sigma_g^2]}{\sigma_v^2 + c^2\sigma_g^2 + c^2a^2p(n-1)}} \tag{2.52}$$

Note that $k(n)$ must be calculated first from a knowledge of $p(n-1)$ and then $p(n)$ is calculated from

$$\boxed{p(n) = \frac{1}{c}\sigma_v^2k(n)} \tag{2.53}$$

The three equations given as (2.42), (2.52), and (2.53) are the recursions necessary to define the first-order Kalman filter. Unlike the Wiener filter, the Kalman filter gain must be defined in terms of an iteration and may not, therefore, be defined in terms of a universally stable solution.

2.4 VECTOR KALMAN FILTER

It is normally found in practical situations that a first-order autoregressive signal model is insufficient to characterize the physical process adequately. It is more likely that an AR process of order N will be needed. Such is the case, for instance, in the modeling of communications paths.

The first-order Kalman filter equations specified in Section 2.3 may be modified to take higher-order filters into account by replacing the scalars by vectors of order N. This process may best be illustrated by a simple example.

The simplest instance is where the autoregressive process under consideration is of a second-order form defined by

$$x(n) = ax(n-1) + bx(n-2) + g(n-1) \tag{2.54}$$

by defining two state variables $x_1(n)$ and $x_2(n)$ such that

$$x_1(n) = x(n) \quad \text{and} \quad x_2(n) = x(n-1)$$

(2.54) may be rewritten as a pair of state equations:

$$\begin{aligned} x_1(n) &= ax_1(n-1) + bx_2(n-1) + g(n-1) \\ x_2(n) &= x_1(n-1) \end{aligned} \tag{2.55}$$

Writing (2.55) as a matrix equation, we have

$$\begin{bmatrix} x_1(n) \\ x_2(n) \end{bmatrix} = \begin{bmatrix} a & b \\ 1 & 0 \end{bmatrix} \begin{bmatrix} x_1(n-1) \\ x_2(n-1) \end{bmatrix} + \begin{bmatrix} g(n-1) \\ 0 \end{bmatrix}$$

or

$$\mathbf{X}(n) = \mathbf{AX}(n-1) + \mathbf{G}(n-1) \tag{2.56}$$

The Kalman filter equations for the estimation problem are now in vector format but have the same form as for the scalar filter:

$$\hat{\mathbf{X}}(n) = \mathbf{A}\hat{\mathbf{X}}(n-1) + \mathbf{K}(n)[\mathbf{Y}(n) - \mathbf{CA}\hat{\mathbf{X}}(n-1)] \tag{2.57}$$

$$\mathbf{K}(n) = \mathbf{P}_1(n)\mathbf{C}^{\mathrm{T}}[\mathbf{CP}_1(n)\mathbf{C}^{\mathrm{T}} + \mathbf{Z}(n)]^{-1} \tag{2.58}$$

$$\mathbf{P}_1(n) = \mathbf{AP}(n-1)\mathbf{A}^{\mathrm{T}} + \mathbf{Q}(n-1) \tag{2.59}$$

$$\mathbf{P}(n) = \mathbf{P}_1(n) - \mathbf{K}(n)\mathbf{CP}_1(n) \tag{2.60}$$

where the scalar observation noise variance σ_v^2 and the system noise variance σ_g^2 have been replaced by matrices $\mathbf{Z}(n)$ and $\mathbf{Q}(n)$, respectively.

$$\mathbf{Z}(n) = E\{\mathbf{V}(n)\mathbf{V}^{\mathrm{T}}(n)\}$$

$$\mathbf{Q}(n) = E\{\mathbf{G}(n)\mathbf{G}^{\mathrm{T}}(n)\}$$

Similarly, the filter parameter, or Kalman gain, has been replaced by an $N \times N$ matrix $\mathbf{K}(n)$.

The following section, 2.4.1, demonstrates an example of the use of the vector Kalman filter in communications equalization. Therefore, rather than specifying arbitrary generalized forms for the vector estimator, the filter format will be illustrated purely through this example.

2.4.1 Vector Kalman Filter as a Channel Equalizer

The example taken here to illustrate the operation of the Kalman filter is that of channel equalization [Lawrence and Kaufman]. This is required when a random data sequence is transmitted over a distorting communications channel causing intersymbol interference (ISI), and it is then necessary to filter or equalize the channel output in order to recover the original data sequence. This application is covered in some depth, from the adaptive filter viewpoint, in Chapter 8; however, it is used here merely as an illustration of the use of vector Kalman estimation.

The distortion introduced by the channel may be modeled by a linear finite impulse response filter of the same type as that shown in Figure 2.2 (with the weights denoted by c_i instead of h_i) with noise added at the output. This is the model of the observation process of Figure 2.3. The channel input data, $x(n)$, is randomly distributed and the signal generation model may thus be represented by

$$\mathbf{X}(n+1) = \mathbf{A}\mathbf{X}(n) + \mathbf{F}s(n) \tag{2.61}$$

where

$$\mathbf{A} = \begin{bmatrix} 0 & \cdots\cdots\cdots & & & 0 \\ 1 & 0 & \cdots\cdots & & 0 \\ 0 & 1 & 0 & \cdots\cdots & 0 \\ \vdots & & & & \vdots \\ \vdots & & & & \vdots \\ 0 & \cdots\cdots & 0 & 1 & 0 \end{bmatrix} \quad \text{and} \quad \mathbf{F} = \begin{bmatrix} 1 \\ 0 \\ \vdots \\ \vdots \\ \vdots \\ 0 \end{bmatrix} \tag{2.62}$$

The observed channel output at sample time, n, is then given by

$$y(n) = \mathbf{C}^{\mathrm{T}}\mathbf{X}(n) + v(n) \tag{2.63}$$

where $\mathbf{C}$ is an $N \times 1$ vector of the channel coefficients and $v(n)$ is the additive noise at the channel output. Given the unitary nature of $\mathbf{A}$ the prediction $\mathbf{A}\hat{\mathbf{X}}(n-1)$ is simply $\hat{\mathbf{X}}(n-1)$ and the Kalman equalizer is represented by

$$\hat{\mathbf{X}}(n) = \hat{\mathbf{X}}(n-1) + \mathbf{K}(n)[y(n) - \mathbf{C}^{\mathrm{T}}\hat{\mathbf{X}}(n-1)] \tag{2.64}$$

and the iterations for $\mathbf{K}(n)$ are as given in (2.58)–(2.60). It should be noted here that it is implied that the distorting channel coefficients, $\mathbf{C}$, are known a priori.

The filter, or equalizer, described by (2.64) is shown schematically in Figure 2.6(a), showing how the various matrix products are implemented physically to form the prediction and correction. This may be compared with the literal multichannel matrix realization shown in Figure 2.6(b).

A more common use of Kalman estimation techniques is in the calculation of tap-weight values for finite impulse response (FIR) equalizers [Godard]. A dis-

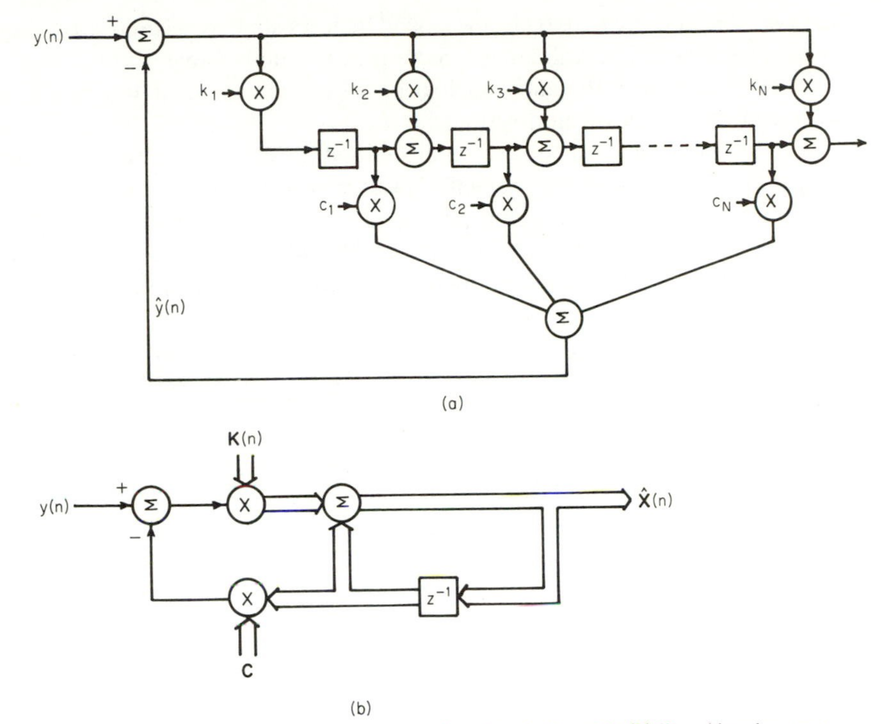

Figure 2.6 (a) Nth-order Kalman equalizer block diagram; (b) literal implementation of the vector Kalman filter.

cussion of the use of Kalman and fast Kalman techniques in this context is reported by [Mueller 1981].

2.5 CONCLUSIONS

Only a limited treatment of the estimation of stochastic signals has been presented in this chapter, making use of linear estimators only. Two types of linear processes have been considered:

1. The finite, or parallel, estimator resulting from the solution of the Wiener–Hopf equation, which produces an estimate as a weighted sum of a finite number of observation samples.
2. The sequential or recursive, Kalman, estimator which makes use of an assumed autoregressive signal generation model to produce an estimate that may be efficiently updated as a result of each new observation in a time series.

The estimators considered find direct application in a number of areas. Wiener filters are most suited to the situation where only a finite data field is actually available and they are typically applied in areas such as seismic surveying and image processing. The Kalman estimator is best suited to analysis of continuous time series and it is therefore applied in areas such as radar tracking and smoothing in inertial navigation systems.

It should be noted that the estimator defined in (2.2) need not be a linear function of $\hat{x}(n)$, and indeed many practical problems require the use of nonlinear estimation procedures (e.g., homomorphic filters in image processing). However, the analysis of the estimation procedure is tractable only in a few special cases for nonlinear functions and no general nonlinear analysis is available [Assefi].

3

ADAPTIVE ALGORITHMS FOR FINITE IMPULSE RESPONSE FILTERS

Benjamin Friedlander

3.1 INTRODUCTION

Adaptive filters generally consist of two distinct parts: a filter, whose structure is designed to perform a desired processing function, and an adaptive algorithm for adjusting the parameters (coefficients) of that filter. The many possible combinations of filter structures and the adaptive laws governing them lead to a sometimes bewildering variety of adaptive filters.

In this chapter we focus on what is, perhaps, the simplest class of filter structure: linear filters with a finite impulse response (FIR). A typical FIR filter, implemented in direct form, is depicted in Figure 3.1. Note that the filter output is a linear combination of a finite number of past inputs. The filter is not recursive (i.e., contains no feedback); the recursive adaptive filter is discussed in Chapter 4. This property leads to particularly simple adaptive algorithms, as will be demonstrated in the following sections.

Having specified the filter structure it is next required to design an adaptive algorithm for adjusting its coefficients. In this chapter we consider adaptive laws whose objective is to minimize the energy of the filter output (i.e., the output variance or the output sum of squares). The need to minimize this particular cost function arises in many applications involving least-squares estimation, such as adaptive noise canceling, adaptive line enhancement, and adaptive spectral estimation. Applications of adaptive filters are discussed in more detail in Chapters 8 and 9.

In the following two sections we present two adaptive algorithms for FIR filters: the recursive least-squares (RLS) algorithm and the Widrow–Hoff least-mean-squares (LMS) algorithm. The LMS algorithm has gained considerable popularity since the early 1960s. Its simplicity makes it attractive for many applications in which computational requirements need to be minimized. The RLS algorithm has been used extensively for system identification and time-series analysis. In spite of its potentially superior performance, its use in signal processing applications has been relatively limited, due to its higher computational requirements. In recent years there has been renewed interest in the RLS algorithm, especially in its "fast" (computationally efficient) versions. The RLS algorithm has been applied to adaptive channel equalization [Godard, Lawrence and Kaufman, Satorius and Pack 1981], adaptive array processing [Monzingo and Miller], and other problems.

A derivation of the RLS algorithm and a summary of its main properties are presented in Section 3.2. The RLS algorithm is used extensively in Chapter 5 in the context of lattice filters, where it is referred to as the least-squares lattice (LSL). In Section 3.3 we derive the LMS algorithm as an approximation of the RLS, and discuss its properties and its performance. Finally, in Section 3.4, we consider the design of an adaptive filter with constrained structure. The filter structure is chosen to guarantee linear phase characteristics.

3.2 RECURSIVE LEAST-SQUARES ALGORITHM

We start by considering the following prototype problem, which provides the basis for many adaptive filtering techniques. Let $y(n)$ be the input to an FIR filter of order N (see Figure 3.1). We denote the output of the filter by $e(n)$, where

$$e(n) = y(n) - h_1 y(n-1) \cdots - h_N y(n-N), \qquad n \geq 0 \tag{3.1}$$

The vector of filter coefficients will be denoted by $\mathbf{H}$,

$$\mathbf{H}^T = [h_1 \quad \cdots \quad h_N] \tag{3.2}$$

The filter output will, of course, depend on the filter coefficients [i.e., $e(n) = e_{\mathbf{H}}(n)$]. We will usually drop the explicit dependence on $\mathbf{H}$ for notational convenience. Next we consider an algorithm designed to minimize the sum of squares of

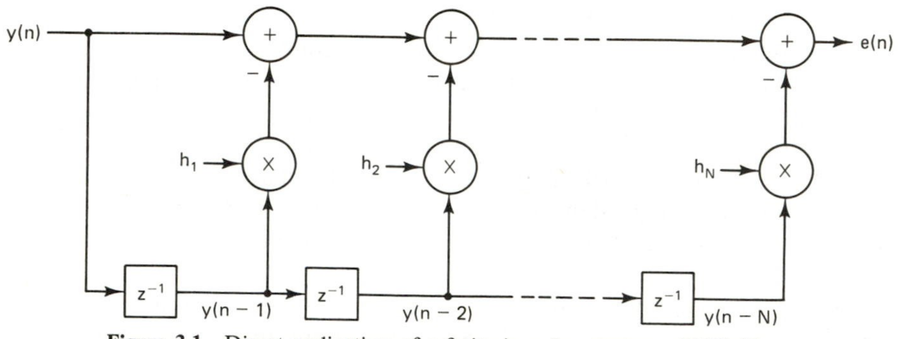

Figure 3.1 Direct realization of a finite impulse response (FIR) filter.

the output,

$$V(n) = \sum_{s=0}^{n} e^2(s) \tag{3.3}$$

The vector of filter coefficients for which this sum is minimized will be denoted by $\hat{\mathbf{H}}(n)$. This type of filter can be interpreted as a least-squares predictor for the input sequence $y(n)$, and the output $e(n)$ as a prediction error. This type of problem arises often in time-series analysis and signal processing. Note that we have made no assumptions about the statistics of the sequence $y(n)$. We are dealing for the moment with a purely deterministic minimization problem. To see the problem more clearly, we rewrite (3.1) in matrix form as

$$\underbrace{\begin{bmatrix} e(0) \\ e(1) \\ \vdots \\ e(n) \end{bmatrix}}_{\mathbf{e}(n)} = \underbrace{\begin{bmatrix} y(0) \\ y(1) \\ \vdots \\ y(n) \end{bmatrix}}_{\mathbf{y}(n)} - \underbrace{\begin{bmatrix} 0 & \cdots & \cdots \\ y(0) & \ddots & 0 \\ & \ddots & y(0) \\ & & \vdots \\ y(n-1) & \cdots & y(n-N) \end{bmatrix}}_{\mathbf{Y}(n)} \underbrace{\begin{bmatrix} h_1 \\ \vdots \\ \vdots \\ h_N \end{bmatrix}}_{\mathbf{H}} \tag{3.4}$$

In forming this equation we assumed that $y(n) = 0$ for $n < 0$. In other words, the data are "prewindowed," that is, multiplied by a window function $w(n)$, where $w(n) = 0$ for $n < 0$ and $w(n) = 1$ for $n \geq 1$. Other window functions can be chosen, leading to different adaptive algorithms. For example, when the data are not windowed, we obtain the "unwindowed" or "covariance" form of these equations:

$$\begin{bmatrix} e(N) \\ \vdots \\ e(n) \end{bmatrix} = \begin{bmatrix} y(N) \\ \\ y(n) \end{bmatrix} - \begin{bmatrix} y(N-1) & \cdots & y(0) \\ \\ y(n-1) & \cdots & y(n-N) \end{bmatrix} \begin{bmatrix} h_1 \\ \vdots \\ h_N \end{bmatrix} \tag{3.5}$$

We will consider first the prewindowed form, which has the simplest structure.

Note that minimizing the cost function $V(n)$ is equivalent to minimizing the squared norm of the left-hand side of (3.4), since

$$V(n) = \| \mathbf{e}(n) \|^2 \triangleq \mathbf{e}^{\mathrm{T}}(n)\mathbf{e}(n) \tag{3.6}$$

The vector that minimizes this norm is given by

$$\hat{\mathbf{H}}(n) = [\mathbf{Y}^{\mathrm{T}}(n)\mathbf{Y}(n)]^{-1}\mathbf{Y}^{\mathrm{T}}(n)\mathbf{y}(n) \tag{3.7}$$

This equation follows from a standard result in linear algebra regarding the solution of overdetermined sets of linear equations. Equation (3.7) provides, in principle, a solution to the adaptive filtering problem. At each time step (3.7) can be evaluated to give the values of the filter coefficients. This involves, however, a considerable amount of computation, since the coefficients are recomputed "from scratch" each time. A more useful form of the algorithm is obtained by developing a recursive method for computing $\hat{\mathbf{H}}(n)$. The recursive least-squares (RLS) algorithm described next is capable of updating the filter coefficients while making full use of the information contained in the current set of coefficients. Only an incremental amount of computation is needed at each time step. This is essentially done by the use of a Kalman-type estimation process described in Section 2.3.

3.2.1 Derivation of the RLS Algorithm

To derive the recursive least-squares algorithm we consider the situation where we make just one extra observation $y(n+1)$. Defining

$$\mathbf{P}(n) = [\mathbf{Y}^{\mathrm{T}}(n)\mathbf{Y}(n)]^{-1} \tag{3.8}$$

we have

$$\begin{aligned}\mathbf{P}(n+1) &= [\mathbf{Y}^{\mathrm{T}}(n) \mid \boldsymbol{\phi}(n+1)]\left[\begin{array}{c}\mathbf{Y}(n)\\ \hline \boldsymbol{\phi}^{\mathrm{T}}(n+1)\end{array}\right]^{-1} \\ &= [\mathbf{Y}^{\mathrm{T}}(n)\mathbf{Y}(n) + \boldsymbol{\phi}(n+1)\boldsymbol{\phi}^{\mathrm{T}}(n+1)]^{-1}\end{aligned} \tag{3.9}$$

where

$$\boldsymbol{\phi}^{\mathrm{T}}(n+1) = [y(n) \quad \cdots \quad y(n-N+1)] \tag{3.10}$$

The matrix inverse in (3.9) can be rewritten as

$$\mathbf{P}(n+1) = \mathbf{P}(n) - \mathbf{P}(n)\boldsymbol{\phi}(n+1)[1 + \boldsymbol{\phi}^{\mathrm{T}}(n+1)\mathbf{P}(n)\boldsymbol{\phi}(n+1)]^{-1}\boldsymbol{\phi}^{\mathrm{T}}(n+1)\mathbf{P}(n) \tag{3.11}$$

This equation follows from the matrix identity

$$[\mathbf{A} + \mathbf{BCD}]^{-1} = \mathbf{A}^{-1} - \mathbf{A}^{-1}\mathbf{B}(\mathbf{C} + \mathbf{DA}^{-1}\mathbf{B})^{-1}\mathbf{DA}^{-1} \tag{3.12}$$

which holds for all matrices **A**, **B**, **C**, and **D** of compatible dimensions (for nonsingular **A**).

Next note that

$$\begin{aligned}\hat{\mathbf{H}}&(n+1)\\ &= \mathbf{P}(n+1)\mathbf{Y}^{\mathrm{T}}(n+1)\mathbf{Y}(n+1)\\ &= \mathbf{P}(n+1)[\mathbf{Y}^{\mathrm{T}}(n)\mathbf{y}(n) + \boldsymbol{\phi}(n+1)\mathbf{y}(n+1)]\\ &= \{\mathbf{P}(n) - \mathbf{P}(n)\boldsymbol{\phi}(n+1)[1 + \boldsymbol{\phi}^{\mathrm{T}}(n+1)\mathbf{P}(n)\boldsymbol{\phi}(n+1)]^{-1}\boldsymbol{\phi}(n+1)\mathbf{P}(n)\}\\ &\quad \cdot\mathbf{Y}^{\mathrm{T}}(n)\mathbf{y}(n) + \mathbf{P}(n+1)\boldsymbol{\phi}(n+1)y(n+1)\\ &= \hat{\mathbf{H}}(n) - \mathbf{P}(n)\boldsymbol{\phi}(n+1)[1 + \boldsymbol{\phi}^{\mathrm{T}}(n+1)\mathbf{P}(n)\boldsymbol{\phi}(n+1)]^{-1}\boldsymbol{\phi}^{\mathrm{T}}(n+1)\hat{\mathbf{H}}(n)\\ &\quad + \mathbf{P}(n+1)\boldsymbol{\phi}(n+1)y(n+1)\end{aligned} \tag{3.13}$$

Define

$$\begin{aligned}\mathbf{K}&(n+1)\\ &= \mathbf{P}(n+1)\boldsymbol{\phi}(n+1)\\ &= \mathbf{P}(n)\boldsymbol{\phi}(n+1) - \mathbf{P}(n)\boldsymbol{\phi}(n+1)[1 + \boldsymbol{\phi}^{\mathrm{T}}(n+1)\mathbf{P}(n)\boldsymbol{\phi}(n+1)]^{-1}\\ &\quad \cdot\boldsymbol{\phi}^{\mathrm{T}}(n+1)\mathbf{P}(n)\boldsymbol{\phi}(n+1)\\ &= \mathbf{P}(n)\boldsymbol{\phi}(n+1)[1 - [1 + \boldsymbol{\phi}^{\mathrm{T}}(n+1)\mathbf{P}(n)\boldsymbol{\phi}(n+1)]^{-1}\boldsymbol{\phi}^{\mathrm{T}}(n+1)\mathbf{P}(n)\boldsymbol{\phi}(n+1)]\\ &= \mathbf{P}(n)\boldsymbol{\phi}(n+1)[1 + \boldsymbol{\phi}^{\mathrm{T}}(n+1)\mathbf{P}(n)\boldsymbol{\phi}(n+1)]^{-1}\end{aligned} \tag{3.14}$$

Using this definition, (3.13) gives

$$\hat{\mathbf{H}}(n+1) = \hat{\mathbf{H}}(n) + \mathbf{K}(n+1)[y(n+1) - \boldsymbol{\phi}^{\mathrm{T}}(n+1)\hat{\mathbf{H}}(n)] \tag{3.15}$$

Equations (3.11), (3.14), and (3.15) comprise the RLS algorithm. The initial value of $\mathbf{P}$ may be obtained either by evaluating $[\mathbf{Y}^{\mathrm{T}}(0)\mathbf{Y}(0)]^{-1}$ from an initial block of data or simply by letting $\mathbf{P}(0) = \sigma\mathbf{I}$, where σ is a large (positive) number; typically, $\sigma = 100/\mathrm{Var}\{y(0)\}$. Similarly, the initial value of $\mathbf{H}$ can be obtained by evaluating $\hat{\mathbf{H}}(0) = -\mathbf{P}(0)\mathbf{Y}^{\mathrm{T}}(0)\mathbf{y}(0)$, or simply by letting $\hat{\mathbf{H}}(0) = 0$.

3.2.2 Exponentially Weighted RLS

The adaptive algorithm derived above has an infinite memory. The values of the filter coefficients are functions of all past inputs. As will be discussed later, it is often useful to introduce a "forgetting factor" into the algorithm, so that recent data are given greater importance than are old data. One way of accomplishing this is to replace the sum-of-squares cost function, by an exponentially weighted sum of squares of the output:

$$\bar{V}(n) = \sum_{s=0}^{n} \lambda^{n-s} e^2(s) \tag{3.16}$$

where $0 \leq \lambda \leq 1$ is a constant determining the effective memory of the algorithm. When $\lambda = 1$, the algorithm will have infinite memory, as before. When $\lambda < 1$, the algorithm will have an effective memory of $\tau = -1/\log \lambda \cong 1/(1-\lambda)$ data points. To see this, consider for the moment the case where $e^2(s) = 1$. Then

$$\bar{V}(n) = \sum_{s=0}^{n} \lambda^{n-s} = \frac{1-\lambda^{n+1}}{1-\lambda} \tag{3.17}$$

We will define the effective memory length as the value $n + 1 = \tau$ for which $\bar{V}(\tau)/\bar{V}(\infty) \simeq 0.9$. Thus $1 - \lambda^{\tau} = 0.9$ or $\tau = \log 0.1/\log \lambda$. When λ is very close to unity, $\log \lambda \simeq \lambda - 1$.

To see the effect of the forgetting factor on the recursive algorithm, note that the new error criterion can be written as

$$\bar{V}(n) = \mathbf{e}^{\mathrm{T}}(n)\boldsymbol{\Lambda}(n)\mathbf{e}(n) \tag{3.18}$$

where

$$\boldsymbol{\Lambda}(n) = \mathrm{diag}\{\lambda^n, \lambda^{n-1}, \ldots, \lambda, 1\} \tag{3.19}$$

Premultiplying (3.4) by $\boldsymbol{\Lambda}^{1/2}(n)$ and solving for $\hat{\mathbf{H}}(n)$, as before, will give [see (3.7)]

$$\hat{\mathbf{H}}(n) = [\mathbf{Y}^{\mathrm{T}}(n)\boldsymbol{\Lambda}(n)\mathbf{Y}(n)]^{-1}\mathbf{Y}^{\mathrm{T}}(n)\boldsymbol{\Lambda}(n)\mathbf{y}(n) \tag{3.20}$$

Next we redefine $\mathbf{P}(n)$ as

$$\mathbf{P}(n) = [\mathbf{Y}^{\mathrm{T}}(n)\boldsymbol{\Lambda}(n)\mathbf{Y}(n)]^{-1} \tag{3.21}$$

from which it follows that

$$\begin{aligned}\mathbf{P}^{-1}(n+1) &= \lambda\mathbf{Y}^{\mathrm{T}}(n)\boldsymbol{\Lambda}(n)\mathbf{Y}(n) + \boldsymbol{\phi}(n+1)\boldsymbol{\phi}^{\mathrm{T}}(n+1) \\ &= \lambda\mathbf{P}^{-1}(n) + \boldsymbol{\phi}(n+1)\boldsymbol{\phi}^{\mathrm{T}}(n+1)\end{aligned} \tag{3.22}$$

Inverting (3.22) gives

$$\mathbf{P}(n+1) = \frac{\mathbf{P}(n) - \mathbf{P}(n)\boldsymbol{\phi}(n+1)[\lambda + \boldsymbol{\phi}^{\mathrm{T}}(n+1)\mathbf{P}(n)\boldsymbol{\phi}(n+1)]^{-1}\boldsymbol{\phi}^{\mathrm{T}}(n+1)\mathbf{P}(n)}{\lambda} \tag{3.23}$$

Next note that

$$\begin{aligned}\hat{\mathbf{H}}(n+1) &= \mathbf{P}(n+1)\mathbf{Y}^{\mathrm{T}}(n+1)\boldsymbol{\Lambda}(n+1)y(n+1) \\ &= \mathbf{P}(n+1)[\lambda\mathbf{Y}^{\mathrm{T}}(n)\boldsymbol{\Lambda}(n)\mathbf{y}(n) + \boldsymbol{\phi}(n+1)y(n+1)]\end{aligned} \tag{3.24}$$

Following the same steps as before it is straightforward to show that (3.15) still holds, except that the new definition of $\mathbf{P}(n+1)$ is used in computing $\mathbf{K}(n+1)$:

$$\mathbf{K}(n+1) = \mathbf{P}(n+1)\boldsymbol{\phi}(n+1) = \mathbf{P}(n)\boldsymbol{\phi}(n+1)[\lambda + \boldsymbol{\phi}^{\mathrm{T}}(n+1)\mathbf{P}(n)\boldsymbol{\phi}(n+1)]^{-1} \tag{3.25}$$

The complete algorithm is summarized in Table 3.1. The structure of the RLS adaptive filter is depicted in Figure 3.2.

TABLE 3.1 RLS Algorithm (exponentially weighted, prewindowed)

- Initialization:

$$\hat{\mathbf{H}}(0) = 0, \quad \mathbf{P}(0) = \sigma I, \quad \boldsymbol{\phi}(0) = 0$$

- At each time step do the following:

$$e(n) = y(n) - \boldsymbol{\phi}^{\mathrm{T}}(n)\hat{\mathbf{H}}(n-1) \quad [= e_{\hat{\mathbf{H}}(n-1)}(n)]$$

Compute filter output

$$\mathbf{P}(n) = \frac{\mathbf{P}(n-1) - \mathbf{P}(n-1)\boldsymbol{\phi}(n)[\lambda + \boldsymbol{\phi}^{\mathrm{T}}(n)\mathbf{P}(n-1)\boldsymbol{\phi}(n)]^{-1}\boldsymbol{\phi}^{\mathrm{T}}(n)\mathbf{P}(n-1)}{\lambda}$$

Gain update

$$\mathbf{K}(n) = \mathbf{P}(n-1)\boldsymbol{\phi}(n)[\lambda + \boldsymbol{\phi}(n)\mathbf{P}(n-1)\boldsymbol{\phi}(n)]^{-1}$$

$$\hat{\mathbf{H}}(n) = \hat{\mathbf{H}}(n-1) + \mathbf{K}(n)e(n)$$

Update filter coefficients

$$\boldsymbol{\phi}(n+1) = \mathbf{Z}\boldsymbol{\phi}(n) + [y(n+1), 0, \ldots, 0]^{\mathrm{T}}$$

Update the states of the filter

$$\mathbf{Z} = \begin{bmatrix} 0 & & & 0 \\ 1 & \ddots & & \\ & \ddots & \ddots & \\ 0 & & 1 & 0 \end{bmatrix}$$

Downshift operator

3.2.3 Computational Complexity

The adaptive algorithm described above performs well in many situations. The main reason that limits the usefulness of this technique in some applications is its computational complexity. Counting the number of operations involved in performing one step of the algorithm in Table 3.1 provides a measure of its computational complexity. Using the fact that $\mathbf{P}(n)$ is a symmetric matrix, it is possible to implement the algorithm so that it will require $2.5N^2 + 4N$ multiplications and additions per time step (N being the filter order). For high-order filters, a compu-

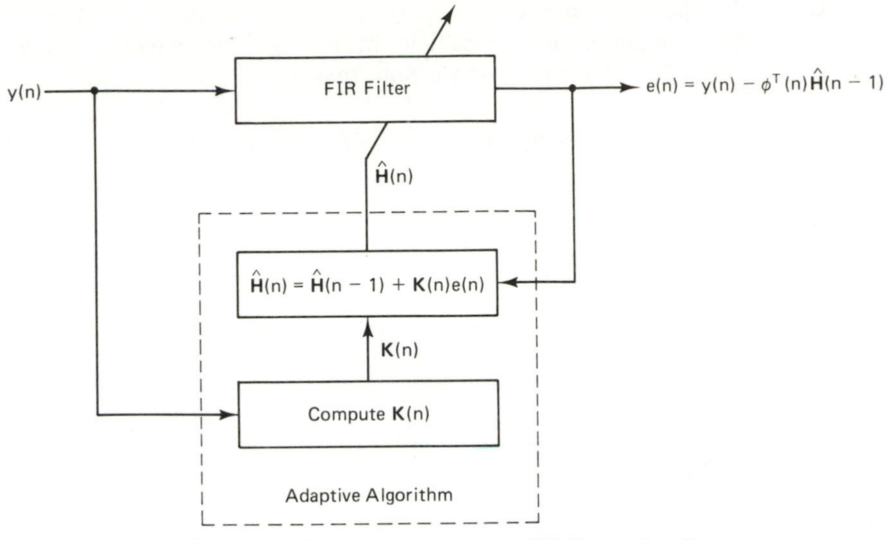

Figure 3.2 Recursive least-squares (RLS) adaptive filter.

tational complexity that increases in proportion to the square of the filter order is often unacceptable. Note that updating the gain vector $\mathbf{K}(n)$ requires most of the computations. Given this gain vector, the rest of the algorithm involves only $2N$ multiplications and additions. To reduce the complexity of the algorithm it is necessary, therefore, to simplify the gain update. Using the special structure of the matrix $\mathbf{P}^{-1}(n)$, it is possible to derive a gain update formula with complexity proportional to N rather than N^2 operations. This so-called "fast" algorithm (fast Kalman) is described in more detail in [Ljung et al. 1978]. An even simpler update formula can be derived by approximating the least-squares algorithm by a gradient search technique, as discussed in Section 3.3. In this algorithm the gain vector is proportional to the data vector $\boldsymbol{\phi}(n)$ and thus requires no additional computations.

3.2.4 Stochastic Interpretation

The RLS algorithm was derived without making any assumptions about the input data. Its defining property is that it will minimize the sum of squares of its output at every time step. This fact is in itself very useful. However, it is often desirable to know more about various aspects of the filter behavior. To explore in more detail the properties of this filter, we need to make some assumptions regarding the statistics of the data. It is important to remember, however, that the algorithm will perform its stated function [minimization of $V(n)$ or $\bar{V}(n)$] for arbitrary input sequences. Consider the case where the data $y(n)$ comprise a stationary zero-mean stochastic process with correlation $r(\tau)$,

$$r(\tau) = E\{y(n)y(n-\tau)\} \tag{3.26}$$

A fundamental property of stationary stochastic processes is ergodicity. Stated in a somewhat simplified manner this implies that time averages can be replaced in the limit by expected values. In particular we note that

$$\lim_{n\to\infty} \frac{1}{n} \sum_{s=\tau}^{n} y(n)y(n-\tau) = E\{y(n)y(n-\tau)\} = r(\tau) \tag{3.27}$$

It follows that

$$\lim_{n\to\infty} \frac{1}{n} \mathbf{Y}^{\mathrm{T}}(n)\mathbf{Y}(n) = \mathbf{R} \triangleq \begin{bmatrix} r(0) & \cdots\cdots & r(N-1) \\ \vdots & \ddots & \vdots \\ r(N-1) & \cdots\cdots & r(0) \end{bmatrix} \tag{3.28}$$

and

$$\lim_{n\to\infty} \frac{1}{n} \mathbf{Y}^{\mathrm{T}}(n)\mathbf{y}(n) = \mathbf{r} \triangleq \begin{bmatrix} r(1) \\ \vdots \\ r(N) \end{bmatrix} \tag{3.29}$$

and therefore

$$\lim_{n\to\infty} n\mathbf{P}(n) = \lim_{n\to\infty} \left[\frac{1}{n}\mathbf{Y}^{\mathrm{T}}(n)\mathbf{Y}(n)\right]^{-1} = \mathbf{R}^{-1} \tag{3.30}$$

and

$$\mathbf{H}_{\mathrm{opt}} \triangleq \lim_{n\to\infty} \hat{\mathbf{H}}(n) = \lim\, [\mathbf{Y}^{\mathrm{T}}(n)\mathbf{Y}(n)]^{-1}\mathbf{Y}^{\mathrm{T}}(n)\mathbf{y}(n) = \mathbf{R}^{-1}\mathbf{r} \tag{3.31}$$

In other words, in the limit, the filter coefficient vector converges to the solution of the Yule–Walker equations [Yule, Walker]:

$$\mathbf{R}\mathbf{H}_{\mathrm{opt}} = \mathbf{r} \tag{3.32}$$

and the properly normalized matrix $\mathbf{P}(n)$ converges to the inverse of the covariance matrix. The discussion above is meant only as a plausibility argument and not as a formal proof. However, these facts can be proven completely rigorously under some mild regularity conditions.

It remains to interpret the limiting filter, that is, the filter with coefficient vector $\mathbf{H}_{\mathrm{opt}}$. Recall that the RLS algorithm was designed to minimize the sum of squares of the output. Not surprisingly, the limiting filter can be shown to minimize the variance of the output process. To see this we consider the following linear least-squares estimation problem. Let $\hat{y}(n|n-1)$ denote the Nth-order one-step-ahead predictor of the process $y(n)$,

$$\hat{y}(n|n-1) = \sum_{i=1}^{N} a(i)y(n-i) \tag{3.33}$$

The predictor coefficients $a(i)$ are chosen so as to minimize the variance of the least-squares prediction error $\epsilon(n)$,

$$\epsilon(n) = y(n) - \hat{y}(n|n-1) = y(n) - \sum_{i=1}^{N} a(i)y(n-i) \tag{3.34}$$

A well-known property of least-squares estimators is the orthogonality property: The prediction error $\epsilon(n)$ is uncorrelated with all past data,

$$E\{\epsilon(n)y(s)\} = 0 \qquad \text{for } s < n \tag{3.35}$$

Postmultiplying equation (3.34) by $y(s)$ and taking expected values, we get

$$0 = r(n-s) - \sum_{i=1}^{N} a(i)r(n-s-i) \tag{3.36}$$

Setting $s = n-1, \ldots, n-N$, we can rewrite this equation in matrix form as

$$\underbrace{\begin{bmatrix} r(0) & \cdots & r(N-1) \\ \vdots & \ddots & \\ r(N-1) & \cdots & r(0) \end{bmatrix}}_{\mathbf{R}} \underbrace{\begin{bmatrix} a(1) \\ \vdots \\ a(N) \end{bmatrix}}_{\mathbf{A}} = \underbrace{\begin{bmatrix} r(1) \\ \vdots \\ r(N) \end{bmatrix}}_{\mathbf{r}} \tag{3.37}$$

Comparison with (3.32) leads immediately to the conclusion that $\mathbf{H}_{\text{opt}} = A$. In other words, for stationary input data, the RLS adaptive filter will converge to the linear least-squares one-step-ahead prediction filter (sometimes called the Wiener filter, referred to in more detail in Chapter 2).

Note that in the discussion above we considered the case where $\lambda = 1$. The convergence of the filter coefficients requires, in essence, their dependence on an infinite amount of data. Some statements regarding the finite memory case $\lambda \neq 1$ will be made in the following section.

3.2.5 Asymptotic Accuracy of Least-Squares Estimates

The least-squares estimator defined by (3.7) belongs to a more general class of prediction error estimators which has been studied intensively in the system identification literature. The theory of prediction error estimators has many applications to the analysis of adaptive filters.

Here we consider one particular result regarding the asymptotic accuracy of such estimators. This result is based on a more general theorem whose precise statement and proof can be found in [Ljung 1981, Ljung and Söderström 1983]. Specialized to the problem treated in this chapter, the theorem can be stated as follows.

Theorem. Subject to mild stationarity and regularity conditions, the least-squares estimator $\hat{\mathbf{H}}(n)$ defined by (3.37) is asymptotically normally distributed in the sense that

$$\sqrt{n}(\hat{\mathbf{H}}(n) - \mathbf{H}_{\text{opt}}) \longrightarrow \mathbf{x} \tag{3.38}$$

where

$$x \sim N(0, \sigma_\epsilon^2 \mathbf{R}^{-1}) \tag{3.39}$$

$$\sigma_\epsilon^2 \triangleq E\{e^2(n)\} = \text{prediction error variance} \tag{3.40}$$

$\mathbf{H}_{\text{opt}}$ is defined by (3.32), and $\mathbf{R}$ is the data covariance matrix (3.28). It follows from this theorem that for large enough values of n, the covariance matrix of the parameter estimation error vector

$$\tilde{\mathbf{H}}(n) \triangleq \hat{\mathbf{H}}(n) - \mathbf{H}_{\text{opt}} \tag{3.41}$$

is given approximately by

$$E\{\tilde{\mathbf{H}}(n)\tilde{\mathbf{H}}^{\mathrm{T}}(n)\} \triangleq \frac{\sigma_\epsilon^2}{n}\mathbf{R}^{-1} \tag{3.42}$$

This theorem therefore provides a useful measure of the accuracy of the estimator based on long data sequences [by accuracy we mean here the deviation of $\hat{\mathbf{H}}(n)$ from its limiting value $\mathbf{H}_{\text{opt}}$]. For short data sequences the covariance of $\tilde{\mathbf{H}}(n)$ may be much larger than that predicted by (3.42). Thus the matrix $(\sigma_\epsilon^2/n)\mathbf{R}^{-1}$ can be interpreted as a lower bound on the estimation error variance.

3.2.6 Asymptotic Properties of the Adaptive Filter

The theorem presented above also provides a useful measure of the "noisiness" of the adaptive filter coefficients as a function of the number of data points and the statistics of the data. When the adaptive algorithm operates with infinite memory ($\lambda = 1$), the variance of the filter coefficients will reduce in inverse proportion to the number of data points. Eventually, the coefficients become, for all practical purposes, constant. In the case where $\lambda < 1$, the coefficients will always be noisy, with variance depending on the effective window length τ. The variance of the filter coefficients will be bounded from below by $(\sigma_\epsilon^2/\tau)\mathbf{R}^{-1}$ [according to (3.42), with the running time to replace by the effective window length τ]. When the window is sufficiently long, the variance of the filter coefficients will be predicted quite well by the asymptotic bound. For short windows, the actual variance may be much larger than the predicted by the bound. At this time there are no analytical results for determining the window length beyond which the asymptotic results hold. Simulation studies are needed to answer this question.

A useful measure of the adaptive filter performance is provided by the variance of the output. The output of a filter with an arbitrary coefficient vector $\mathbf{H}$ is given by [see (3.1)]

$$e_{\mathbf{H}}(n) = y(n) - \boldsymbol{\phi}^{\mathrm{T}}(n)\mathbf{H} \tag{3.43}$$

The output can be always decomposed as the sum of the "optimal" prediction error and another term,

$$e_{\mathbf{H}}(n) = \underbrace{y(n) - \boldsymbol{\phi}^{\mathrm{T}}(n)\mathbf{H}_{\text{opt}}}_{\epsilon(n)} + \boldsymbol{\phi}^{\mathrm{T}}(n)[\mathbf{H}_{\text{opt}} - \mathbf{H}] \tag{3.44}$$

It follows that

$$\begin{aligned} E\{e_{\mathbf{H}}^2(n)\} = E\{\epsilon^2(n)\} &+ E\{\mathbf{H} - \mathbf{H}_{\text{opt}}\}^{\mathrm{T}}\boldsymbol{\phi}(n)\boldsymbol{\phi}^{\mathrm{T}}(n)(\mathbf{H} - \mathbf{H}_{\text{opt}})\} \\ &+ 2E\{\epsilon(n)\boldsymbol{\phi}^{\mathrm{T}}(n)[\mathbf{H}_{\text{opt}} - \mathbf{H}]\} \end{aligned} \tag{3.45}$$

The last term is zero, since the prediction error is uncorrelated with past data. Thus

$$E\{e_{\mathbf{H}}^2(n)\} = \sigma_\epsilon^2 + (\mathbf{H} - \mathbf{H}_{\text{opt}})^{\mathrm{T}}\mathbf{R}(\mathbf{H} - \mathbf{H}_{\text{opt}}) \tag{3.46}$$

Equation (3.46) can be used to evaluate the output variance for any fixed coefficient vector $\mathbf{H}$. As expected, the output variance will be minimized if and only if $\mathbf{H} = \mathbf{H}_{\text{opt}}$. The minimum variance σ_ϵ^2 can be evaluated as follows:

$$\begin{aligned} E\{\epsilon^2(n)\} &= E\{[y(n) - \boldsymbol{\phi}^{\mathrm{T}}(n)\mathbf{H}_{\text{opt}}]^2\} \\ &= E\{y^2(n)\} + E\{\mathbf{H}_{\text{opt}}^{\mathrm{T}}\boldsymbol{\phi}(n)\boldsymbol{\phi}^{\mathrm{T}}(n)\mathbf{H}_{\text{opt}}\} - 2E\{\mathbf{H}_{\text{opt}}^{\mathrm{T}}\boldsymbol{\phi}(n)y(n)\} \\ &= r(0) + \mathbf{H}_{\text{opt}}^{\mathrm{T}}\mathbf{R}\mathbf{H}_{\text{opt}} + 2\mathbf{H}_{\text{opt}}^{\mathrm{T}}\mathbf{r} \end{aligned} \tag{3.47}$$

Using the definition of $\mathbf{H}_{\text{opt}}$ (3.32), we get

$$\sigma_\epsilon^2 = r(0) - \mathbf{H}_{\text{opt}}^{\mathrm{T}}\mathbf{r} = r(0) - \mathbf{r}^{\mathrm{T}}\mathbf{R}^{-1}\mathbf{r} \tag{3.48}$$

During the operation of the adaptive algorithm, the filter coefficients are themselves random. To evaluate the output variance in this case we must use (3.45) with $\mathbf{H}$ replaced by $\hat{\mathbf{H}}(n-1)$:

$$\begin{aligned} E\{e_{\hat{\mathbf{H}}(n-1)}^2(n)\} &= \sigma_\epsilon^2 + E\{\tilde{\mathbf{H}}^{\mathrm{T}}(n-1)\boldsymbol{\phi}(n)\boldsymbol{\phi}^{\mathrm{T}}(n)\tilde{\mathbf{H}}(n-1)\} \\ &= \sigma_\epsilon^2 + \operatorname{tr} E\{[\boldsymbol{\phi}(n)\boldsymbol{\phi}^{\mathrm{T}}(n)][\tilde{\mathbf{H}}(n-1)\tilde{\mathbf{H}}^{\mathrm{T}}(n-1)]\} \\ &= \sigma_\epsilon^2 + \operatorname{tr}\{E[\boldsymbol{\phi}(n)\boldsymbol{\phi}^{\mathrm{T}}(n)]E[\tilde{\mathbf{H}}(n-1)\tilde{\mathbf{H}}^{\mathrm{T}}(n-1)]\} \\ &\simeq \sigma_\epsilon^2 + \operatorname{tr}\left\{\mathbf{R}\frac{\sigma_\epsilon^2}{n}\mathbf{R}^{-1}\right\} = \sigma_\epsilon^2 + \frac{N\sigma_\epsilon^2}{n} \end{aligned} \tag{3.49}$$

assuming $\mathbf{H}$ and $\boldsymbol{\phi}$ to be uncorrelated. In other words, the increase in the output variance due to the noisiness of the coefficients is given by

$$M = \frac{E\{e_{\hat{\mathbf{H}}(n-1)}^2(n)\} - \sigma_\epsilon^2}{\sigma_\epsilon^2} = \frac{N}{n} \tag{3.50}$$

In the signal processing literature this ratio is sometimes called the "misadjustment factor" of the adaptive filter. The result above was derived for the infinite memory case ($\lambda = 1$). In the finite memory case, (3.50) will be used with n replaced by the effective window length $\tau \simeq 1/(1-\lambda)$, and will be interpreted as a lower bound on the actual misadjustment factor.

3.2.7 Square-Root Implementation

A crucial part of the RLS algorithm is the computation of the gain vector $\mathbf{K}(n)$, which involves updating of the error covariance matrix $\mathbf{P}(n)$. The straightforward update of $\mathbf{P}(n)$ using the difference equation (3.11) or (3.23) can lead to numerical problems, especially when a finite-precision machine is used. The main source of trouble is the fact that this update formula does not guarantee the positive definiteness of the (inverse) covariance matrix $\mathbf{P}(n)$. This difficulty is circumvented by developing update equations for the square root $\mathbf{P}^{1/2}(n)$ of the covariance matrix, defined as any matrix with the property $(\mathbf{P}^{1/2}(n))(\mathbf{P}^{1/2}(n))^{\mathrm{T}} = \mathbf{P}(n)$.

Square-root implementations of least-squares estimation algorithm have been studied extensively in the literature [Bierman]. Table 3.2 summarizes a square-root algorithm for updating $\mathbf{P}(n)$ and the gain vector $\mathbf{K}(n)$. The matrices $\tilde{\mathbf{U}}$

TABLE 3.2 Square-Root Form of the RLS Algorithm

- Initialization:

$\tilde{\mathbf{U}} = \mathbf{I}, \qquad \tilde{\mathbf{D}} = \text{diag}\,[\sigma]$

- First step:

$\mathbf{f} = \tilde{\mathbf{U}}^{\mathrm{T}}\boldsymbol{\phi}(n)$, where $\mathbf{f}^{\mathrm{T}} = [f_1 \quad \cdots \quad f_N]$

$\mathbf{v} = \tilde{D}f$ (i.e., $v_i = \tilde{d}_i f_i$ for $i = 1, \ldots, N$)

$\tilde{d}_1 = \dfrac{\tilde{d}_1}{\alpha_1}$, where $\alpha_1 = \lambda_t + v_1 f_1$

$\mathbf{k}_1^{\mathrm{T}} = [v_1, 0, \ldots, 0]$

- Main loop:

for $j = 2, \ldots, N$ do

$\alpha_j = \alpha_{j-1} + v_j f_j$

$\hat{d}_j = \dfrac{\tilde{d}_j \alpha_{j-1}}{\alpha_j \lambda(n)}$

$\hat{u}_j = \tilde{u}_j + a_j \bar{k}_{j-1}$, where $a_j = \dfrac{-f_i}{\alpha_{j-1}}$

$\bar{k}_j = \bar{k}_{j-1} + v_j \tilde{u}_j$

- We use the notation u_i, u_i for the columns of U, U; that is,

$\tilde{\mathbf{U}} = [\tilde{u}_1 \quad \cdots \quad \tilde{u}_N], \hat{\mathbf{U}} = [\hat{u}_1 \quad \cdots \quad \hat{u}_N]$

- Compute the new gain:

$K(n) = \dfrac{\bar{k}_N}{\alpha_N}$

and $\hat{\mathbf{U}}$ are diagonal matrices, where

$$\mathbf{P}(n-1) = \tilde{\mathbf{U}}\tilde{\mathbf{D}}\tilde{\mathbf{U}}^{\mathrm{T}} = \text{old covariance matrix}$$

$$\mathbf{P}(n) = \hat{\mathbf{U}}\hat{\mathbf{D}}\hat{\mathbf{U}}^{\mathrm{T}} = \text{new covariance matrix} \tag{3.51}$$

The computational complexity of this square-root algorithm is comparable to that of the standard RLS. For a derivation and analysis of this type of algorithms see [Bierman, Lawson and Hanson].

3.2.8 Sliding Window Form of the RLS

The essential feature of the exponentially weighted RLS algorithm is that the estimate at time n is based on all the past data, with more recent data weighted more heavily than "older" data. In some adaptive applications it is desired to have the estimate $\hat{\mathbf{H}}(n)$ depend only on a finite number of past data. This can be achieved by a version of the covariance form introduced in (3.5), in which the estimate is a function of data over a (rectangular) window of length L: $\{y(n-L+1), \ldots, y(n)\}$. The derivation of a recursive algorithm for updating $\hat{\mathbf{H}}(n)$ follows the same procedure as in the prewindowed case and is left as an exercise to the reader.

The sliding window algorithm consists of two distinct steps. First, a new data point $y(n)$ is added. Then an old data point $y(n-L)$ is discarded, thus keeping the number of active points equal to L. The first step will be the same as in (3.11), (3.14), and (3.15) (with an obvious change in notation):

$$\mathbf{P}^1(n) = \mathbf{P}(n-1) - \mathbf{P}(n-1)\boldsymbol{\phi}(n)[1 + \boldsymbol{\phi}^T(n)\mathbf{P}(n-1)\boldsymbol{\phi}(n)]^{-1}\boldsymbol{\phi}^T(n)\mathbf{P}(n-1) \tag{3.52a}$$

$$\hat{\mathbf{H}}^1(n) = \hat{\mathbf{H}}(n-1) + \mathbf{K}(n)[y(n) - \boldsymbol{\phi}^T(n)\mathbf{P}(n-1)] \tag{3.52b}$$

$$\mathbf{K}(n) = \mathbf{P}(n-1)\boldsymbol{\phi}(n)[1 + \boldsymbol{\phi}^T(n)\mathbf{P}(n-1)\boldsymbol{\phi}(n)]^{-1} \tag{3.52c}$$

where $\mathbf{P}^1(n)$ and $\hat{\mathbf{H}}^1(n)$ denote quantities based on $(L+1)$ data points.

Next, we discard the data point $y(n-L)$:

$$\mathbf{P}(n) = \mathbf{P}^1(n) + \mathbf{P}^1(n)\boldsymbol{\phi}(n-L)[1 + \boldsymbol{\phi}^T(n-L)\mathbf{P}^1(n)\boldsymbol{\phi}(n-L)]^{-1}\boldsymbol{\phi}^T(n-L)\mathbf{P}^1(n) \tag{3.53a}$$

$$\hat{\mathbf{H}}(n) = \hat{\mathbf{H}}^1(n) - \mathbf{K}^1(n)[y(n-L) - \boldsymbol{\phi}^T(t-L)\hat{\mathbf{H}}^1(n)] \tag{3.53b}$$

$$\mathbf{K}^1(n) = \mathbf{P}^1(n)\boldsymbol{\phi}(n-L)[1 + \boldsymbol{\phi}^T(n-L)\mathbf{P}^1(n)\boldsymbol{\phi}(n-L)]^{-1} \tag{3.53c}$$

This algorithm completely discards old data points and is more effective than the exponentially weighted version in tracking sudden changes in the statistics of the data $y(n)$. Also, a "bad" data point [e.g., a very large value of $y(n)$ due to a noise impulse] will be completely forgotten after L time points, while in the exponentially weighted case its effect may linger on for a long time.

The foregoing algorithm requires that the last $L + N$ data points be stored. For a large window size this may represent a problem. In addition, the computational complexity of the algorithm is about double that of the one presented in Table 3.1. For these reasons the exponentially weighted form of the RLS is often preferred in practice.

3.3 LEAST-MEAN-SQUARES ADAPTIVE ALGORITHM

The least-mean-squares (LMS) algorithm developed by Widrow and Hoff [Widrow and Hoff 1960] is a predecessor of the RLS adaptive filter. The LMS algorithm has been used for many signal processing applications. Its simplicity and ease of implementation make this algorithm an attractive solution for many practical problems.

The main disadvantages of the LMS algorithm are related to its convergence properties. As will be shown in this section, this algorithm is based on a gradient search technique for minimizing a quadratic performance function, whereas the

RLS algorithm is a Newton–Raphson type of procedure. It is known from the theory of iterative optimization procedures that the Newton–Raphson technique will generally converge faster than a gradient search technique [Fletcher].

In this section we derive the LMS algorithm and study some of its properties.

3.3.1 Iterative Computation of the Optimal Coefficient Vector

To introduce the LMS algorithm we first rewrite the variance of the filter output in yet another form [see (3.46)]:

$$\begin{aligned} J(\mathbf{H}) &\triangleq E\{e_{\mathbf{H}}^2(n)\} = E\{[y(n) - \boldsymbol{\phi}^{\mathrm{T}}(n)\mathbf{H}]^2\} \\ &= E\{y^2(n)\} - 2E\{y(n)\boldsymbol{\phi}^{\mathrm{T}}(n)\}\mathbf{H} + \mathbf{H}^{\mathrm{T}}E\{\boldsymbol{\phi}(n)\boldsymbol{\phi}^{\mathrm{T}}(n)\}\mathbf{H} \\ &= r(0) - 2\mathbf{r}^{\mathrm{T}}\mathbf{H} + \mathbf{H}^{\mathrm{T}}\mathbf{R}\mathbf{H} \end{aligned} \tag{3.54}$$

Note that $J(\mathbf{H})$ is a quadratic function of the entries of the coefficient vector $\mathbf{H}$. The method of steepest descent is an optimizatioh technique that uses the gradients of this surface to seek its minimum. The gradient at any point on the surface is obtained by differentiating $J(\mathbf{H})$ with respect to the parameter vector $\mathbf{H}$:

$$\frac{\partial J(\mathbf{H})}{\partial \mathbf{H}} \triangleq \begin{bmatrix} \dfrac{\partial J(\mathbf{H})}{\partial h(1)} \\ \vdots \\ \dfrac{\partial J(\mathbf{H})}{\partial h(N)} \end{bmatrix} = 2(\mathbf{R}\mathbf{H} - \mathbf{r}) \tag{3.55}$$

Alternatively [see (3.46)],

$$\frac{\partial J(\mathbf{H})}{\partial \mathbf{H}} = 2\mathbf{R}(\mathbf{H} - \mathbf{H}_{\mathrm{opt}}) \tag{3.56}$$

Setting the gradient vector to zero defines the optimal vector $\mathbf{H}_{\mathrm{opt}}$, as in (3.32). The method of steepest descent is an iterative procedure for computing this optimal parameter vector. We start with an arbitrary vector $\mathbf{H}(0)$ and at each step change the current vector by an amount proportional to the negative of the gradient vector:

$$\mathbf{H}(k+1) = \mathbf{H}(k) + \mu\left[-\frac{\partial J(\mathbf{H})}{\partial \mathbf{H}}\right]_{\mathbf{H}=\mathbf{H}(k)} \tag{3.57}$$

The scalar parameter μ is a convergence factor that controls stability and rate of adaptation.

To study the properties of this difference equation, consider the error vector

$$\tilde{\mathbf{H}}(k) = \mathbf{H}(k) - \mathbf{H}_{\mathrm{opt}} \tag{3.58}$$

From (3.56)–(3.58) we obtain the following difference equation for this error vector:

$$\tilde{\mathbf{H}}(k+1) = (\mathbf{I} - 2\mu\mathbf{R})\tilde{\mathbf{H}}(k) \tag{3.59}$$

By choosing μ to be sufficiently small it is always possible to make the matrix $(\mathbf{I} - 2\mu\mathbf{R})$ stable, guaranteeing that the error will go asymptotically to zero.

To see more clearly the convergence behavior of the error vector, we transform (3.59) into a set of N decoupled scalar equations. To do this we first note that the covariance matrix $\mathbf{R}$, being symmetric and positive definitive, can be decomposed as

$$\mathbf{R} = \mathbf{U}\mathbf{D}\mathbf{U}^{\mathrm{T}} \tag{3.60}$$

where $\mathbf{U}$ is an orthonormal matrix $(\mathbf{U}^{\mathrm{T}}\mathbf{U} = \mathbf{U}\mathbf{U}^{\mathrm{T}} = \mathbf{I})$ and $\mathbf{D}$ is a diagonal matrix containing the eigenvalues of $\mathbf{R}$,

$$\mathbf{D} = \mathrm{diag}\{\lambda_1 \quad \cdots \quad \lambda_N\} \tag{3.61}$$

Inserting (3.60) into (3.59) and premultiplying by U^{T}, we get

$$\mathbf{U}^{\mathrm{T}}\tilde{\mathbf{H}}(k+1) = (I - 2\mu\mathbf{D})(\mathbf{U}^{\mathrm{T}}\tilde{\mathbf{H}}(k)) \tag{3.62}$$

Denoting the transformed error vector by

$$\mathbf{X}(k) = \begin{bmatrix} x_1(k) \\ \vdots \\ x_N(k) \end{bmatrix} = \mathbf{U}^{\mathrm{T}}\tilde{\mathbf{H}}(k) \tag{3.63}$$

we get

$$x^i(k+1) = (1 - 2\mu\lambda_i)x^i(k) \tag{3.64}$$

For convergence, it is necessary to have $|1 - 2\mu\lambda_i| < 1$, for all the eigenvalues λ_i. This condition will be certainly satisfied if

$$0 < \mu < \frac{1}{\lambda_{\max}} \tag{3.65}$$

where $\lambda_{\max}$ is the largest eigenvalue of $\mathbf{R}$. Note also that as long as $\mu < 1/2\lambda_{\max}$, the convergence rates of all the modes will increase as μ increases. Once $\mu = 1/2\lambda_{\max}$, however, the mode corresponding to $\lambda_{\max}$ will begin to slow down again. Thus the choice of μ in the vicinity of $1/2\lambda_{\max}$ will generally yield the best convergence rate for the difference equation (3.59).

Equation (3.64) also provides an indication of the rate at which the various modes of the error equation decay to zero. Defining the time constant of the ith mode by τ_i, we get (assuming that $\mu\lambda_i \ll 1$)

$$1 - 2\mu\lambda_i = e^{-1/\tau_i} = 1 - \frac{1}{\tau_i} \tag{3.66}$$

or

$$\tau_i \simeq \frac{1}{2\mu\lambda_i} \tag{3.67}$$

Thus the longest time constant involved in the error system is given by

$$\tau_{\max} = \frac{1}{2\mu\lambda_{\min}} \tag{3.68}$$

where $\lambda_{\min}$ is the smallest eigenvalue of the covariance matrix $\mathbf{R}$. From (3.65) and (3.68) we conclude that

$$\tau_{\max} > \frac{\lambda_{\max}}{2\lambda_{\min}} \tag{3.69}$$

In other words, the larger the eigenvalue spread of the covariance matrix, the longer it will take for the steepest descent method to converge.

3.3.2 LMS Algorithm

The iterative procedure described above required knowledge of the exact gradient of the cost function $J(\mathbf{H})$. In practice, the gradient will not be known a priori and needs to be estimated from the data. The gradient estimate used by the LMS is the gradient of a single squared-error sample:

$$\frac{\partial e_{\mathbf{H}}^2(n)}{\partial \mathbf{H}} = 2e_{\mathbf{H}}(n)\frac{\partial e_{\mathbf{H}}(n)}{\partial \mathbf{H}} = -2e_{\mathbf{H}}(n)\boldsymbol{\phi}(n) \tag{3.70}$$

The LMS algorithm can, therefore, be written as

$$\mathbf{H}(n) = \mathbf{H}(n-1) + 2\mu e_{\mathbf{H}(n-1)}(n)\boldsymbol{\phi}(n) \tag{3.71}$$

The summation of many terms of the form $e(n)\boldsymbol{\phi}(n)$ makes the crude single-sample gradient approach the true gradient, and the expected value of the coefficient vector $\mathbf{H}(n)$ will approach the optimal coefficient vector $\mathbf{H}_{\text{opt}}$. Note that

$$E\{-2e_{\mathbf{H}}(n)\boldsymbol{\phi}(n)\} = 2[E\{\boldsymbol{\phi}(n)\boldsymbol{\phi}^{\mathrm{T}}(n)\}\mathbf{H} - E\{y(n)\boldsymbol{\phi}(n)\}] = 2[\mathbf{RH} - \mathbf{r}] = \frac{\partial J(\mathbf{H})}{\partial \mathbf{H}} \tag{3.72}$$

Thus the expected value of the approximate gradient equals the true gradient. More precise proofs of convergence will be discussed shortly. Note the simplicity of the LMS update equation. Each filter coefficient is updated by the addition of a weighted error term:

$$h_i(n) = h_i(n-1) + 2\mu y(n-i)e(n) \tag{3.73}$$

The error $e(n)$ is common to all the coefficients, while the weighting $[2\mu y(n-i)]$ is proportional to the data currently stored in the ith filter section (see Figure 3.1). The computation of $[2\mu e(n)]$ requires $N + 1$ operations (multiplications and additions). Thus each step of the filter update will require a total of $(2N + 1)$ operations (compared to $2.5N^2 + 4N$ operations for the RLS).

3.3.3 Convergence of the LMS Algorithm

A key issue in the analysis of any stochastic algorithm is the question of convergence. The convergence properties of the LMS algorithm have been studied extensively, at various levels of mathematical rigor. We start by writing down the difference equation obeyed by the parameter error vector $\tilde{\mathbf{H}}(n) = \mathbf{H}(n) - \mathbf{H}_{\text{opt}}$.

Inserting (3.44) into (3.71) and subtracting $\mathbf{H}_{\text{opt}}$ from both sides of the equation, we get

$$\begin{aligned}\tilde{\mathbf{H}}(n) &= \tilde{\mathbf{H}}(n-1) + 2\mu\boldsymbol{\phi}(n)[\epsilon(n) - \boldsymbol{\phi}^{\mathrm{T}}(n)\tilde{\mathbf{H}}(n-1)] \\ &= [\mathbf{I} - 2\mu\boldsymbol{\phi}(n)\boldsymbol{\phi}^{\mathrm{T}}(n)]\tilde{\mathbf{H}}(n-1) + 2\mu\phi(n)\epsilon(n)\end{aligned} \tag{3.74}$$

Note that $\tilde{\mathbf{H}}(n)$ is now a random variable and not a deterministic function as was the case in (3.59). Let us denote the expected value of this random vector by $\bar{\mathbf{H}}(n)$.

Taking the expected value of (3.74) gives

$$\bar{\mathbf{H}}(n) = (\mathbf{I} - 2\mu\mathbf{R})\bar{\mathbf{H}}(n-1) \tag{3.75}$$

since $\epsilon(n)$ and $\boldsymbol{\phi}(n)$ are uncorrelated. Thus we get back (3.59), indicating that the mean of the error vector behaves exactly as if the true gradient vector were known.

Next we want to compute the covariance matrix of the error vector $\tilde{\mathbf{H}}(n)$. From (3.74) it follows that

$$\begin{aligned}\tilde{\mathbf{H}}(n)\tilde{\mathbf{H}}^{\mathrm{T}}(n) &= [\mathbf{I} - 2\mu\boldsymbol{\phi}(n)\boldsymbol{\phi}^{\mathrm{T}}(n)]\tilde{\mathbf{H}}(n-1)\tilde{\mathbf{H}}^{\mathrm{T}}(n-1)[\mathbf{I} - 2\boldsymbol{\phi}(n)\boldsymbol{\phi}^{\mathrm{T}}(n)] \\ &\quad + 4\mu^2\epsilon^2(n)\boldsymbol{\phi}(n)\boldsymbol{\phi}^{\mathrm{T}}(n) + 2\mu\epsilon(n)[\mathbf{I} - 2\boldsymbol{\phi}(n)\boldsymbol{\phi}^{\mathrm{T}}(n)]\tilde{\mathbf{H}}(n-1)\boldsymbol{\phi}^{\mathrm{T}}(n) \\ &\quad + 2\mu\epsilon(n)\boldsymbol{\phi}(n)\boldsymbol{\phi}^{\mathrm{T}}(n)\tilde{\mathbf{H}}^{\mathrm{T}}(n-1)[\mathbf{I} - 2\mu\boldsymbol{\phi}(n)\boldsymbol{\phi}^{\mathrm{T}}(n)]\end{aligned} \tag{3.76}$$

To see the evolution of the error covariance matrix we take the expected value of both sides of (3.76).

Note that $\epsilon(n)$ is uncorrelated with $\boldsymbol{\phi}(n)$ [since $\epsilon(n)$ is the optimal prediction error]. We also assume that $\boldsymbol{\phi}(n)$ and $\tilde{\mathbf{H}}(n)$ are uncorrelated. Since $E\{\epsilon(n)\} = 0$, the expected value of the last two terms in (3.76) will be zero. Thus

$$\begin{aligned}&E\{\tilde{\mathbf{H}}(n)\tilde{\mathbf{H}}^{\mathrm{T}}(n)\} \\ &\quad = E\{[\mathbf{I} - 2\mu\boldsymbol{\phi}(n)\boldsymbol{\phi}^{\mathrm{T}}(n)]\tilde{\mathbf{H}}(n-1)\tilde{\mathbf{H}}^{\mathrm{T}}(n-1)[\mathbf{I} - 2\mu\boldsymbol{\phi}(n)\boldsymbol{\phi}(n)]\} + 4\mu^2\sigma_\epsilon^2\mathbf{R}\end{aligned} \tag{3.77}$$

In earlier work on the convergence of the LMS [Widrow et al. 1975(2) and 1976(1)] it was assumed that (3.77) can be replaced by

$$\begin{aligned}&E\{\tilde{\mathbf{H}}(n)\tilde{\mathbf{H}}^{\mathrm{T}}(n)\} \\ &\quad = [\mathbf{I} - 2\mu\mathbf{R}]E\{\tilde{\mathbf{H}}(n-1)\tilde{\mathbf{H}}^{\mathrm{T}}(n-1)\}[\mathbf{I} - 2\mu\mathbf{R}] + 4\mu^2\sigma_\epsilon^2\mathbf{R}\end{aligned} \tag{3.78}$$

A more precise evaluation of (3.77) was presented recently in [Horowitz and Senne], as will be discussed shortly. But first we derive the asymptotic properties of the LMS algorithm which follow from (3.78), as given in [Widrow et al. 1975(2) and 1976(1)]. To do this note that the error covariance matrix in (3.78) is the same as if $\mathbf{H}(n)$ were generated by the stochastic difference equation

$$\tilde{\mathbf{H}}(n) = [\mathbf{I} - 2\mu\mathbf{R}]\tilde{\mathbf{H}}(n-1) + 2\mu\epsilon(n)\boldsymbol{\phi}(n) \tag{3.79}$$

which is simply the "noisy" version of (3.75).

Next we premultiply (3.79) by $\mathbf{U}^{\mathrm{T}}$ [see (3.60)] to get

$$[\mathbf{U}^{\mathrm{T}}\tilde{\mathbf{H}}(n)] = [\mathbf{I} - 2\mu\mathbf{D}][\mathbf{U}^{\mathrm{T}}\tilde{\mathbf{H}}(n-1)] + 2\mu\epsilon(n)[\mathbf{U}^{\mathrm{T}}\boldsymbol{\phi}(n)] \tag{3.80}$$

Since the elements of $\mathbf{U}^{\mathrm{T}}\boldsymbol{\phi}_t$ are mutually uncorrelated and the matrix $(\mathbf{I} - 2\mathbf{D})$ is diagonal, it follows that the elements of $[\mathbf{U}^{\mathrm{T}}\tilde{\mathbf{H}}(n)]$ are mutually uncorrelated.

By pre- and postmultiplying (3.78) by $\mathbf{U}^{\mathrm{T}}$ and $\mathbf{U}$, respectively, or directly from (3.79), we get

$$E\{[\mathbf{U}^{\mathrm{T}}\tilde{\mathbf{H}}(n)][\mathbf{U}^{\mathrm{T}}\tilde{\mathbf{H}}(n)]^{\mathrm{T}}\} = [\mathbf{I} - 2\mu\mathbf{D}]E\{[\mathbf{U}^{\mathrm{T}}\tilde{\mathbf{H}}(n-1)][\mathbf{U}^{\mathrm{T}}\tilde{\mathbf{H}}(n-1)]^{\mathrm{T}}\}[\mathbf{I} - 2\mu\mathbf{D}] + 4\mu^2\sigma_\epsilon^2\mathbf{D} \tag{3.81}$$

Asymptotically, the statistics of $[\mathbf{U}^{\mathrm{T}}\tilde{\mathbf{H}}(n)]$ and $[\mathbf{U}^{\mathrm{T}}\tilde{\mathbf{H}}(n-1)]$ are the same, and therefore we can rearrange (3.81) to give

$$(\mathbf{I} - \mu\mathbf{D})E\{[\mathbf{U}^{\mathrm{T}}\tilde{\mathbf{H}}(n)][\mathbf{U}^{\mathrm{T}}\mathbf{H}(n)]^{\mathrm{T}}\} = \mu\sigma_\epsilon^2\mathbf{I} \tag{3.82}$$

For a convergent algorithm it is reasonable to assume that

$$\mu\mathbf{D} \ll \mathbf{I} \tag{3.83}$$

and thus we finally get

$$E\{[\mathbf{U}^{\mathrm{T}}\tilde{\mathbf{H}}(n)][\mathbf{U}^{\mathrm{T}}\tilde{\mathbf{H}}(n)]^{\mathrm{T}}\} \simeq \mu\sigma_\epsilon^2\mathbf{I} \tag{3.84a}$$

or

$$E\{\tilde{\mathbf{H}}(n)\tilde{\mathbf{H}}^{\mathrm{T}}(n)\} \simeq \mu\sigma_\epsilon^2\mathbf{I} \tag{3.84b}$$

Equation (3.84) provides a measure of the asymptotic noisiness of the filter coefficients in the LMS algorithm. We will use this equation next to compute the asymptotic variance of the filter output $e(n)$.

3.3.4 Learning Curve

As mentioned in Section 3.1, the variance of the filter output is often used as a measure of its performance. Consider the behavior of the output variance associated with the iterative computation of the coefficient vector. From (3.46), (3.58), (3.60), and (3.63) it immediately follows that

$$\begin{aligned} J[\mathbf{H}(k)] &= \sigma_\epsilon^2 + \tilde{\mathbf{H}}^{\mathrm{T}}(k)\mathbf{R}\tilde{\mathbf{H}}(k) = \sigma_\epsilon^2 + \mathbf{X}^{\mathrm{T}}(k)\mathbf{D}\mathbf{X}(k) \\ &= \sigma_\epsilon^2 + \sum_{i=1}^{N} \lambda_i[x^i(k)]^2 \end{aligned} \tag{3.85}$$

The elements of $\mathbf{X}(k)$ were shown to approach zero at exponential rates, with time constants given by (3.68). Since $J[\mathbf{H}(k)]$ involves a sum of the squares elements of $\mathbf{X}(k)$, it will decay at an exponential rate that is twice the rate of decay $\mathbf{X}(k)$. In other words, the time constants associated with $J[\mathbf{H}(k)]$ will be half of those associated with $\mathbf{X}(k)$. Thus, the time constant associated with the ith mode is

$$\tau_i^J = \frac{1}{2}\tau_i = \frac{1}{4\mu\lambda_i} \tag{3.86}$$

The plot of $J[\mathbf{H}(k)]$ versus the number of iterations is called in the literature the "learning curve." The plot starts at

$$J[\mathbf{H}(0)] = \sigma_\epsilon^2 + \tilde{\mathbf{H}}^{\mathrm{T}}(0)\mathbf{R}\tilde{\mathbf{H}}(0) = \sigma_\epsilon^2 + \mathbf{X}^{\mathrm{T}}(0)\mathbf{D}\mathbf{X}(0) \tag{3.87}$$

and decays toward its final value

$$J[\mathbf{H}(\infty)] = \sigma_\epsilon^2 \tag{3.88}$$

In the LMS algorithm the actual output variance will be larger than predicted by (3.85), due to the noisiness of the coefficient error vector $\tilde{\mathbf{H}}(n)$. From (3.49) and the fact that $\tilde{\mathbf{H}}(n-1)$ and $\boldsymbol{\phi}(n)$ are uncorrelated, it follows that

$$E\{e^2_{\mathbf{H}(n-1)}(n)\} = \sigma^2_\epsilon + \text{tr}\,[\mathbf{R}E\{\tilde{\mathbf{H}}(n-1)\tilde{\mathbf{H}}^{\mathrm{T}}(n-1)\}] \tag{3.89}$$

Inserting their asymptotic error covariance from (3.84) into (3.89), we get

$$E\{e^2_{\mathbf{H}(n-1)}(n)\} = \sigma^2_\epsilon + \mu\sigma^2_\epsilon\,\text{tr}\,\{\mathbf{R}\} = \sigma^2_\epsilon + \mu \sum_{i=1}^{N} \lambda_i \tag{3.90}$$

Using the terminology introduced in the adaptive filtering literature, we define the misadjustment factor M,

$$M = \frac{E\{e^2_{\mathbf{H}(n-1)}(n)\} - \sigma^2_\epsilon}{\sigma^2_\epsilon} = \mu \sum_{i=1}^{N} \lambda_i = \mu\,\text{tr}\,\{\mathbf{R}\} \tag{3.91}$$

This formula works well for small values of misadjustment ($M \leq 0.25$) since it requires that $\mathbf{H}(n)$ be sufficiently close to the optimal value $\mathbf{H}_{\text{opt}}$. For stationary input data it is always possible to achieve a small misadjustment by choosing a sufficiently small value for μ.

Recall that the misadjustment factor of the RLS algorithm is given by $N/\tau \simeq (1-\lambda)N$ [see (3.50)]. The misadjustment factor of the LMS algorithm is given by $\mu N\,\text{Var}\{y(n)\}$ [see (3.91)]. The two algorithms can be made to operate with the same misadjustment factor by choosing the parameters μ and λ so that $(1-\lambda) = \mu\,\text{Var}\{y(n)\}$. If the input data are normalized to have unit variance, we have $(1-\lambda) = \mu$. This equation is useful in making performance comparisons of these two algorithms.

For design purposes it is useful to express the misadjustment factor in terms of the filter order and the speed of adaptation. Note, for example, that

$$M = \mu \sum_{i=1}^{N} \lambda_i = \mu N \lambda_{\text{avg}} \tag{3.92}$$

where λ_{avg} is the average of the eigenvalues of the covariance matrix. From (3.86) it follows that

$$\lambda_i = \frac{1}{4\mu\tau_i^J} \tag{3.93}$$

Defining an average time constant

$$\frac{1}{\tau^J_{\text{avg}}} = \frac{1}{N}\sum_{i=1}^{N} \frac{1}{\tau_i^J} \tag{3.94}$$

we get

$$\lambda_{\text{avg}} = \frac{1}{4\mu\tau^J_{\text{avg}}} \tag{3.95}$$

Thus

$$M = \frac{N}{4\tau^J_{\text{avg}}} \tag{3.96}$$

This equation relates the misadjustment factor to the average settling time of the learning curve and the number of filter coefficients.

Another design constraint that needs to be taken into account is related to the stability of the difference equation (3.80) or the equivalent deterministic equation (3.62). To ensure the stability of the matrix $(\mathbf{I} - 2\mu\mathbf{D})$ it is necessary that $0 < \mu < 1/\lambda_{max}$. To avoid computation of the eigenvalues of the sample covariance matrix, the following rule of thumb is often used in practice:

$$\frac{1}{\text{tr}\,\mathbf{R}} > \mu > 0 \tag{3.97}$$

This condition follows from the fact that for positive-definite matrices

$$\text{tr}\,\mathbf{R} = \sum_{i=1}^{N} \lambda_i > \lambda_{max} \tag{3.98}$$

and therefore $\mu < 1/\text{tr}\,\mathbf{R}$ guarantees that $\mu < 1/\lambda_{max}$. The advantage of using (3.97) is that $\text{tr}\,\mathbf{R} = Nr(0) = N\,\text{Var}\{y(n)\}$ can be easily computed (with $\text{Var}\{y(n)\}$ estimated from the data).

As in the deterministic case [see (3.69)] the convergence rate is inherently limited by the eigenvalue spread of the data covariance matrix.

3.3.5 Recent Convergence Results

In the past few years there has been renewed activity in the analysis of the LMS convergence properties. An exact evaluation of the evolution equation (3.77) for the error covariance matrix was presented in [Horowitz and Senne] and later in [Weinstein 1983]. It was found that the gain μ must be restricted to an interval significantly smaller than the domain commonly stated in the literature. It was shown that stability is ensured if and only if $0 \leq \mu \leq 1/3\lambda_{max}$ and

$$f(\mu) \triangleq \sum_{i=1}^{N} \frac{\mu\lambda_i}{1 - 2\mu\lambda_i} < 1 \tag{3.99}$$

The misadjustment factor was shown to be

$$M(\mu) = \frac{f(\mu)}{1 - f(\mu)} \tag{3.100}$$

When $\sum_{i=1}^{n} \mu\lambda_i \ll 1$, we have

$$f(\mu) \simeq \sum_{i=1}^{n} \mu\lambda_i \ll 1 \tag{3.101}$$

and

$$M(\mu) \simeq \mu\,\text{tr}\,\{\mathbf{R}\} \tag{3.102}$$

which is the same as the misadjustment factor derived earlier (3.92). Note that in this case (3.99) holds and therefore the stability region is $0 < \mu < 1/3\lambda_{max}$, which differs from the result stated earlier.

3.3.6 LMS Algorithm as a Stochastic Approximation Method

Another way of looking at the LMS adaptive algorithm is to consider it as a special case of the stochastic approximation method. Stochastic approximation is the name given to a general class of recursive algorithms for solving equations of the form

$$g(\mathbf{H}) = E\{G(y, \mathbf{H})\} = 0 \tag{3.103}$$

where G is a known function of the data y and $\mathbf{H}$ is an unknown parameter vector. The original stochastic approximation algorithm developed by [Robbins and Monroe] uses the following update formula to solve for $\mathbf{H}$ given observation of the time series $y(n)$:

$$\hat{\mathbf{H}}(n+1) = \hat{\mathbf{H}}(n) + \gamma(n) G[y(n+1), \hat{\mathbf{H}}(n)] \tag{3.104}$$

where $\{\gamma(n)\}$ is a suitably chosen scalar sequence. It can be shown [Blum] that the algorithm above converges to the true parameter value $\mathbf{H}$, provided that certain conditions are satisfied by $\{\gamma(n)\}$, G, and the data $y(n)$. The usual conditions on $\{\gamma(n)\}$ are

$$\sum_{n=1}^{\infty} \gamma(n) = \infty \tag{3.105}$$

$$\sum_{n=1}^{\infty} \gamma^p(n) < \infty \qquad \text{for some } p > 1 \tag{3.106}$$

A common choice for the sequence $\{\gamma(n)\}$ is $\gamma(n) = 1/n$. This choice will result in almost sure convergence of $\hat{\mathbf{H}}(n)$ to the true value of $\mathbf{H}$ [Ljung 1977].

In the adaptive filtering problem considered here we have

$$g(\mathbf{H}) = \frac{\partial J(\mathbf{H})}{\partial \mathbf{H}} = E\{2\phi(n)e(n)\} \tag{3.107}$$

in which case (3.104) gives

$$\hat{\mathbf{H}}(n+1) = \hat{\mathbf{H}}(n) + 2\gamma(n)\phi(n)e(n) \tag{3.108}$$

In many applications it is desirable to prevent $\gamma(n)$ from becoming too small to enable the adaptive algorithm to respond to changes in the statistics of the input data $y(n)$. A simple way of achieving this is to set $\gamma(n) = \mu$ where μ is some small positive number. The filter coefficients will not converge in this case but keep fluctuating around some value. This situation corresponds to the finite memory case of the RLS algorithm, while the choice $\gamma(n) = 1/n$ corresponds to the infinite memory case.

Some examples. To illustrate the behavior of the RLS and LMS algorithms, we performed the following experiment. Stationary data were generated by passing a white noise sequence through a second-order all-pole filter $1/A(z)$ with parameters:

$$\text{Case 1:}\quad A(z) = 1 - 1.6z^{-1} + 0.95z^{-2}$$

$$\text{Case 2:}\quad A(z) = 1 - 1.8z^{-1} + 0.95z^{-2}$$

The output data were normalized to have unit variance. The RLS algorithm was initialized with $\mathbf{H}(0) = 0, \mathbf{P}(0) = 100\mathbf{I}$, and the forgetting factor was chosen so that $1 - \lambda = \mu$. The squared-error sequence generated by each algorithm was averaged over 200 independent simulation runs, and plotted as function of time (see Figures 3.3 and 3.5). Each figure depicts the learning curves for both the RLS and LMS algorithms.

Figures 3.3 and 3.4 depict the behavior of the algorithms for case 1, where $\lambda_{max}/\lambda_{min} = 10$, for two different choices of the convergence parameter μ (and the forgetting factor λ). Figure 3.3 depicts the fastest possible convergence of the LMS algorithm. The parameter μ was increased almost to the point of instability. Choosing a smaller value for μ slows down the convergence rate of the LMS algorithm, as depicted in Figure 3.4. Figure 3.5 depicts the convergence of the algorithms for case 2.

Examination of these figures leads to the following observations:

1. The initial convergence of the RLS is considerably faster than the convergence of the LMS.
2. The convergence of the LMS is clearly affected by the eigenvalue ratio (case 1 versus case 2), whereas the RLS is insensitive to it.
3. The convergence rate of the LMS is strongly affected by the choice of μ. The forgetting factor λ has little effect on the initial convergence rate of the RLS. Of course, λ determines the tracking behavior of the algorithm (after initial convergence) and the noisiness of the filter coefficients.

Figures 3.6 and 3.7 depict the trajectory of the filter coefficient in a single run of the algorithm. The parameters of the RLS algorithm are shown to converge very quickly to their true values, while the LMS parameters take a much longer time to converge.

It should be noted that after the initial convergence phase, the tracking behavior of the RLS algorithm will be dominated by the effective window length $\tau \simeq 1/(1 - \lambda)$. Its response to sudden changes in the process parameters will generally be slower than during the initial phase, since the update gain $\mathbf{K}(n)$ will be smaller than the initial gain $\mathbf{K}(0)$. The LMS algorithm, on the other hand, operates with a fixed update gain and its behavior in tracking mode is essentially the same as during initial convergence.

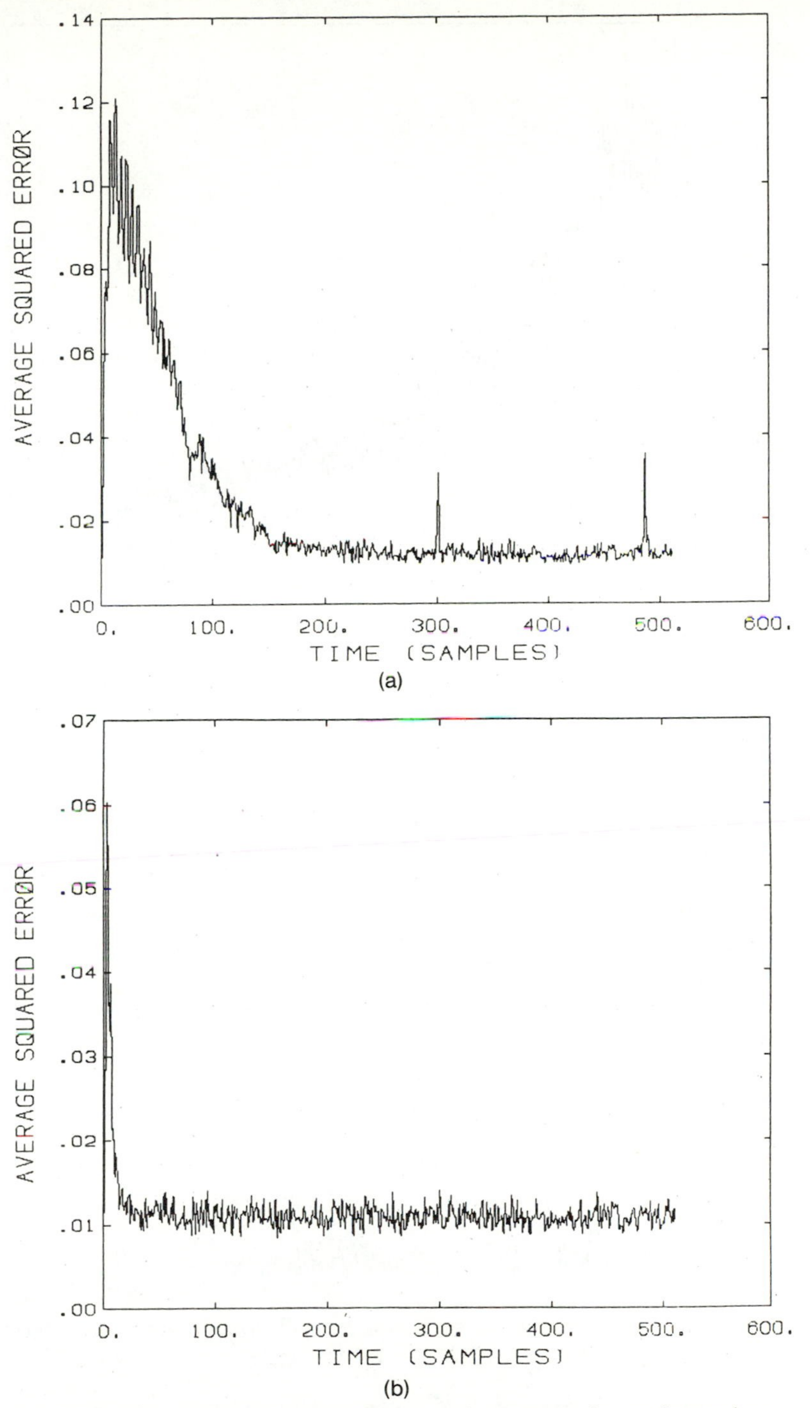

Figure 3.3 Learning curves for (a) a LMS and (b) a RLS adaptive predictor using a two-pole autoregressive input signal generation model (specified as case 1 in the text) with convergence coefficient $\mu = 1 - \lambda = 0.05$.

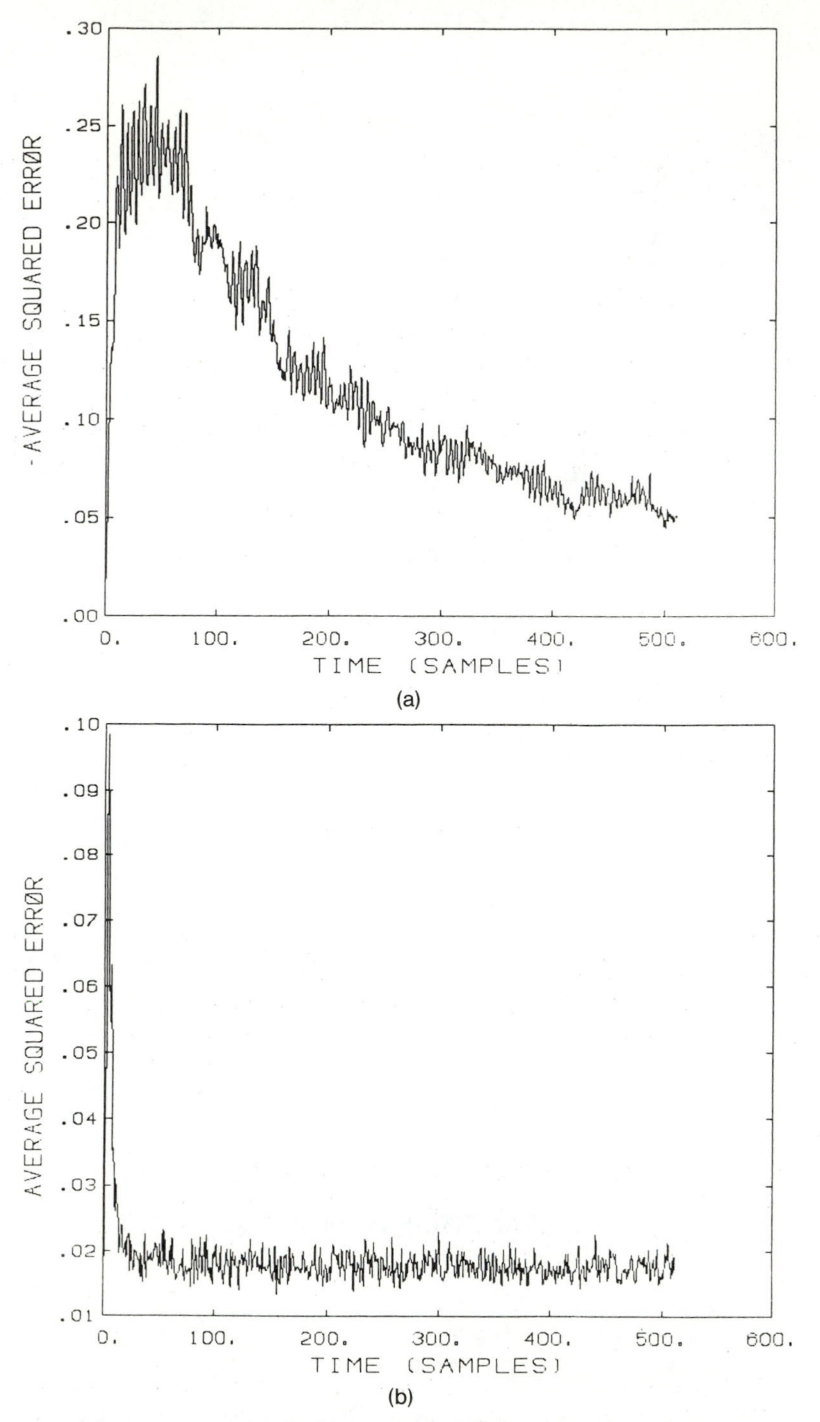

Figure 3.4 Learning curves for (a) a LMS and (b) a RLS adaptive predictor using a two-pole autoregressive input signal generation model (specified as case 1 in the text) with convergence coefficient $\mu = 1 - \lambda = 0.005$.

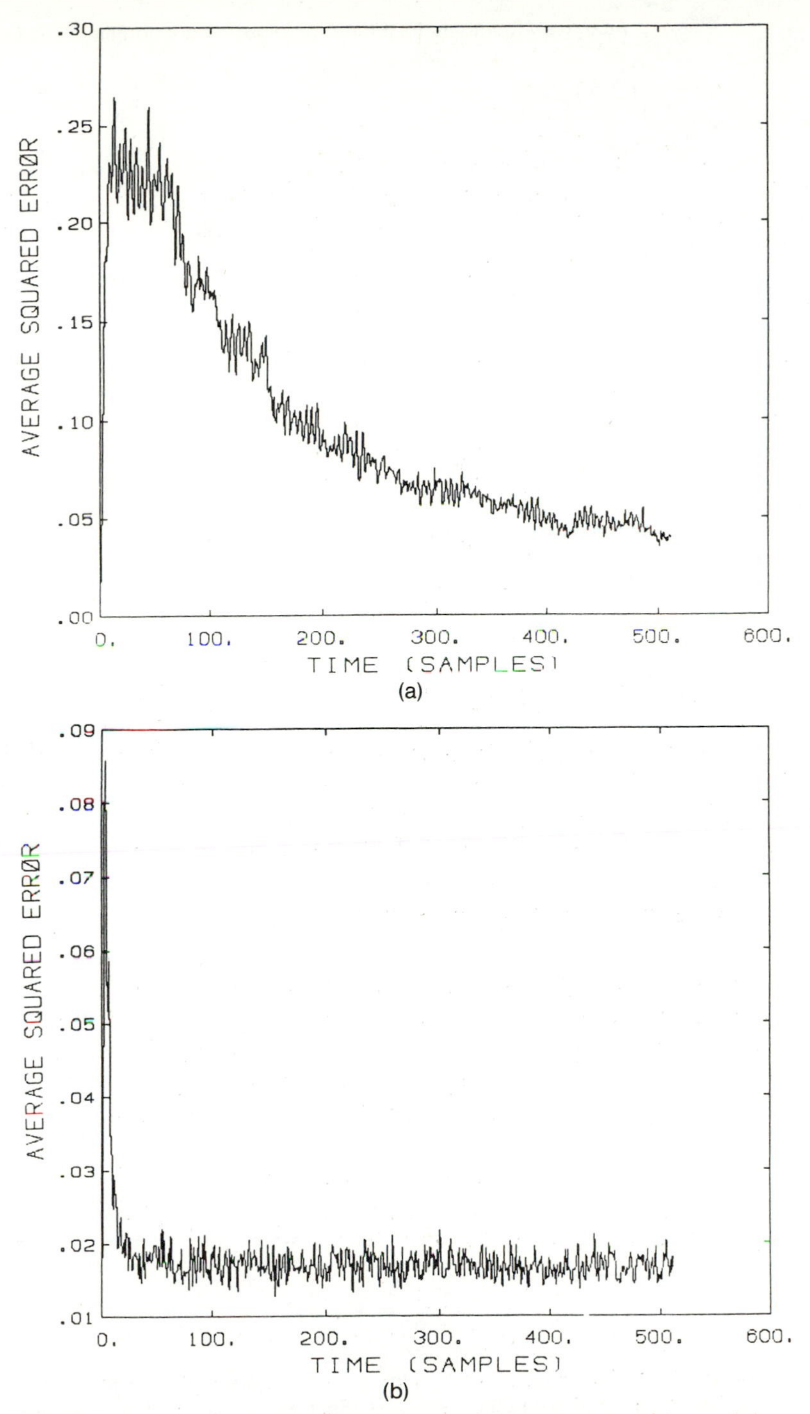

Figure 3.5 Learning curves for (a) a LMS and (b) a RLS adaptive predictor using a two-pole autoregressive input signal generation model (specified as case 2 in the text) with convergence coefficient $\mu = 1 - \lambda = 0.05$.

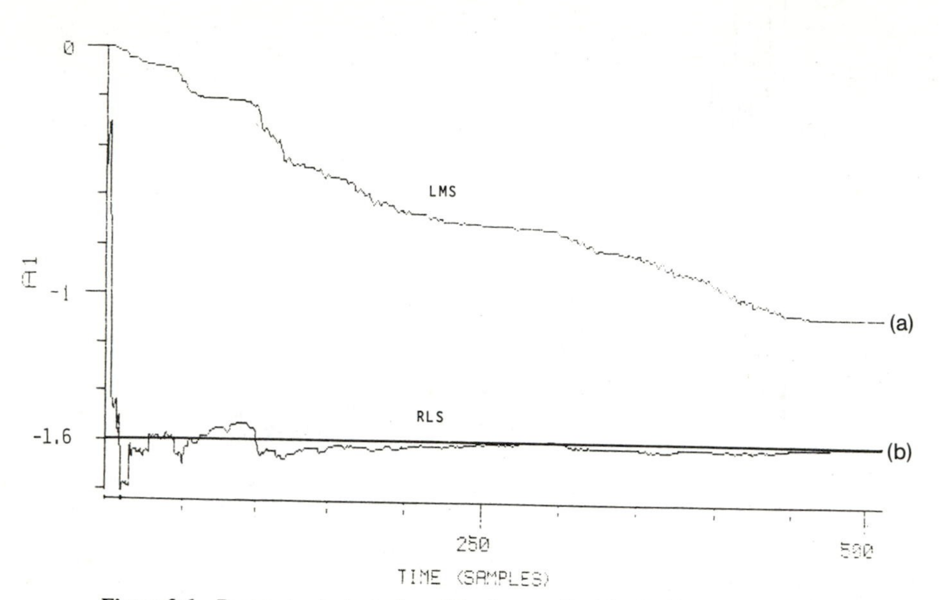

Figure 3.6 Parameter trajectories of the first tap for (a) a LMS and (b) a RLS adaptive predictor using a two-pole autoregressive input signal generation model (specified as case 1 in the text) with convergence coefficient $\mu = 1 - \lambda = 0.005$.

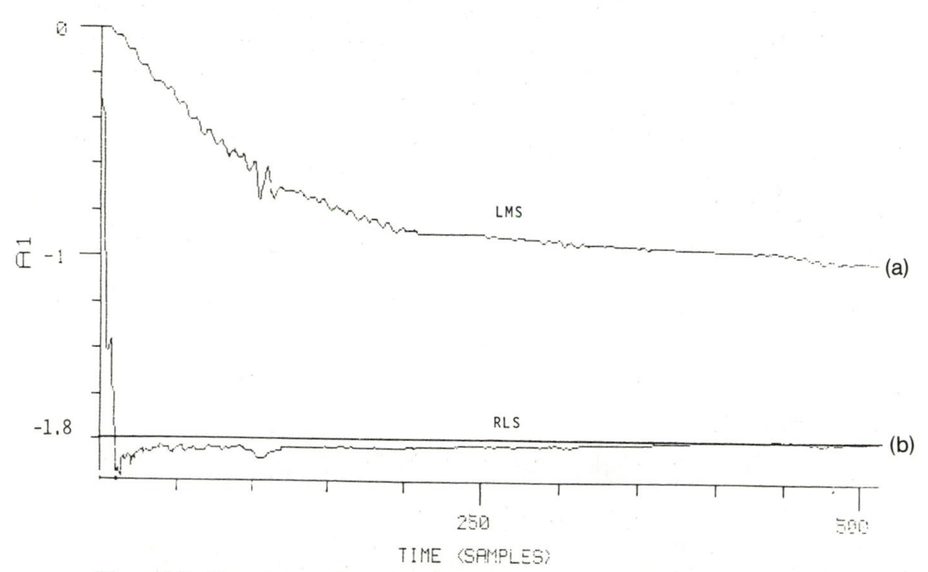

Figure 3.7 Parameter trajectories of the first tap for (a) a LMS and (b) a RLS adaptive predictor using a two-pole autoregressive input signal generation model (specified as case 2 in the text) with convergence coefficient $\mu = 1 - \lambda = 0.005$.

3.4 ADAPTIVE FINITE IMPULSE RESPONSE FILTERS WITH LINEAR-PHASE CHARACTERISTICS

An important class of filters commonly used in digital signal processing is the class of finite impulse response (FIR) linear-phase filters. Such filters are important for applications where frequency dispersion due to nonlinear phase is harmful (e.g., speech processing and data transmission). Signals in the passband of linear-phase filters are reproduced exactly at the filter output, except for a delay corresponding to the slope of the phase versus frequency plot. An extensive literature exists on the design of linear-phase filters with a desired frequency response, and various efficient design algorithms have been developed [Oppenheim and Schafer, Rabiner and Gold 1975]. Most of this literature deals, however, with nonadaptive processing, that is, designing the filter to have a known frequency response (or impulse response). Relatively little work seems to have been done on adaptive linear-phase filters, that is, filters whose characteristics are determined on the basis of an observed time series and not on a priori specifications.

An FIR linear-phase filter is characterized by a symmetric impulse response. Consider the general FIR filter specified in (3.1). The filter will be linear phase if

$$h_i = h_{M-i}, \qquad i = 0, \ldots, N \tag{3.109}$$

where

$$h_0 = -h_M = 1 \tag{3.110}$$

In the following we will assume for simplicity that $M = 2N + 1$ (i.e., M is an odd number). The filter output can then be written as

$$\begin{aligned} e(n) = {} & [y(n) + y(n-2N-1)] \\ & - h_1[y(n-1) + y(n-2N)] - \cdots - h_N[y(n-N) + y(n-N-1)] \end{aligned} \tag{3.111}$$

The structure of the filter is depicted in Figure 3.8. As before, we define the coefficient vector $\mathbf{H} = [h_1 \ \cdots \ h_N]^{\mathrm{T}}$ and denote the corresponding output by $e(n)$. The optimal (least-squares) coefficient vector will be denoted by $\mathbf{H}_{\text{opt}} = [a_1, \ldots, a_n]^{\mathrm{T}}$ and the corresponding output by $\epsilon(n) = e_{\mathbf{H}_{\text{opt}}}(n)$. The objective is to design an adaptive filter that will minimize the output sum of squares, or the output variance.

We will develop this filter in three steps. First, we derive the equations for the optimal filter coefficients for the case where the data covariance matrix is known. Then we discuss the case where only the input data are known, but not the statistics. Finally, we present recursive update equations for the filter coefficients for both the RLS and LMS versions.

3.4.1 Stochastic Case

Recall again the orthogonality property of the least-squares estimator, which states that $E\{\epsilon(n)y(s)\} = 0$ for all $s < n$. Postmultiplying (3.111) by $y(s)$ and taking

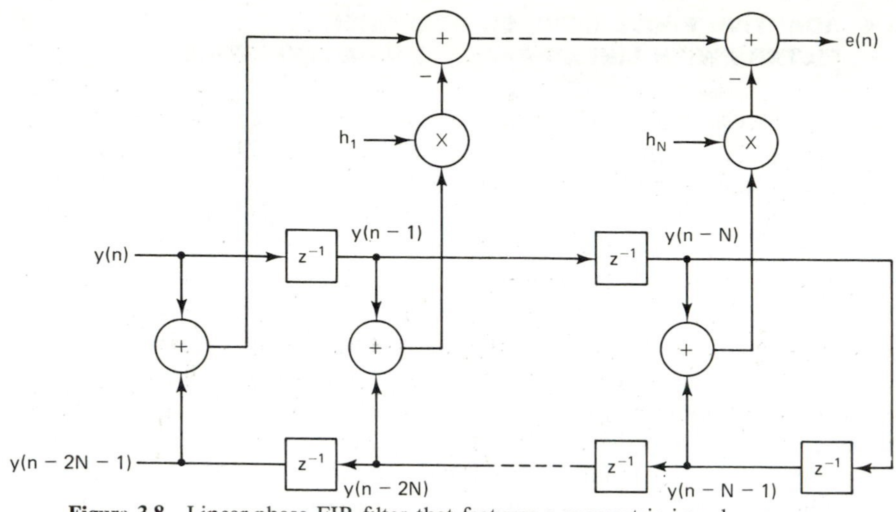

Figure 3.8 Linear-phase FIR filter that features a symmetric impulse response.

expected values gives

$$0 = r(n-s) + r(n-s-2N-1) - a_1[r(n-s-1) + r(t-s-2N)] \\ \cdots - a_n[r(n-s-N) + r(n-s-N-1)] \qquad (3.112)$$

This equation can be rewritten in matrix form (for $n - s = 1, \ldots, N$) as

$$\left\{ \underbrace{\begin{bmatrix} r(0) & r(1) & \cdots & r(N-1) \\ r(1) & \ddots & & \vdots \\ \vdots & & \ddots & r(1) \\ r(N-1) & & r(1) & r(0) \end{bmatrix}}_{\mathbf{R}} + \underbrace{\begin{bmatrix} r(2N-1) & \cdots & r(N+1) & r(N) \\ \vdots & & & \vdots \\ r(N+1) & & & \vdots \\ r(N) & r(N-1) & \cdots & r(1) \end{bmatrix}}_{\tilde{\mathbf{R}}} \right\} \underbrace{\begin{bmatrix} a_1 \\ \vdots \\ \vdots \\ a_N \end{bmatrix}}_{\mathbf{H}}$$

$$= \underbrace{\begin{bmatrix} r(1) \\ \vdots \\ \vdots \\ r(N) \end{bmatrix}}_{\mathbf{r}} + \underbrace{\begin{bmatrix} r(2N) \\ \vdots \\ \vdots \\ r(N+1) \end{bmatrix}}_{\tilde{\mathbf{r}}} \qquad (3.113)$$

Thus

$$\mathbf{H}_{\text{opt}} = [\mathbf{R} + \tilde{\mathbf{R}}]^{-1}[\mathbf{r} + \tilde{\mathbf{r}}] \qquad (3.114)$$

Another way of deriving this result is to define a new set of variables:

$$z(n, i) = y(n-i) + y(n-2N-1+i) \qquad (3.115)$$

Using this notation we can write

$$e(n) = z(n) - h_1 z(n, 1) - \cdots h_N z(n, N) = z(n) - \bar{\boldsymbol{\phi}}^{\mathrm{T}}(n) \qquad (3.116)$$

where

$$\bar{\phi}(n) = [z(n,1) \quad \cdots \quad z(n,N)]^{\mathrm{T}} \tag{3.117}$$

By comparison to the least-squares problem treated in Section 3.2 it is clear that

$$\mathbf{H}_{\mathrm{opt}} = E\{\bar{\phi}(n)\bar{\phi}^{\mathrm{T}}(n)\}^{-1}E\{z(n)\bar{\phi}(n)\} \tag{3.118}$$

It is straightforward to check that

$$E\{\bar{\phi}(n)\bar{\phi}^{\mathrm{T}}(n)\} = 2[\mathbf{R} + \tilde{\mathbf{R}}] \tag{3.119}$$

$$E\{z(n)\bar{\phi}(n)\} = 2[\mathbf{r} + \tilde{\mathbf{r}}] \tag{3.120}$$

from which (3.114) follows. Transforming this constrained filtering problem to a standard (unconstrained) filtering problem for the process $z(n)$ is very useful in deriving the adaptive algorithm. Note that the covariance matrix $E\{\bar{\phi}(n)\bar{\phi}^{\mathrm{T}}(n)\}$ is nonstationary, even though the original process $y(n)$ was stationary.

The output variance of the optimal linear-phase filter can be computed as in (3.47):

$$\begin{aligned}
\sigma_\epsilon^2 &= E\{\epsilon^2(n)\} \\
&= E\{z^2(n)\} + E\{\mathbf{H}_{\mathrm{opt}}^{\mathrm{T}}\bar{\phi}(n)\bar{\phi}^{\mathrm{T}}(n)\mathbf{H}_{\mathrm{opt}} - 2E\{\mathbf{H}_{\mathrm{opt}}^{\mathrm{T}}\bar{\phi}(n)z(n)\} \\
&= 2[r(0) + r(2N+1)] + 2\mathbf{H}_{\mathrm{opt}}[\mathbf{R} + \tilde{\mathbf{R}}]\mathbf{H}_{\mathrm{opt}} - 4\mathbf{H}_{\mathrm{opt}}^{\mathrm{T}}[\mathbf{r} + \tilde{\mathbf{r}}] \\
&= 2[[r(0) + r(2N+1)] - (\mathbf{r} + \tilde{\mathbf{r}})^{\mathrm{T}}[\mathbf{R} + \tilde{\mathbf{R}}]^{-1}(\mathbf{r} + \tilde{\mathbf{r}})]
\end{aligned} \tag{3.121}$$

The output of the linear-phase filter with an arbitrary set of coefficients is given by [see (3.46)]

$$E\{e^2(n)\} = \sigma_\epsilon^2 + (\mathbf{H} - \mathbf{H}_{\mathrm{opt}})^{\mathrm{T}}[\mathbf{R} + \tilde{\mathbf{R}}]^{-1}(\mathbf{H} - \mathbf{H}_{\mathrm{opt}}) \tag{3.122}$$

which follows from the observation [see (3.44)] that

$$e(n) = z(n) - \bar{\phi}^{\mathrm{T}}(n)\mathbf{H} = \underbrace{z(n) - \bar{\phi}^{\mathrm{T}}(n)\mathbf{H}_{\mathrm{opt}}}_{\epsilon(n)} + \bar{\phi}^{\mathrm{T}}(n)[\mathbf{H}_{\mathrm{opt}} - \mathbf{H}] \tag{3.123}$$

and the fact that $\epsilon(n)$ is uncorrelated with past data [e.g., $E\{\epsilon(n)\bar{\phi}(n)\} = 0$]. Equation (3.122) again verifies the fact that the output variance will be minimized for $\mathbf{H} = \mathbf{H}_{\mathrm{opt}}$. Note that $\mathbf{R}^z = \mathbf{R} + \tilde{\mathbf{R}}$ is a symmetric positive-definite matrix. Its positive definiteness follows from that fact that $\mathbf{R}^z$ is the covariance matrix of $\bar{\phi}(n)$.

3.4.2 RLS Algorithm

Next we consider the case where the covariance matrix is not known. Here we want to compute a set of filter coefficients that will minimize the output sum of squares. Following the same step as in Section 3.2 we get

$$\underbrace{\begin{bmatrix} e(0) \\ \vdots \\ e(n) \end{bmatrix}}_{\mathbf{e}(n)} = \underbrace{\begin{bmatrix} z(0) \\ \vdots \\ z(n) \end{bmatrix}}_{\mathbf{z}(n)} - \underbrace{\begin{bmatrix} 0 & 0 & \cdots & \cdot \\ z(0) & \ddots & & \\ \vdots & & \ddots & z(0) \\ z(n-1) & \cdots & & z(n-N) \end{bmatrix}}_{\mathbf{Z}(n)} \begin{bmatrix} h_1 \\ \vdots \\ h_N \end{bmatrix} \tag{3.124}$$

The coefficient vector that minimizes the error norm $\mathbf{e}^T(n)\mathbf{e}(n)$ will be denoted by $\hat{\mathbf{H}}(n)$ and is given by

$$\hat{\mathbf{H}}(n) = [\mathbf{Z}^T(n)\mathbf{Z}(n)]^{-1}\mathbf{Z}^T(n)\mathbf{z}(n) \tag{3.125}$$

Based on the similarity of this equation to (3.7) we can immediately transform various results derived earlier to the case considered here. For example, the asymptotic properties of this estimator are

$$\lim_{n\to\infty} \frac{1}{n}\mathbf{Z}^T(n)\mathbf{Z}(n) = 2[\mathbf{R} + \tilde{\mathbf{R}}] \tag{3.126}$$

$$\lim_{n\to\infty} \frac{1}{n}\mathbf{Z}^T(n)\mathbf{z}(n) = 2[\mathbf{r} + \tilde{\mathbf{r}}] \tag{3.127}$$

Also, the RLS algorithm summarized in Table 3.1 can be used to update the coefficient vector $\hat{\mathbf{H}}(n)$ recursively. The only differences are that $\boldsymbol{\phi}(n)$ will be changed into $\bar{\boldsymbol{\phi}}(n)$ everywhere, and the update equation for $e(n)$ and $\boldsymbol{\phi}(n)$ will be changed into (3.115)–(3.117). It is important to note that $\bar{\boldsymbol{\phi}}(n)$ no longer has the shift property that was used to update $\boldsymbol{\phi}(n)$ in Table 3.1. Thus (3.118) must be used to evaluate $\bar{\boldsymbol{\phi}}(n)$ at each time step. Table 3.3 summarizes the algorithm. A more detailed

TABLE 3.3 Linear-Phase RLS Adaptive Filter

- Initialization:

$\hat{\mathbf{H}}(0) = 0, \quad \mathbf{P}(0) = \sigma\mathbf{I}, \quad \bar{\boldsymbol{\phi}}(n) = 0$

- At each time step do the following:

$e(n) = y(n) + y(n - 2N + 1) - \bar{\boldsymbol{\phi}}^T(n)\hat{\mathbf{H}}(n-1)$ Compute filter output

$$\mathbf{P}(n) = \frac{\mathbf{P}(n-1) - \mathbf{P}(n-1)\bar{\boldsymbol{\phi}}(n)[\lambda + \bar{\boldsymbol{\phi}}^T(n)\mathbf{P}(n-1)\bar{\boldsymbol{\phi}}(n)]^{-1}\bar{\boldsymbol{\phi}}^T(n)\mathbf{P}(n-1)}{\lambda}$$

Gain update

$\mathbf{K}(n) = \mathbf{P}(n-1)\bar{\boldsymbol{\phi}}(n)[\lambda + \bar{\boldsymbol{\phi}}^T(n)\mathbf{P}(n-1)\bar{\boldsymbol{\phi}}(n)]^{-1}$

$\hat{\mathbf{H}}(n) = \hat{\mathbf{H}}(n-1) + \mathbf{K}(n)e(n)$ Update filter coefficients

$\bar{\boldsymbol{\phi}}(n) = [y(n-1), y(n-2), \ldots, y(n-N)]^T + [y(n-2N), \ldots, y(n-N-1)]^T$ Update the states of the filter

description of the linear-phase RLS algorithm and some of its variations are given in [Friedlander and Morf 1982(2)].

3.4.3 LMS Algorithm

Following the deviation in Section 3.3 it is straightforward to write down the LMS version of the RLS algorithm discussed above. The LMS algorithm is given by

$$\mathbf{H}(n) = \mathbf{H}(n-1) + 2\mu\bar{\boldsymbol{\phi}}(n)e(n) \tag{3.128}$$

with $e(n)$, $\bar{\phi}(n)$ as defined in (3.116) and (3.117). The convergence properties of this algorithm will depend on the eigenvalues of the covariance matrix $\mathbf{R}^z = \mathbf{R} + \tilde{\mathbf{R}}$. For example, the expected value of the error vector $\tilde{\mathbf{H}}(n) = \mathbf{H}(n) - \mathbf{H}_{\text{opt}}$ will obey the difference equation [see (3.75)]

$$\tilde{\mathbf{H}}(n) = (\mathbf{I} - 2\mu\mathbf{R}^z)\tilde{\mathbf{H}}(n-1) \tag{3.129}$$

Note that by choosing a sufficiently small value for μ it is always possible to guarantee stability of the state matrix $(\mathbf{I} - 2\mu\mathbf{R}^z)$. The time constants associated with the exponential decay of $\tilde{\mathbf{H}}(n)$ are given by [see (3.67)]

$$\tau_i = \frac{1}{2\mu\lambda_i} \tag{3.130}$$

where λ_i are the eigenvalues of $\mathbf{R}^z$. All the other results in Section 3.3 can be similarly translated to the (constrained) linear-phase case.

The linear-phase filter discussed in this section is an example of an adaptive filter with a constrained structure. Filters with other types of constraints have been designed for different applications. In array processing the direction of the main beam imposes a linear constraint on the coefficients of an adaptive beamformer.

Other interesting variations of FIR adaptive filters are obtained by different choices of the cost function to be minimized by the algorithm. The quadratic error function $\mathbf{E}\{\epsilon^2(n)\}$ is the most common choice and is the easiest to analyze. However, other cost functions such as the fourth power of the error $E\{\epsilon^4(n)\}$, or the absolute value function $E\{|\epsilon(n)|\}$, are useful in some applications. The function $J = E\{(\epsilon^2(n) - 1)^2\}$ has been used in problems where the filter output is required to have a constant modulus [Treichler and Agee 1983]. The development and evaluation of specialized FIR adaptive filters is an area of active investigation.

4

ADAPTIVE ALGORITHMS FOR INFINITE IMPULSE RESPONSE FILTERS

John R. Treichler

4.1 INTRODUCTION

4.1.1 General Scope

The concept of adaptation in digital filtering has proven to be a powerful and versatile means of signal processing in applications where precise a priori filter design is impractical. For the most part, such signal processing applications have relied on the well-known adaptive finite impulse response (FIR) filter configuration. Yet, in practice, situations commonly arise wherein the nonrecursive nature of this adaptive filter results in a heavy computational load. Consequently, in recent years active research has attempted to extend the adaptive FIR filter into the more general feedback or infinite impulse response (IIR) configuration. The immediate reward lies in the substantial decrease in computation that a feedback filter can offer over an FIR filter. This computational improvement comes at certain costs, however. In particular, the presence of feedback makes filter stability an issue and can impact adversely on the algorithm's convergence time and the general numerical sensitivity of the filter. Even so, the largest obstacle to the wide use of adaptive IIR filters is the lack of robust and well-understood algorithms for adjusting the required filter gains. This chapter explores the classes of algorithms currently under development, those based on minimum mean-square-error concepts, and another which has its roots in nonlinear stability theory. The basic derivation of each will be presented and certain aspects of performance examined.

Other key design concerns, such as the fact that certain algorithms require the use of specific filter structures, will also be illuminated.

4.1.2 Why Use IIR Adaptive Filters?

As discussed at length in Chapter 3, the adaptive FIR digital filter is well understood and several reliable algorithms exist for choosing the proper filter coefficients. Why, then, explore the alternative of using an IIR filter? The answer can be seen through two practical examples. First, consider the problem of correcting or compensating for signal distortions introduced into a radio signal which encounters multipath propagation effects. The signal sensed at the receiving antenna is, under reasonable assumptions, a linear combination of variously delayed versions of the transmitted signal. Moreover, it can be shown that for large enough filter order M, the transmission path between transmitter and receiver can be accurately represented as an FIR filter. The objective of the adaptive filter in this case is to process the received signal in such a way as to compensate or correct for the propagation-induced signal degradation. The simplified block diagram of this situation is shown in Figure 4.1. The transfer function of the compensation filter, $H(z)$, should be chosen so that $P(z)H(z) = 1$. If this can be done, the compensation is complete and $\hat{s}(n)$, the output estimate, exactly equals $s(n)$, the transmitted signal. But if the propagation path is well represented by an FIR filter, we might expect the optimal correction filter $H(z)$ to be an IIR filter. This is exactly the case [Sussman], and moreover, if $H(z)$ is restricted to being an FIR filter, the order M' required to attain a given compensation performance is generally much higher than that of the proper IIR filter. Since higher order implies high computation

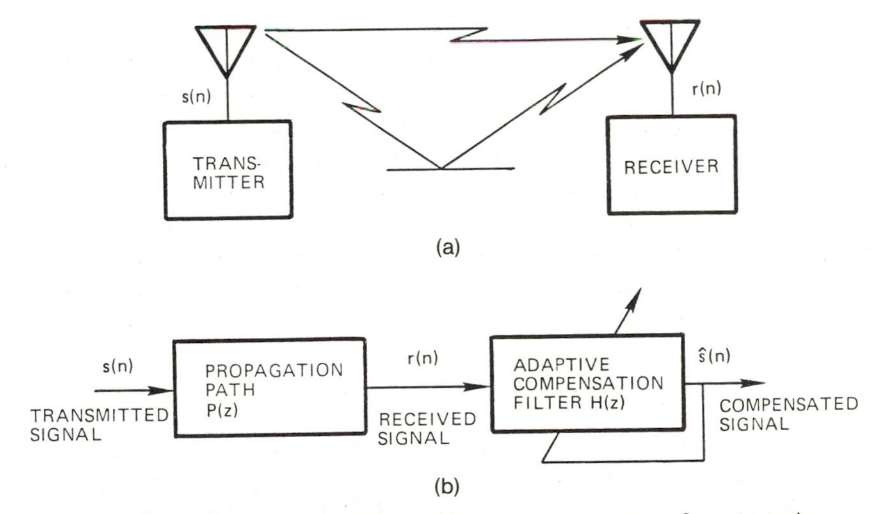

Figure 4.1 One use for an adaptive IIR filter: compensation for propagation effects: (a) multipath propagation in a radio channel; (b) simplified block diagram of signal processing steps.

costs, the IIR filter appears to be preferable. We return to this example in Section 4.4.

Another problem of practical interest is that of modeling systems whose outputs contain sinusoid or near-sinusoid signals. This situation is shown in Figure 4.2. The input excitation $u(n)$ is applied to the system under test, producing $s(n)$. The same excitation is applied to an IIR filter whose coefficients are adjusted with the goal of making its output, $\hat{s}(n)$, equal $s(n)$ for all possible choices of $u(n)$. If this can be achieved, the IIR filter's coefficients form an accurate model of the actual system. Again, the attraction here to an IIR filter is based on the a priori knowledge that the system has sinusoidal and near-sinusoidal outputs, implying the presence of highly resonant modes. An FIR filter could be used to approximate the impulse response, but only at the cost of a very high filter order with the associated computation costs.

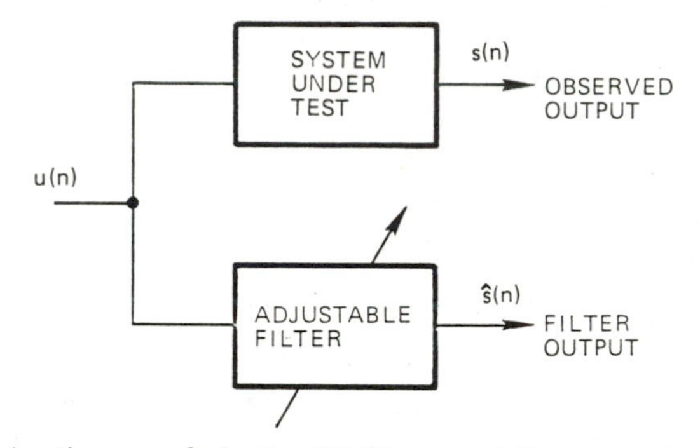

Figure 4.2 Another use of adaptive IIR filters: modeling a system with strong resonances.

The situation shown in Figure 4.2 is generally referred to as the "direct modeling" problem, while that in Figure 4.1 is called the "inverse modeling" problem (see Section 1.1.2). We may conclude that IIR filters form the canonical, and therefore, least computationally expensive, filter form for inverse modeling with "transmission zeros," and direct modeling problems with strong resonances. The computational advantages of IIR filters can be dramatic. A single strong resonance in an otherwise noisy background can be modeled well by a second-order IIR filter while an FIR filter hundreds of taps long would be required to produce equal output quality.

4.1.3 Problem Formulation

The basic IIR adaptive filtering problem is shown in Figure 4.3. The input $x(n)$ is applied to the IIR filter, resulting in the output $y(n)$. The output is subtracted from the desired output $d(n)$ to yield $e(n)$, the error between the two. This error, just with the FIR adaptive algorithms discussed in Chapter 3, is supplied to

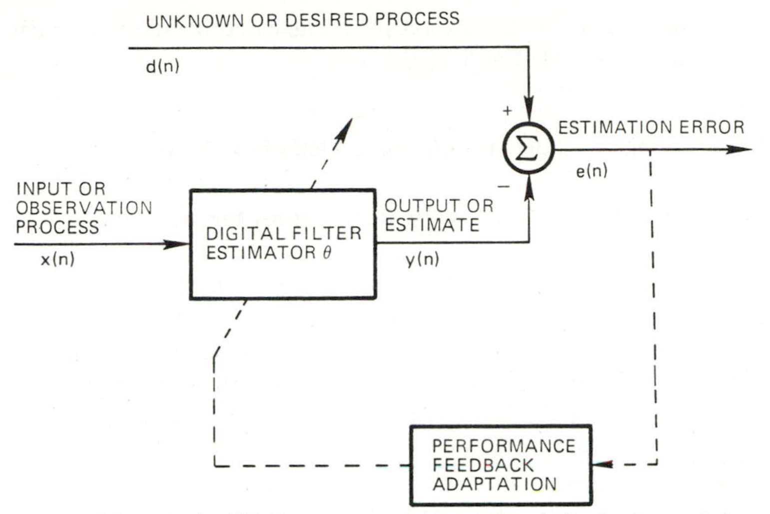

Figure 4.3 Adaptive IIR filter structure and notation. (After Larimore et al., copyright © 1980 IEEE.)

an algorithm that selects the filter coefficients properly. Given this formulation, the analytic objective can be stated: Given $x(n)$ and $d(n)$, choose the coefficients of the filter in such a way as to minimize some measure of the error $e(n)$.

4.1.4 Implications of Feedback

Before examining specific IIR adaptive algorithms, it is useful to consider the general implication of using an IIR filter as the basic kernel of the adaptive structure. As observed earlier, the key difference is the availability of feedback in the filter itself. This, of course, is the advantage of an IIR filter; that is, feedback allows a very low order filter to have a long-duration impulse response and with it the potential for sharp frequency response. In the context of adaptation, however, the feedback has several additional implications:

1. Unlike FIR filters, the IIR filter kernel is not stable for all choices of coefficients. This implies that the stability of the kernel may not be guaranteed explicitly as a part of the adaptive algorithm's design.
2. It will be seen that the feedback makes functions such as the gradient of $y(n)$ substantially more complex than for the FIR case.
3. High-resonance poles of the filter can have very long time constants, sometimes longer than the adaptive modes of algorithm. This factor complicates convergence analysis which, for FIR adaptive algorithms, such as LMS, is quite straightforward (see Section 3.3).

In the light of these observations, then, we examine two classes of adaptive IIR filter algorithms. In general, our goal is the development of a computationally simple IIR adaptive filter which converges to a stable optimal solution under very general specifications on $x(n)$ and $d(n)$.

4.2 MINIMUM MEAN-SQUARE-ERROR TECHNIQUES

4.2.1 Developing Necessary Conditions for a Solution

We consider, first, algorithms based on minimizing the mean-square value of the error $e(n)$. These techniques bear the closest relationship to the approaches used in Chapters 2 and 3; to that end, we assumed that both $x(n)$ and $d(n)$ are zero-mean stochastic processes which are stationary in the wide sense. For a nonadaptive filter, this would imply that both $y(n)$ and $e(n)$ are zero-mean wide-sense stationary stochastic processes as well.

For convenience, we assume at this point that the particular IIR filter kernel is constructed in direct form; that is, the output $y(n)$ is given by the difference equation

$$y(n) = \sum_{i=1}^{N} \hat{a}_i y(n-i) + \sum_{j=0}^{M} \hat{b}_j x(n-j) \tag{4.1}$$

where the $[\hat{b}_j]$ and the $[\hat{a}_i]$ are feedforward and the feedback coefficients, respectively, which will be adaptively modified. For later convenience, it is useful to define an aggregate coefficient vector $\hat{\boldsymbol{\theta}}$ and the aggregate data vector $\mathbf{X}(n)$ by

$$\hat{\boldsymbol{\theta}} = [\hat{a}_1 \cdots \hat{a}_N \quad \hat{b}_0 \cdots \hat{b}_M]^{\mathrm{T}} \tag{4.2}$$

and

$$\mathbf{X}(n) = [y(n-1) \cdots y(n-N) x(n) \cdots x(n-M)]^{\mathrm{T}} \tag{4.3}$$

Given this notation, the cost function J can be written as

$$J = \tfrac{1}{2}E[e(n)]^2 = \tfrac{1}{2}\cdot E\{d(n) - y(n)\}^2 \tag{4.4}$$

When $\hat{\boldsymbol{\theta}}$ is fixed, it is well known that the optimal filter occurs for $\hat{\boldsymbol{\theta}} \equiv \boldsymbol{\theta}_{\mathrm{opt}}$, where

$$\nabla_\theta J_{\boldsymbol{\theta}_{\mathrm{opt}}} = \tfrac{1}{2}E\{\nabla_\theta e^2(n)\}|_{\boldsymbol{\theta}_{\mathrm{opt}}} = E\{e(n)\nabla_\theta e(n)\}|_{\boldsymbol{\theta}_{\mathrm{opt}}} = 0 \tag{4.5}$$

Since $d(n)$ is not a function of $\hat{\boldsymbol{\theta}}$, this necessary condition reduces to a set of scalar equations

$$E\left\{e(n)\frac{\partial e(n)}{\partial \theta_i}\right\}\Bigg|_{\boldsymbol{\theta}_{\mathrm{opt}}} = -E\left\{(d(k) - y(n))\frac{\partial y(n)}{\partial \theta_i}\right\}\Bigg|_{\boldsymbol{\theta}_{\mathrm{opt}}} = 0 \tag{4.6}$$

We now consider the partial derivatives of $y(n)$. Examining (4.1), it is clear that the partial derivative of $y(n)$ with respect to $\hat{a}_1$ must contain $y(n-1)$; but is that all? In fact, no; evaluating (4.1) for $n' = n - 1$ shows that $y(n-1)$ is also a function of $\hat{a}_1$. In general, it can be seen that this partial derivative will depend on all values of $y(n)$ still in the "delay line." With this in mind, it can be shown that the orthogonality condition (4.5) results in the following scalar conditions:

$$E\left[e(n)\left\{y(n-i)+\sum_{j=1}^{N}\hat{a}_i\frac{\partial y(n-j)}{\partial \hat{a}_i}\right\}\right]\Bigg|_{\theta_{\text{opt}}}=0, \qquad 1\leq i\leq N \tag{4.7}$$

and

$$E\left[e(n)\left\{x(n-j)+\sum_{i=1}^{N}\hat{a}_i\frac{\partial y(n-i)}{\partial \hat{b}_j}\right\}\right]\Bigg|_{\theta_{\text{opt}}}=0, \qquad 0\leq j\leq M \tag{4.8}$$

4.2.2 Solution Techniques

Direct solution. The most obvious method of finding the optimal vector of filter parameters is by direct solution of the $M + N + 1$ simultaneous equations shown in (4.7) and (4.8). In principle, this is straightforward. The auto- and cross-correlation functions of $x(n)$ and $d(n)$ may be assumed to be known and, for any M and N, an equation for $y(n)$, and hence its auto- and cross-correlation properties can be developed. These functions may be plugged in (4.7) and (4.8) and, again in principle, a solution produced. Note also that (4.7) and (4.8) are, in effect, the "normal equations" developed in Chapter 2. As a practical matter, however, this approach is completely intractable for denominator orders greater than 2. Even for $N = 1$, the equations become nonlinear in the filter coefficients. The nonlinearity not only complicates the solution but allows the possibility of nonunique and nonexistent solutions. These analytical difficulties, plus the fact that the auto- and cross-correlation functions of $x(n)$ and $d(n)$ are rarely known in practice, have stemmed any continuing interest in the direct solution approach.

Iterative solution. Given the analytical difficulties found in direct solution of the IIR minimum mean-square-error (MMSE) problem, attention focused on developing iterative techniques for solution. Not only may there be no real alternative, but iterative techniques have several strong practical advantages:

1. No complicated numerical techniques (such as matrix inversion) are required.
2. The correlation functions for $x(n)$ and $d(n)$ need not be known.
3. "Slow" nonstationarities of $x(n)$ and $d(n)$ can be tracked.

These facts have, in fact, encouraged the development of iterative techniques for finding the optimal coefficient vector.

The most common iterative technique is based on gradient search of the MMSE performance surface. If the parameter vector at time n is denoted as $\hat{\boldsymbol{\theta}}(n)$, the updating algorithm is given by

$$\hat{\boldsymbol{\theta}}(n+1)=\hat{\boldsymbol{\theta}}(n)-\mu(n)\nabla_{\boldsymbol{\theta}}J(\hat{\boldsymbol{\theta}}(n)) \tag{4.9}$$

where $\mu(n)$ is a scalar sequence and $\nabla_{\boldsymbol{\theta}}J$ is the gradient of the cost function with respect to the parameter vector $\boldsymbol{\theta}$. For proper choice of sequence $\mu(n)$ convergence can be assured; however, only when the cost function is unimodal with respect to $\boldsymbol{\theta}$ will convergence be to the globally optimal design $\boldsymbol{\theta}_{\text{opt}}$ for arbitrary initial choice of $\boldsymbol{\theta}(0)$. We revisit this point shortly.

Returning to (4.5), we see that the gradient of the performance surface, for $\boldsymbol{\theta}$ fixed, is

$$\nabla_\theta J(\boldsymbol{\theta}) = E\{e(n)\nabla_\theta y(n)\}|_{\boldsymbol{\theta}} \tag{4.10}$$

The components of the gradient of $y(n)$ are, again with $\boldsymbol{\theta}$ fixed,

$$\frac{\partial y(n)}{\partial \hat{a}_i} = y(n-i) + \sum_{j=1}^{N} \hat{a}_j \frac{\partial y(n-j)}{\partial \hat{a}_i}, \qquad 1 \le i \le N \tag{4.11}$$

$$\frac{\partial y(n)}{\partial \hat{b}_j} = x(n-j) + \sum_{i=1}^{N} \hat{a}_i \frac{\partial y(n-i)}{\partial \hat{b}_j}, \qquad 1 \le j \le M \tag{4.12}$$

Recalling the definition of $\mathbf{X}(n)$ and $\boldsymbol{\theta}$, we note that for $\boldsymbol{\theta}$ fixed, (4.11) and (4.12) simplify into a convenient recursive formula; that is,

$$\nabla_\theta y(n) = \mathbf{X}(n) + \sum_{i=1}^{N} \hat{a}_i \nabla_\theta y(n-i) \tag{4.13}$$

This equation shows that the gradient of $y(n)$ and hence of J depends explicitly on past values of the gradient as reflected through the feedback coefficients of the filter.

The true gradient search algorithm is obtained by substituting (4.10) into (4.9), but clearly this is not a practical formulation since it involves the expectation operator. When the true gradient is unavailable, unbiased estimates will suffice in some cases; and to that end, the following adaptive algorithm has been suggested and employed:

For each n:

$$y(n) = \hat{\boldsymbol{\theta}}^{\mathrm{T}}(n)\mathbf{X}(n) \qquad \text{filter output} \tag{4.14a}$$

$$\hat{\nabla}_\theta y(n) = \mathbf{X}(n) + \sum_{i=1}^{N} \hat{a}_i(n)\hat{\nabla}_\theta y(n-i) \qquad \text{gradient estimate} \tag{4.14b}$$

$$\hat{\boldsymbol{\theta}}(n+1) = \hat{\boldsymbol{\theta}}(n) + \mu(n)\{d(n) - y(n)\}\hat{\nabla}_\theta y(n) \qquad \text{coefficient update} \tag{4.14c}$$

For each value of n, we first produce the filter output and then an estimate of the gradient of $y(n)$. The two are then used to compute the next set of coefficients $\hat{\boldsymbol{\theta}}(n+1)$.

Equation (4.14) represents a simplification from the ideal in two respects. First, the performance function gradient is approximated by an estimate rather than the expectation itself. Second, the gradient of $y(n)$ is approximated by allowing the coefficients of $\hat{\boldsymbol{\theta}}(n)$ to vary in time. This time variation affects (4.14b) since the recursion employs $\hat{a}_i(n)$ and not fixed values for the $\{\hat{a}_i\}$. In practice, both of these problems can be ameliorated by making $\mu(n)$ very small. When this is true, the vector $\hat{\boldsymbol{\theta}}(n)$ changes very slowly and (4.14b) closely approximates (4.13). This slow movement implies an effective time averaging of the performance gradient as well. If averaged long enough, ergodicity arguments imply that the actual value of the gradient will be arbitrarily close to the expected value. Both of these conditions influence the proper choice of the sequence $\mu(n)$. Note that the first of the two effectively implies that convergence be slow enough so that all time constants of the filter kernel be allowed to decay. In effect, (4.14b) carries with it past history of the output gradient. To avoid influencing the evolving new gradient information with the old, the filter must change slowly enough to allow the old information to die out.

Extensive computation simulations have been used to validate this adaptive algorithm and they have shown, generally speaking, that with adequately small $\mu(n)$, and attention to the kernel's stability, the algorithm will be reliably convergent to a minimum in the MMSE performance surface. However, since the gradient equations are intrinsically nonlinear, their solution, iterative or direct, can in general be satisfied by multiple extrema.

Consequently, seeking the optimal design $\boldsymbol{\theta}_{\text{opt}}$ by a gradient search of the performance function will not necessarily be successful. Convergence to local minima will invariably occur for certain initial values of parameters. Furthermore, the computation of the gradient, as indicated by (4.14b), is itself a recursive process, and can represent significant computation [Parikh and Ahmed].

Summarizing, the general problem of designing IIR filters, adaptive or not, lacks the mathematical advantages that have made least-squares performance useful for estimation. Also, the general gradient search framework is less suitable to develop an adaptation strategy to seek IIR filter design.

4.2.3 Historical Perspective

The IIR adaptive filtering problem has been studied for many years, mostly in the guise of IIR filter synthesis [Rabiner and Gold 1975] and system identification [Landau 1974, Mendel]. Although mathematically similar in many ways, neither area treated a problem with as much generality as that shown in Figure 4.3, nor did they address the need for simple on-line techniques with a minimum of computation. The development of IIR adaptive filtering algorithms were spurred, to a large degree, by the practical success that many workers were having with FIR adaptive filters based on MMSE concepts. As discussed in Section 4.1, the long FIR filters required for certain practical problems drove researchers toward the IIR approach.

Much early work on the MMSE approach was done by Stephen Horvath [Horvath and earlier works], who developed (4.11) to (4.14) for the specific application of telephone channel equalization. Not only were the equations developed and computer simulations conducted, but Horvath considered the stability of the kernel and, to that end, developed gradient update equations for several IIR filter forms, including the cascade and lattice structures.

At about the same time, work in the United States was being published; [White 1975] developed the MMSE gradient algorithm (4.14) for application to a speech analyzer and synthesizer, while Stearns and coworkers [Stearns and Elliott 1976] evolved the concept further. All these algorithms and variants employed the gradient recursion (4.14b) and therefore involved a reasonably large amount of computation per iteration. In 1976, Feintuch suggested a simplified algorithm given by

For every n:

$$y(n) = \hat{\boldsymbol{\theta}}^{\mathrm{T}}(n)\mathbf{X}(n) \tag{4.15a}$$

$$\hat{\nabla}_{\theta} y(n) = \mathbf{X}(n) \tag{4.15b}$$

$$\hat{\boldsymbol{\theta}}(n+1) = \hat{\boldsymbol{\theta}}(n) + \mu(n)\{d(n) - y(n)\}\hat{\nabla}_{\theta} y(n) \tag{4.15c}$$

These equations are identical to (4.14) except that the recursion based on old output gradients is deleted in (4.15b). This has the effect of reducing the number of multiplications required at each iteration by $N \cdot (M + N + 1)$, a healthy contribution considering that only about $2 \cdot (M + N + 1)$ are required for all other aspects of the filter computation. It might also be recognized that (4.15) represents the direct application of the FIR LMS algorithm to the IIR filter kernel.

Insomuch as (4.15b) is a substantially poorer approximation to the output gradient than (4.14b), the algorithm was shown to converge to false minima [Johnson and Larimore 1977] even when the algorithm in (4.14) converged to an actual minimum of the MSE surface [Parikh and Ahmed]. Even so, Feintuch's work sparked the search for simple, reliable IIR adaptive algorithms. Since 1976, most of the work on MMSE IIR adaptive filters has focused on developing an understanding of their convergence behavior [Stearns 1981, David].

4.3 TECHNIQUES BASED ON NONLINEAR STABILITY THEORY

4.3.1 Problem Formulation

Consider a less general, more structured problem. Let $d(n)$ itself be a bounded autoregressive moving-average (ARMA) process driven by $x(n)$,

$$d(n) = \sum_{j=0}^{M_d} b_j x(n-j) + \sum_{i=1}^{N_d} a_i d(n-i) \tag{4.16}$$

where the generating parameters a_i and b_j are constant. Assume further that sufficient variables are provided in the adaptive filter to span the parameter space generating $d(n)$ (i.e., $M \geq M_d$ and $N \geq N_d$). Without loss of generality, the assumption that $M = M_d$ and $N = N_d$ is permitted, with any excess generating parameters equaling zero. The error process $e(n)$ becomes

$$e(n) = \sum_{j=0}^{M} \{b_j - \hat{b}_j(n)\}x(n-j) + \sum_{i=1}^{N} \{a_i d(n-i) - \hat{a}_i(n)y(n-i)\} \tag{4.17}$$

It is sufficient to choose $\hat{b}_j \equiv b_j$ and $\hat{a}_i \equiv a_i$ to minimize J. If so, then

$$e(n) = \sum_{i=1}^{N} a_i e(n-i) \tag{4.18}$$

and due to the bounded-input bounded-output (BIBO) stability of (4.16)

$$\lim_{n\to\infty} e(n) = 0 \tag{4.19}$$

Once steady state has occurred, this choice of parameters, as would be expected, yields a minimal squared error.

With the filter design problem stated in this fashion it is a restatement of the output error identification problem shown in Figure 4.2. In this situation, an unknown ARMA plant has input $x(n)$ and output $d(n)$, perhaps measured in the presence of noise. It is desired to estimate or identify its internal parameters a_i and b_j in an unbiased fashion. This can be done by using a parallel model [Landau 1974], driven by the same input, the output $y(n)$ of which is compared with the plant output $d(n)$. On the basis of this output error, the parameter estimates are formed. However, there is an important distinction: In the identification problem, a performance measure based on the error is used only as a means of attaining small parameter error; in the filtering problem, a small output error is instead the desired end. In certain cases, a filter can tolerate substantial parameter error and still perform satisfactorily.

4.3.2 Hyperstable Adaptive Recursive Filter

Landau (1976) introduced an unbiased output error procedure for seeking parameter estimates of an ARMA plant, the hyperstable output error identifier. From this, [Johnson 1979(2)] developed an IIR adaptive filtering algorithm based on this hyperstable output error identifier. This technique, the hyperstable adaptive recursive filter (HARF), embodies two modifications of Landau's algorithm making it more suitable for the signal processing context. First, Landau's technique calls for diminishing adaptation gain factors, ultimately converging to zero. When identifying a plant with constant parameters, such an algorithm is acceptable; however, it has been recognized that for most adaptive filtering problems, adaptation must remain active to track changes in the signal environment [Widrow et al. 1976(2)]. Second, the hyperstable identifier is not strictly causal, requiring the current output sample $d(n)$ to form the current parameter values $\hat{b}_j(n)$ and $\hat{a}_i(n)$. Again, while this condition is reasonable for parameter estimation, it is undesirable for real-time filtering.

The HARF algorithm represents the first technique proposed for adaptive IIR filtering which has provable convergence properties. Despite its moderate computational complexity, a careful study of its behavior allows simplifications to be made, preserving most of HARF's desirable properties while reducing the required computation.

The HARF implementation is shown in Figure 4.4 for use in the following discussion. Notice from the figure that in addition to the principal adaptive filter that forms the output $y(n)$ there is an auxiliary process generated:

$$f(n) = \sum_{i=1}^{N} \hat{a}_i(n+1)f(n-i) + \sum_{j=0}^{M} \hat{b}_j(n+1)x(n-j) \tag{4.20}$$

This ARMA process is used both in forming the output

$$y(n) = \sum_{i=1}^{N} \hat{a}_i(n)f(n-i) + \sum_{j=0}^{M} \hat{b}_j(n)x(n-j) \tag{4.21}$$

and in the adaptive algorithm, to appear shortly. The parameters in these two equations are separated in time by one sample; that is, weighting coefficients used in (4.20) have undergone one additional update versus those in (4.21). If convergence should occur, $\hat{a}_i(n+1) = \hat{a}_i(n)$ and $\hat{b}_j(n+1) = \hat{b}_j(n)$, and $y(n)$ asymptotically converges to $f(n)$. However, in the transient stages of adaptation, the distinction between $y(n)$ and $f(n)$ proves necessary.

Suppose that at each sample the adaptive filter coefficients are updated

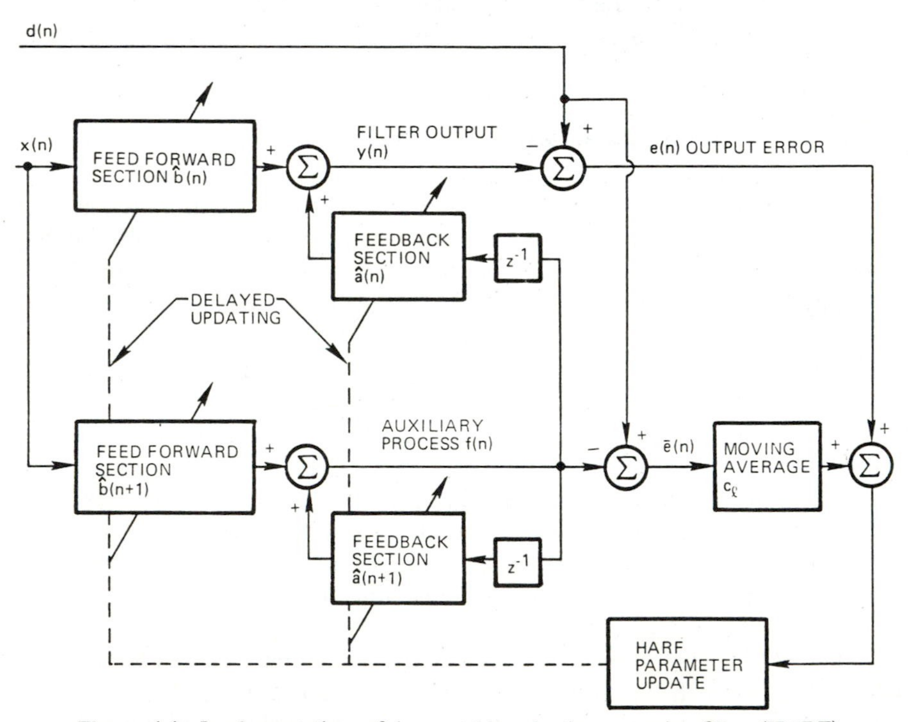

Figure 4.4 Implementation of hyperstable adaptive recursive filter (HARF). (After Larimore et al., copyright © 1980 IEEE.)

according to the formulas

$$\hat{a}_i(n) = \hat{a}_i(n-1) + \frac{\mu_i}{q(n)} f(n-i-1)\Bigg(\{d(n-1) - y(n-1)\} + \sum_{l=1}^{P} c_l\{d(n-l-1) - f(n-l-1)\}\Bigg), \qquad 1 \leq i \leq N \tag{4.22a}$$

$$\hat{b}_j(n) = \hat{b}_j(n-1) + \frac{\rho_j}{q(n)} x(n-j-1)\Bigg(\{d(n-1) - y(n-1)\} + \sum_{l=1}^{P} c_l\{d(n-l-1) - f(n-l-1)\}\Bigg), \qquad 0 \leq j \leq M \tag{4.22b}$$

where $q(n)$ is a normalizing factor greater than unity,

$$q(n) = 1 + \sum_{l=1}^{N} \mu_l f^2(n-l-1) + \sum_{l=0}^{M} \rho_l x^2(n-l-1) \tag{4.23}$$

and μ_l and ρ_l are arbitrary positive constants. In addition, the P constants c_l are chosen by the designer so that the discrete transfer function

$$G(z) = \frac{1 + \sum_{l=1}^{P} c_l z^{-1}}{1 - \sum_{i=1}^{N} a_i z^{-i}} \tag{4.24}$$

is strictly positive real (SPR) [Hitz and Anderson]. The implications of this requirement are discussed in Section 4.3.3.

Under these conditions, proof can be given [Johnson 1979(2)] that the moving-average quantity

$$v(n) = (d(n) - f(n)) + \sum_{l=1}^{P} c_l(d(n-l) - f(n-l)) \tag{4.25}$$

converges to zero and as a result

$$y(n) \longrightarrow f(n) \longrightarrow d(n) \tag{4.26}$$

which is the desired performance.

Before proceeding to a discussion of the SPR assumption necessary for this hyperstable formulation, consider briefly a heuristic description of the HARF updating algorithm. In (4.22) it can be seen that, aside from assorted positive scaling factors, the update to each coefficient is basically a product of two components. First is the value of the signal corresponding to the given weight in the output equation (4.21). For example, the update to $\hat{a}_l(n)$ depends on $f(n-l)$, and their product $\hat{a}_l(n)f(n-l)$ appears in (4.21). The second factor [appearing in brackets in (4.22)] is dependent on the instantaneous performance of the filter, disguised by a moving-average expression. Therefore, for a given quality of performance, largest adjustment is made to the coefficients contributing the most to the output via (4.21). These features are shared by a family of adaptation procedures. Readers will recognize a similarity to the gradient-based adaptation procedures. However, as noted, the complexity of an IIR structure results in specific differences that cannot presently be accounted for by means of a gradient descent approach.

4.3.3 Hyperstability and Adaptive Filtering

The concept of system hyperstability, upon which this adaptive IIR filtering algorithm is based, was developed by [Popov] and provides a generalized description of output stability in time-varying or nonlinear cases. Use of this analysis has occurred primarily in the control theory literature. For example, it has been applied to output error identification via the model reference adaptive system structure [Landau 1976]. In this section the hyperstability theorem is stated and a heuristic description of its conditions and implications is given. In addition, its relationship to adaptive filtering algorithms is shown.

Hyperstability is defined for the discrete-time case as follows [Anderson 1968]. Let $G(z)$ be a rational scalar transfer function for a linear time-invariant system driven by $u(n)$ and responding with $y(n)$. The system is said to be hyperstable if its state vector remains bounded over time for all driving sequences $u(n)$ satisfying jointly with the output

$$\sum_{l=0}^{N_0} u(l)y(l) < K^2 \qquad \forall\ N_0 \tag{4.27}$$

The present discussion deals with a stronger variation of this definition, rigorously denoted as asymptotic hyperstability, which requires that the state vector decay to zero for the same class of input sequences; references to hyperstability in this section will actually imply this latter definition. Note that if $G(z)$ has a proper rational form, the output $y(n)$ will likewise decay to zero.

The hyperstability theorem [Popov] is a simple statement: The system described above is hyperstable if and only if its transfer function $G(z)$ is strictly positive real (SPR) [Hitz and Anderson], that is,

$$\text{Re}[G(z)] > 0, \qquad z = e^{j\theta} \tag{4.28}$$

In other words, an SPR system will have a bounded output when driven by any input (including certain divergent sequences) satisfying (4.27).

The hyperstability theorem has an interesting interpretation in terms of system passivity [Anderson 1968, Landau 1976]. A familiar physical example arises in network theory, in a continuous-time context. It is well known that the driving-point impedance $Z(s)$ of a passive network is SPR and relates driving current to response voltage at a network port. Consider a state realization where each state variable corresponds to an energy storage component. For any current such that the energy delivered into the network is bounded, that is,

$$E = \int_0^T v(t)i(t)\, dt < K^2 \qquad \forall\ T \tag{4.29}$$

then the energy stored internally must be dissipated [i.e., $\|x(t)\| \longrightarrow 0$]. By analogy, any system that is SPR can be thought of as dissipative in a mathematical sense.

It is in the closed-loop configuration that system hyperstability becomes useful in adaptive filtering; parameter adaptation of digital filters can be restated in such a configuration [Landau 1976]. Let $u(n)$ be a sequence derived as a nonlinear time-varying function of the output, denoted as a general feedback element $\mathscr{F}$ in Figure 4.5(a). Had the feedback been linear, $F(z)$, as indicated in Figure 4.5(b), BIBO stability is easily checked in the frequency domain, by locating the zeros of

$$1 + F(z)G(z) \tag{4.30}$$

A zero not inside the unit circle implies instability. A physical interpretation, of course, requires that loop gain never be -1 at any frequency (i.e., give rise to 180° phase shift). A sufficient but unnecessary condition would be to restrict $F(z)$ and

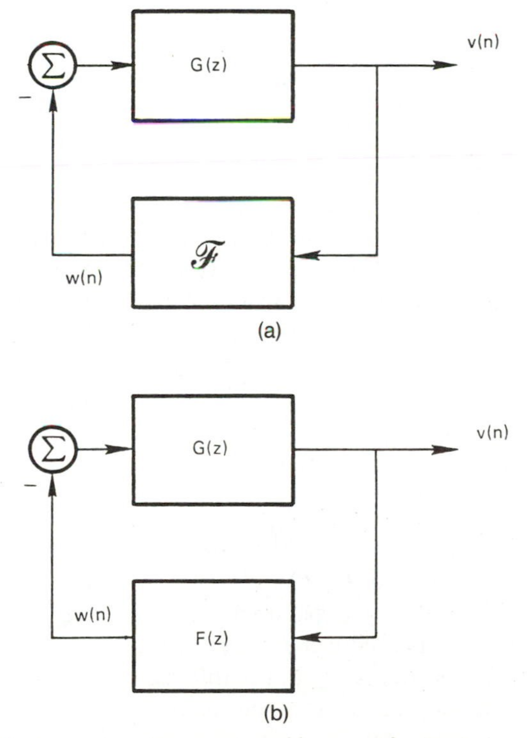

Figure 4.5 Undriven closed-loop system: (a) general feedback element; (b) linear feedback element. (After Larimore et al., copyright © 1980 IEEE.)

$G(z)$ each to contribute less than ± 90 degrees at all frequencies. That is, if both are SPR, the closed loop is guaranteed to be stable.

Still, such a condition is not useful when analyzing cases involving nonlinear time-varying feedback. Instead, a condition must be stated in terms of the time-domain behavior of $\mathscr{F}$. Note from Figure 4.5(a) that the feedback element $\mathscr{F}$ is driven by $v(k)$, and responds with $w(n) = -u(n)$. Then if

$$\sum_{l=0}^{N_0} v(l)w(l) \geq -\gamma_0^2 \qquad \forall\ N_0 > 0 \tag{4.31}$$

it follows that

$$\sum_{l=0}^{N_0} u(l)y(l) \leq \gamma_0^2 \qquad \forall\ N_0 > 0 \tag{4.32}$$

Consequently, if $G(z)$ is SPR, the closed loop is hyperstable. This represents a means of generalizing the positive reality concept to the nonlinear time-varying feedback element $\mathscr{F}$. For linear elements, it can be shown by using eigenfunction analysis that this condition indeed assures that phase response at all frequencies not exceed $\pm 90°$. From a more physical standpoint, if the energy delivered into the feedback element $\mathscr{F}$

$$\sum_{l=0}^{N_0} v(l)w(l)$$

is bounded below as in (4.31), $\mathscr{F}$ is dissipative feedback. (This is analogous to the positive restriction on the physical energy delivered to a passive network.) Alternatively, (4.31) requires that the "sign" of the feedback element "on the average," implied by the summation, should be bounded below.

The hyperstability theorem guarantees stability for a class of intrinsically nonlinear time-varying systems. Clearly, the requirement given by (4.27) represents a sufficiency condition, and as such is overly restrictive. In the simple linear case cited above, both the strict positive reality of the forward path and the positive reality of the feedback path are unnecessary for stability. Consequently, in the general case, one would expect the conditions to be unnecessary. Although a less restrictive criterion for nonlinear systems is of interest, the formulation given here is satisfactory for use in analysis of this class of adaptive filtering algorithms.

To demonstrate the relationship of hyperstability to adaptive filtering, first define an auxiliary error quantity

$$\bar{e}(n) \triangleq d(n) - f(n)$$

where $d(n)$ and $f(n)$ are given by (4.16) and (4.20). Note that $\bar{e}(n)$ is closely related to $e(n)$; as noted, $f(n) \longrightarrow y(n)$ as parameter convergence progresses, implying that $\bar{e}(n) \longrightarrow e(n)$. For practical convergence rates, the error quantities are virtually identical; the distinction is necessary in the interest of rigor. An equation for the auxiliary error can be formed in the same manner as (4.17).

$$\bar{e}(n) = \sum_{j=0}^{M} [b_j - \hat{b}_j(n+1)]x(n-j) + \sum_{i=1}^{N} [a_i d(n-i) - \hat{a}_i(n+1)f(n-i)] \tag{4.33}$$

By adding and subtracting the term

$$\sum_{i=1}^{N} a_i f(n-i) \tag{4.34}$$

and rearranging, we have

$$\begin{aligned} \bar{e}(n) &= \sum_{i=1}^{N} a_i\{d(n-i) - f(n-i)\} + \sum_{i=1}^{N} \{a_i - \hat{a}_i(n+1)\} f(n-i) \\ &\quad + \sum_{j=0}^{M} \{b_j - \hat{b}_j(n+1)\} x(n-j) \\ &= \sum_{i=1}^{N} a_i \bar{e}(n-i) - w(n) \end{aligned} \tag{4.35}$$

where

$$w(n) = -\left\{\sum_{i=1}^{N} [a_i - \hat{a}_i(n+1)] f(n-i) + \sum_{j=0}^{M} [b_j - \hat{b}_j(n+1)] x(n-j)\right\} \tag{4.36}$$

Thus the auxiliary error $\bar{e}(n)$ is an Nth-order autoregressive process whose poles are identical to those of the unknown ARMA plant. The driving function $w(n)$ is a function of the parameter errors $(b_j - \hat{b}_j(n+1))$ and $(a_i - \hat{a}_i(n+1))$. This relation is shown diagrammatically in Figure 4.6(a), where

$$A(z) = 1 - \sum_{i=1}^{N} a_i z^{-i} \tag{4.37}$$

Note that the filter input $x(n)$ enters into the computation of $w(n)$ only as a time-varying factor.

In the adaptive case, the parameter estimates $\hat{a}_i(n)$ and $\hat{b}_j(n)$ are updated using performance feedback. This effectively closes the loop, producing Figure 4.6(b), where the update algorithm determines the functional form of $\mathscr{F}$. Note that the feedback is in general nonlinear and time-varying; therefore, choice of an algorithm that satisfies the conditions of the hyperstability theorem embodied in (4.33) is suitable to assure convergence of the auxiliary error.

To meet the hyperstability conditions, the forward element must be SPR. In general, the simple autoregressive form $1/A(z)$ will fail. A means of augmenting it to force SPR can be achieved by separation of $\mathscr{F}$ into a linear preprocessor $C(z)$ followed by a general element, as in Figure 4.6(c). The output of the $C(z)$ is an auxiliary process $v(n)$, that is, (4.25),

$$v(n) = \bar{e}(n) + \sum_{l=1}^{P} c_l \bar{e}(n-l) \tag{4.38}$$

In this case, $v(n)$ is simply a weighted average or smoothed version of the output error. Rearranging the system gives Figure 4.6(d), with a forward composite linear element

$$G(z) = \frac{C(z)}{A(z)} = \frac{1 + \sum_{l=1}^{P} c_l z^{-l}}{1 - \sum_{i=1}^{N} a_i z^{-i}} \tag{4.39}$$

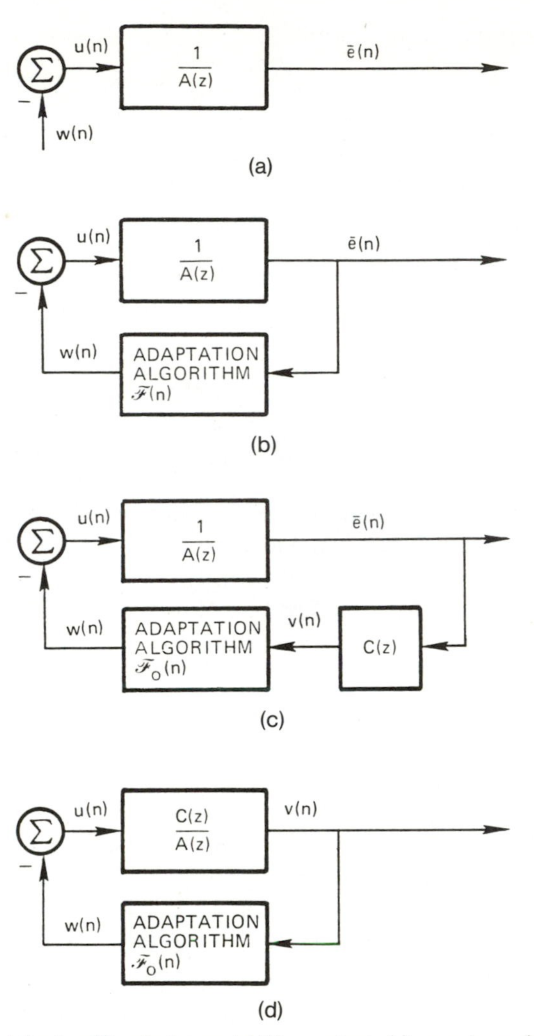

Figure 4.6 Adaptive filter in hyperstability context: (a) open-loop; (b) adaptation feedback; (c) linear preprocessor; and (d) hyperstable adaptation. (After Larimore et al.; copyright © 1980 IEEE.)

In this form, the closed-loop output becomes $v(n)$, a moving-averaged version of the auxiliary error. This error enters into the function $\mathscr{F}_0$ for updating the adaptive filter weights $\hat{a}_i(n)$ and $\hat{b}_j(n)$, and generating the driving sequence $w(n)$. If the values of c_l are chosen to assure the SPR of $G(z)$, and the algorithm using $v(n)$ satisfies the relation (4.31), the closed-loop system is hyperstable and $v(n) \longrightarrow 0$. The error quantity $\bar{e}(n)$, expressible as an internal state variable of $G(z)$, must also converge to zero. The HARF update algorithm described in the preceding section is shown in [Johnson 1979(2)] to satisfy the hyperstability conditions.

The coefficients c_l that form a moving average of the output error represent a set of P design parameters, chosen to assure the SPR of $G(z)$ in (4.39). The denominator, determined by the unknown ARMA process $d(n)$, contributes a phase that must be tempered by zero placement (i.e., choice of c_l) to bound the net phase by ± 90 degrees. As an example, consider Figure 4.7. Assume that $d(n)$ is produced by a second-order filter having complex poles. If the transfer function

$$G(z) = \frac{1}{1 - a_1 z^{-1} - a_2 z^{-2}} \tag{4.40}$$

were analyzed, for only certain conjugate pole pairs would it be SPR; the SPR region is shown in relation to the unit circle as an unshaded oval. Thus, by effectively eliminating the flexibility of a numerator for $G(z)$, only certain pole pairs allow the use of the hyperstability formulation. It is interesting to note that this excludes the vicinity near $z = 1$, the region where poles of an oversampled continuous process will cluster.

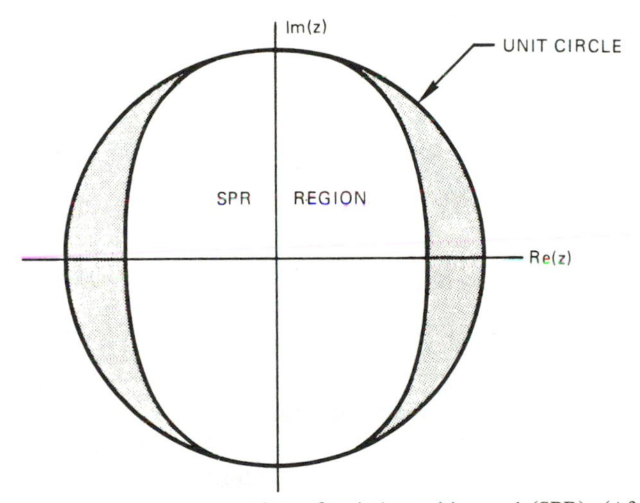

Figure 4.7 Second-order system, region of strictly positive real (SPR). (After Larimore et al., copyright © 1980 IEEE.)

Once the numerator of (4.39) is introduced, the region of SPR pole location can be purposely deformed to encompass parts of the unit circle within which the unknown poles are likely to be found. Figure 4.8 demonstrates this effect for several values of a single smoothing coefficient c_1. Note that $c_1 = 0$ produces the oval region as before; as c_1 becomes negative and $G(z)$ gives a zero on the positive real axis, the SPR region is deformed toward $z = 1$. Finally, for $c_1 = -1$ the region becomes circular, tangential with the unit circle at $z = 1$, and encompasses the low-frequency pole locations. The introduction of the c_1 parameters not only tailors the region of SPR but also influences convergence behavior [Treichler et al. 1978], as shown in a later section.

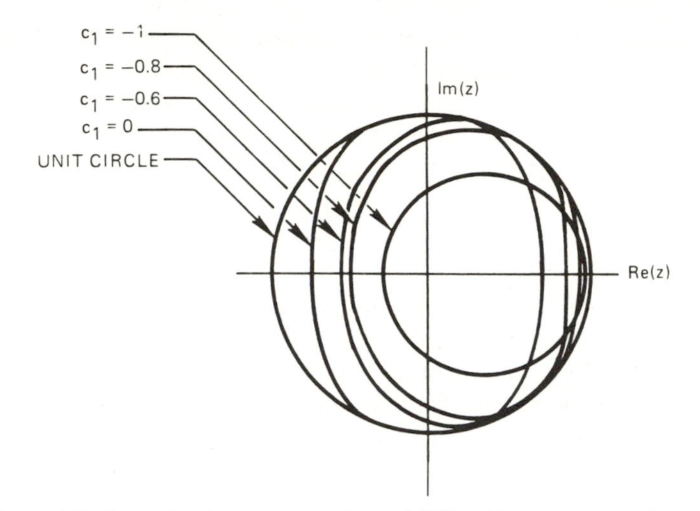

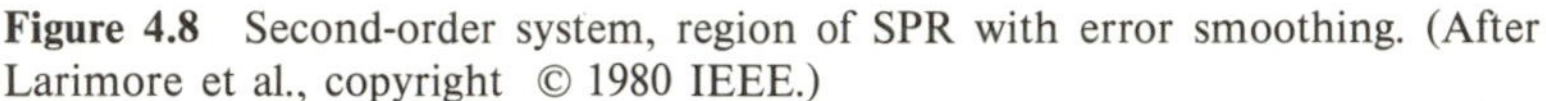

Figure 4.8 Second-order system, region of SPR with error smoothing. (After Larimore et al., copyright © 1980 IEEE.)

The most serious practical consideration in choosing the c_l coefficients to guarantee SPR of (4.39) is that the denominator is unknown. Given some a priori knowledge about the dynamics of the process $d(n)$, a reasonable choice of a numerator for $G(z)$ is simply an estimate of the denominator. Clearly, a perfect estimate produces a total cancellation of numerator and denominator, giving a degenerate SPR result, and a deterministic algorithm equivalent to an equation error variant (e.g., LMS). Inaccurate estimates, although not canceling the dynamics of $G(z)$, may serve to contain the phase angle. For example, a biased estimate for the denominator parameters derived via the equation error technique has been found to give strict positive reality of (4.39) for certain SNR levels [Johnson et al. 1979(1)]. As a rule, placement of a zero in the vicinity of each pole provides a reasonable set of coefficients c_l. It is also possible to develop an algorithm for adjusting the c_l coefficients, paralleling certain results in output error identification [Landau 1978].

4.3.4 Simple Hyperstable Recursive Filter

While the hyperstability formulation of the adaptive IIR filtering problem provides a useful perspective, the resulting HARF algorithm suffers from two significant sources of computational complexity. First, examination of (4.22) indicates that an auxiliary ARMA process $f(n)$ is necessary to compute not only the filter output, but the weight updates as well. Second, the HARF algorithm includes a normalizing scale factor $q(n)$, computed for each iteration. Both of these components of the algorithm substantially increase algorithm cost in terms of hardware and/or sampling rate.

To make the adaptive IIR filtering algorithm more amenable to real-time processing, certain reasonable simplifications can be made. By specifying the rate constants μ and ρ to be sufficiently small as in successful gradient approximation procedures [Parikh and Ahmed], the weights change very little from iteration to iteration; therefore,

$$\begin{aligned} \hat{a}_i(n+1) &\simeq \hat{a}_i(n) \\ \hat{b}_j(n+1) &\simeq \hat{b}_j(n) \end{aligned} \tag{4.41}$$

A comparison of (4.20) and (4.21) indicates that

$$f(n) \simeq y(n) \tag{4.42}$$

The output equation (4.21) becomes

$$y(n) = \sum_{i=1}^{N} \hat{a}_i(n)y(n-i) + \sum_{j=0}^{M} \hat{b}_j(n)x(n-j) \tag{4.43}$$

and the moving-average process $v(n)$ in (4.25) becomes

$$\begin{aligned} v(n) &\simeq \{d(n) - y(n)\} + \sum_{l=1}^{P} c_l\{d(n-l) - y(n-l)\} \\ &= e(n) + \sum_{l=1}^{P} c_l e(n-l) \end{aligned} \tag{4.44}$$

a simple moving average of the output error. Finally, note that $q(n)$ in (4.23) is a simple normalizing factor that controls the instantaneous adaptation rate, reducing the effective step size for large values of filter input and output. Once again, assuming that μ and ρ are sufficiently small,

$$q(n) = 1$$

Using these approximations, (4.22) becomes

$$\hat{a}_i(n) = \hat{a}_i(n-1) + \mu_i y(n-1-i)v(n-1), \qquad 1 \leq i \leq N \tag{4.45a}$$

$$\hat{b}_j(n) = \hat{b}_j(n-1) + \rho_j x(n-1-j)v(n-1), \qquad 0 \leq j \leq M \tag{4.45b}$$

The set (4.43)–(4.45) has been denoted the simple hyperstable adaptive recursive filter, SHARF [Treichler et al. 1978, Larimore et al.].

Note that significant reduction in computation and storage has been realized; the update to each weight requires only knowledge of the smoothed output error process $v(n)$. This computational saving was accomplished at the cost of no longer rigorously satisfying the hyperstability condition (4.31), so that convergence is no longer guaranteed for arbitrary positive μ and ρ. For practical purposes, however, slow adaptation maintains close approximation to a hyperstable structure.

It is interesting to note that certain earlier attempts at adaptive IIR filtering, notably Feintuch's algorithm [Feintuch], are clearly special cases of (4.45). Its update equations, extrapolating from similar equations used in adaptive FIR filtering, are

$$\hat{a}_i(n+1) = \hat{a}_i(n) + \mu e(n) y(n-i) \tag{4.46a}$$

$$\hat{b}_j(n+1) = \hat{b}_j(n) + \rho e(n) x(n-j) \tag{4.46b}$$

where

$$e(n) = d(n) - y(n) \tag{4.47}$$

Note that this is equivalent to the constraint that $c_l = 0$ for $l = 1$ to P in (4.44), that is,

$$v(n) = e(n) \tag{4.48}$$

According to the hyperstability analysis, convergence is assured only if μ and ρ are small and the autoregressive function

$$G(z) = \frac{1}{1 - \sum_{i=1}^{N} a_i z^{-i}} \tag{4.49}$$

is SPR. As shown in Section 4.3.3, in general, zero placement is necessary for SPR satisfaction.

To demonstrate the behavior of the SHARF, a series of simulations are presented. The desired process $d(n)$ is second order, generated by filtering white noise with

$$\frac{0.057}{1 - 1.645z^{-1} + 0.9025z^{-2}} \tag{4.50}$$

The migrations of the adaptive filter's two poles are shown in Figure 4.9 starting at the imaginary poles, $0.6e^{\pm j90°}$, converging to the poles of (4.50), $0.95e^{\pm j30°}$. The four trajectories show the effects of a single smoothing coefficient c_1, beginning with $c_1 = 0$ and ranging through $c_1 = -2.5$. Pole migration, of course, is a complicated transformation of the adaptive weight locus, and as such provides a distorted view of filter behavior. However, on a qualitative level, it can be seen that the variation of the smoothing parameter not only reduces meanderings but also speeds convergence. This is partly due to the effect that smoothing has on the strength of the error process $v(n)$, in turn reducing the average size of the algorithm's update terms in (4.45).

In this example, it is worth noting that despite violation of the SPR requirement, for example, when $c_1 = 0$, convergence did still occur. This simply indicates that the sufficiency condition was overly restrictive in the example. Consider a second case, involving a second-order process generated by a filter

$$\frac{1}{1 - 1.7z^{-1} + 0.7225z^{-2}} \tag{4.51}$$

having a pair of real poles at 0.85. In this case, the SHARF algorithm was simulated using $c_1 = 0$ and $c_1 \simeq -1$; the first case does not satisfy the SPR requirement, whereas the second case does. To eliminate the ambiguities of convergence rate, the adaptive filter was initialized to have its poles at 0.845, within 0.005 of the true

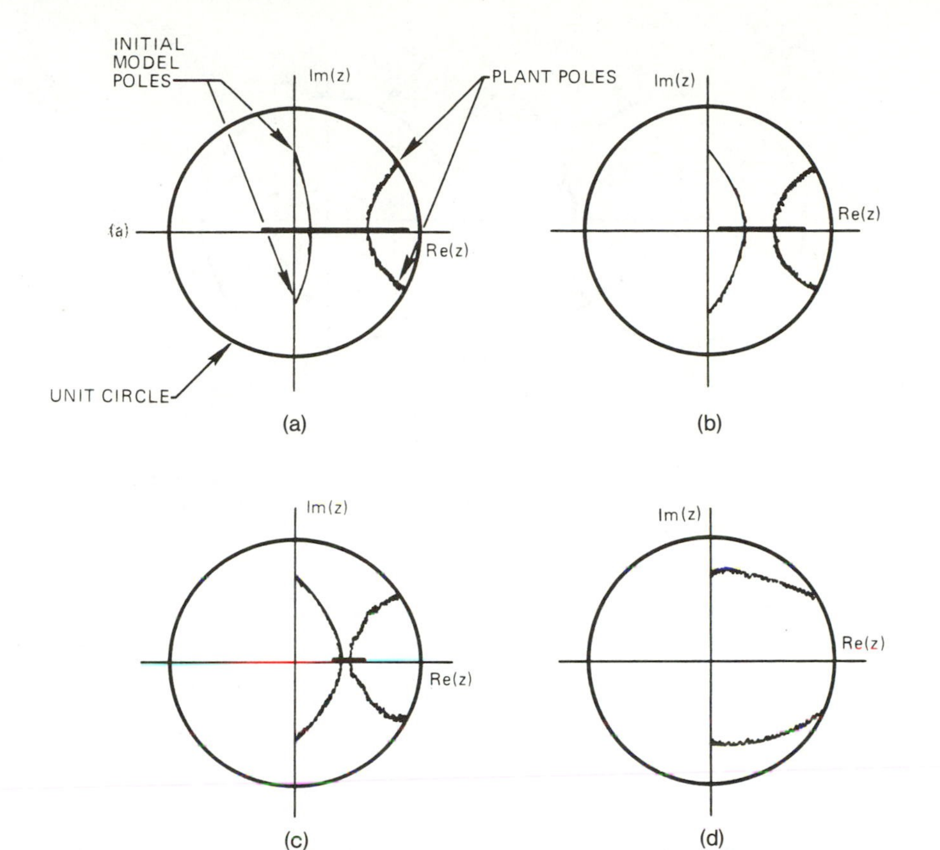

Figure 4.9 Pole trajectories for simulation of the simple hyperstable adaptive recursive filter (SHARF) algorithm: (a) $c_1 = 0$, 160K iterations; (b) $c_1 = -0.8$, 160K iterations; (c) $c_1 = -1$, 120K iterations; and (d) $c_1 = -2.5$, 40K iterations. (After Larimore et al., copyright © 1980 IEEE.)

location. Despite the proximity of the adaptive filter to its desired parameter set, for $c_1 = 0$ [i.e., Feintuch's algorithm (in this nonpositive real case)] the weights quickly readjusted to an alternative configuration, involving a single low-frequency pole, effectively discarding the second degree of freedom. In the second case, convergence was consistent to the unbiased pole estimates (see Figure 4.10).

4.4 CONVERGENCE ANALYSIS

4.4.1 Goals of Convergence Analysis

Once an adaptive filtering algorithm has been defined, it becomes necessary to analyze its convergence properties so that its overall utility can be assessed. Such convergence analysis is done with three nested goals in mind:

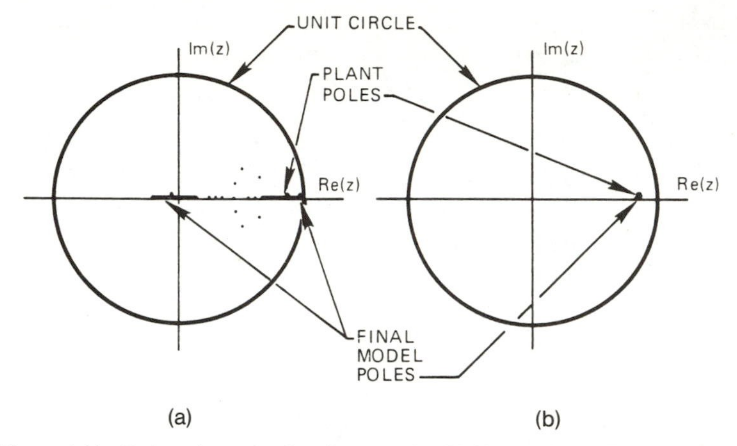

Figure 4.10 Pole trajectories for the recursive LMS and SHARF algorithms: (a) recursive LMS ($c_1 = c_2 = 0$); (b) SHARF ($c_1 = -1.0$, $c_2 = 0$). (After Larimore et al., copyright © 1980 IEEE.)

1. To provide proof that the adaptive process does in fact converge as desired to the proper state
2. To estimate the properties of the convergence process, such as time to convergence
3. To provide some (hopefully simple) model for the behavior of the process as it does converge

Although these seem fairly straightforward, matters become very complicated when the actual performance or convergence criterion is introduced. The biggest distinction comes in the assumptions made on the input processes $x(n)$ and $d(n)$; that is, are they described as stochastic processes or can they realistically be described as deterministic processes?

At least one of two conditions motivates the use of stochastic convergence analysis. The first is the presence of noise on any of the available signal lines. Termed observation noise or plant noise depending on where it appears in the system, it usually forces a stochastic analysis simply because the noise terms are usually most gracefully described as stochastic processes. As a direct result the convergence analysis is usually done in stochastic terms. A second rationale for this approach is when spectral domain analysis is important. This is commonly the case in adaptive filtering applications, where the copious work on Wiener filtering can be drawn upon. In fact, this work is so useful that degenerate stochastic representations are often used for processes that would otherwise allow a deterministic description (e.g., a sinusoid). The results of stochastic analysis follow a standard hierarchy; usually the easiest fact to prove is convergence in the mean, then convergence in the mean square, and finally convergence with probability 1. Progressing up the hierarchy usually requires progressively more knowledge about the statistics of the processes, starting with mean and covariance information, and finally progressing to multivariate probability distributions. This fact accounts for the limited hierarchical progression of most adaptive filtering convergence proofs. However, since most input processes are known by their spectra (hence, covariance properties) but not so commonly by their probability distributions, convergence proofs usually stop at convergence in mean square. In fact, in nonstationary signal environments, it may stop earlier.

In application areas where spectral descriptions are not a dominant concern and where noise is a second-order effect, most convergence analysis centers on "deterministic" approaches. Here the input signals are modeled as fairly arbitrary but bounded processes and the objective is to show, with no equivocation whatsoever, that the adaptive process converges to the desired state. As shall be seen, these techniques basically reduce to showing that the adaptive process is passive in some manner; that is, the system error, defined appropriately, decays to zero as the voltage in a passive circuit would. Clearly, such a result is much stronger than one that indicates the error converges in the mean square. In the past, the price paid for this strength has been the difficulty in extending such proofs to include the effects of observation or plant noise in a graceful fashion. Another concern, as shall be shown in the next section, is that while the deterministic methods yield strong "fact of convergence" results, it is often difficult to obtain convergence behavior (e.g., speed) information from them.

In light of these points, we now examine the limited literature currently available regarding the convergence of IIR adaptive filters.

4.4.2 Approaches

Analysis of MMSE techniques. As discussed earlier in this chapter, most IIR adaptive algorithms that attempt to minimize the mean-square error do so by approximating the gradient of the performance function and iteratively stepping in the specified direction until a stationary point is reached. In light of this, most of the extant analysis regarding MMSE-based IIR adaptive algorithms has focused on examining the properties of the MMSE performance surface itself. In spite of a good deal of work by [Stearns 1981, David], even the basic question has yet to be answered; under what conditions is the MMSE performance function J [see Equation (4.4)] unimodal? If it is unimodal, a gradient based search will not be led astray. Unfortunately, Stearns was able to show unimodality in only certain cases, that is, systems with wideband inputs and with more than enough coefficients to model the desired signal $d(n)$ adequately. Johnson and Larimore have shown that a multimodal performance function can be induced in the "reduced order" case where the filter has insufficient degrees of freedom to accurately model $d(n)$.

Without the basic question answered, other issues, such as convergence speed, have not been seriously addressed, other than through computer simulation. Limiting the utility of this analytic direction even more is the fact that virtually every adaptive algorithm discussed in Section 4.4 only approximates the true gradient of the MMSE performance function. Feintuch's algorithm, for example, has been shown neither to follow the gradient nor converge at a stationary point under certain conditions [Johnson and Larimore 1977]. As a result of these analytical difficulties, little work is currently being done on the convergence properties of the MMSE-based algorithms, even though the techniques discussed in the next section might yield useful results.

Algorithm models. Convergence analysis of adaptive FIR digital filters has historically been done by approximating the filter coefficient recursion expression with a linear time-invariant vector first-order process which can then be analyzed to reveal the time constants associated with the decay of various filter error modes [Widrow et al. 1976(2)]. This procedure is very effective for FIR filters in stationary environments principally because the adaptation of filter parameters does not change the data in the filter delay line. In IIR filters this is definitely not true, however, since the covariance of $y(n-i)$, $1 \leq i \leq N$, is directly affected by the choice of $\hat{a}_i$ and $\hat{b}_j$. Even so, such assertions represent a reasonable starting point to determine the behavior of IIR adaptive filter algorithms, particularly in contrast with that of well-known FIR adaptive filtering algorithms.

We briefly examine three increasingly complex techniques for examining convergence behavior. The SHARF algorithm is used to demonstrate the techniques.

1. Local Linearization about a Stationary Point. This approach uses a linearization of convergence behavior within a small neighborhood of a stationary point of the adaptation process. The approach is motivated by similar linearization associated with iterative numerical optimization procedures [Luenberger].

Recall the coefficient vector

$$\hat{\boldsymbol{\theta}}(n) \triangleq [\hat{a}_1(n) \cdots \hat{a}_N(n) \quad \hat{b}_0(n) \cdots b_M(n)]^{\mathrm{T}} \tag{4.52}$$

and consider the underlying $(N + M + 1)$-dimensional vector space. The SHARF algorithm when driven by a random input u generates a stochastic trajectory in this parameter (or weight) space, but its ensemble mean $E[\boldsymbol{\theta}(n)]$ over all possible input sequences is a deterministic locus. Knowledge of the mean locus as a function of time index n provides an estimate of convergence behavior. If this locus for $E[\boldsymbol{\theta}(n+1)]$ can be defined recursively in terms of $E[\boldsymbol{\theta}(n)]$, it can be approximated by a linear expansion around a stationary point as

$$E[\boldsymbol{\theta}(n+1)] \simeq \boldsymbol{\theta}_{\text{opt}} + G(E[\boldsymbol{\theta}(n)] - \boldsymbol{\theta}_{\text{opt}}) \tag{4.53}$$

or in terms of parameter error,

$$E[\boldsymbol{\phi}(n+1)] \triangleq E[\boldsymbol{\theta}_{\text{opt}} - \boldsymbol{\theta}(n+1)] \simeq GE[\boldsymbol{\phi}(n)] \tag{4.54}$$

The matrix G is a "sensitivity" matrix, evaluated at $\boldsymbol{\theta}_{\text{opt}}$ in accordance with the vector Taylor series. Using the recursive form of (4.54), it follows that

$$E[\boldsymbol{\phi}(n)] \simeq G^n \boldsymbol{\phi}(0) \tag{4.55}$$

That is, given initial parameter locations within a small neighborhood of $\boldsymbol{\theta}_{\text{opt}}$, the expected parameter error vector evolves deterministically according to (4.55).

This interpretation is not unprecedented, as it is commonly used to describe the local behavior of a large class of iterative optimization procedures. A subtle difference is that for SHARF, the matrix G is not necessarily symmetric; this is in contrast to function minimization algorithms operating around a stationary point, where G is a Hessian matrix. Nevertheless, convergence behavior is still governed in a geometric fashion by the eigenvalues of this sensitivity matrix G. The fact that G may be asymmetric simply allows the possibility of complex or negative eigenvalues and nonorthogonal eigenvectors.

To examine the behavior predicted via this approach, consider the simple first-order case of SHARF with (4.43), (4.44), and (4.45) reduced, respectively, to

$$\hat{y}(n) = \hat{a}(n)\hat{y}(n-1) + \hat{b}(n)u(n-1) \tag{4.56a}$$

$$v(n) = y(n) - \hat{y}(n) + c[y(n-1) - \hat{y}(n-1)] \tag{4.56b}$$

$$y(n) = ay(n-1) + bu(n-1) \tag{4.56c}$$

If we assume a white, zero-mean input excitation sequence and assume that μ and ρ are small enough to make filter convergence slow compared to the system time constant $1/(1-a)$, then Larimore has shown [Johnson et al. 1981] that the expected parameter vector error can be written as

$$E[\hat{\boldsymbol{\theta}}(n+1) - \hat{\boldsymbol{\theta}}(n)] = M(E(\hat{\boldsymbol{\theta}}(n)) - \boldsymbol{\theta}_{\text{opt}}) \tag{4.57}$$

where

$$M = \begin{bmatrix} \dfrac{-\mu b^2(1+ac)}{(1-a^2)^2} & -\mu(a+c)\dfrac{b}{1-a^2} \\ 0 & \rho \end{bmatrix} \tag{4.58}$$

Rearranging (4.58), it follows that

$$E[\phi(n+1)] = (I+M)E[\phi(n)] = (I+M)^n\phi(0) \tag{4.59}$$

(i.e., $G = I + M$). Note that this same linearization procedure could be employed at any point in the $\hat{a}$, $\hat{b}$ plane, but only a convergence point results in a free system like (4.59) with a zeroed forcing function.

Two observations can be made from (4.59):

(a) The parameter error vector in this case decays to zero geometrically according to eigenvalues $(1-\rho)$ and $1 - \mu b^2(1+ac)/(1-a^2)^2$. Note that the asymmetry in matrix G means that the eigenvectors are not orthogonal.

(b) If c is chosen as $-a$, the error filtering in (4.56b) by c cancels the effect of the autoregression that arises in $y - \hat{y}$. With this choice, the off-diagonal term in G goes to zero, giving a symmetric result, and gradient-descent behavior typical of FIR adaptive algorithms (e.g., LMS) can be expected.

2. Time-Varying Recursion. While linearization of an algorithm's performance in the vicinity of a stationary point can yield valuable asymptotic convergence results, such an approach does not allow the use of arbitrary (i.e., nonlocal) initial conditions for the filter coefficients. Such arbitrary initial conditions can be permitted by allowing the complexity of a time-varying vector model of the form

$$\mathbf{W}(n+1) = \mathbf{H}(n)\cdot\mathbf{W}(n) + \mathbf{F}(n) \tag{4.60}$$

from which the convergence behavior of the algorithm can be analyzed. The matrix **H** generally describes the local curvature of the functional being "searched" while the vector **F** describes the direction of the forcing function. This approach was used by Treichler [Johnson et al. 1981] to explain the effect that the smoothing filter, described by the $\{c_l\}$, can have on the convergence rate of the SHARF algorithm. By assuming very slow convergence, it was shown that the mean coefficient vector can be written as

$$E[\hat{\boldsymbol{\theta}}(n+1)] = [I - \boldsymbol{\Gamma}\,\mathbf{R}_A(n)\mathbf{S}]\cdot E[\hat{\boldsymbol{\theta}}(n)] + \boldsymbol{\Gamma}\,\mathbf{Q}(n)\mathbf{S}\boldsymbol{\theta}_{\text{opt}} \tag{4.61}$$

where $\boldsymbol{\Gamma}$ is a diagonal matrix containing the adaptation constants, **S** is a rectangular matrix containing the $\{c_l\}$, $\mathbf{R}_A(n)$ is a covariance matrix defined by the adaptive filter's input and output, and $\mathbf{Q}(n)$ is a cross-covariance matrix between the adaptive filter vector and that of the plant being modeled.

Since (4.61) is of the same form as (4.60), it may be used to predict a SHARF's convergence rate around any given operating point, including initial and terminal conditions. Using this form it was shown that by proper choice of the smoothing filter (hence **S**) the eigenvalues of $\mathbf{H}(n)$ can be driven complex, unlike FIR adaptive filtering algorithms which have real nonnegative eigenvalues only. As noted in Chapter 3, the smallest eigenvalue of $\mathbf{H}(n)$ controls the convergence rate, while the largest limits the maximum adaptive gain that can be employed. This, in turn, implies that an FIR adaptive filter (e.g., using LMS) may adapt very slowly when $\lambda_{\max}/\lambda_{\min}$ is large. In contrast, where SHARF can force the eigenvalues into complex pairs, the "eigenvalue spread" can be dramatically reduced (e.g., to 1 for a second-order filter), thus speeding convergence.

3. The ODE Technique. Lennart Ljung (1979, 1980) has developed a third technique for proving convergence and determining convergence speed. This approach, termed the ordinary differential equation (ODE) technique, is directly applicable only to algorithms that employ a time-varying decreasing-gain matrix $\boldsymbol{\Gamma}(n)$. The effect of such a decrescent gain is to slow convergence in the vicinity of a stationary point. By exploiting this fact and converting to a different time scale, the discrete-time, nonlinear coefficient vector recursion [e.g., (4.45)] can be well modeled by an ODE of the form

$$\frac{d\boldsymbol{\theta}(\tau)}{d\tau} = f(\boldsymbol{\theta}(\tau)) \tag{4.62}$$

As Ljung points out, (4.62) can be used to attain most of the goals identified earlier. Proof of convergence of the algorithm consists of demonstrating the stability of

(4.62), which depends on the properties of $\mathbf{f}(\cdot)$, and (4.62) itself can be used as a phase-space model of (4.45).

While decreasing adaptation gains are common in system identification problems, they are not typically used in adaptive filtering work since most applications involve signal waveforms which are at best only quasi-stationary. This lack of stationarity requires that the adaptive filter always be "on," so as to track changes in the signal characteristics. The ODE technique requires decrescent gains and therefore has not been used much in the analysis of either FIR or IIR adaptive filters.

Lyapunov technique. The stochastic methods just discussed can provide a great deal of behavioral information, but usually at the price of representing only "average" and/or local information. Another approach to convergence proofs is based on passivity ideas; that is, the combined system and adaptive process are shown to be passive in some sense with respect to the system errors. This passivity implies that the error will decay to zero in time, and therefore that the algorithm converges. We examine here Lyapunov function analysis. As a general rule, we find that such analysis provides much more definite proof of convergence but less convergence behavior (e.g., speed of convergence) than the stochastic techniques.

Lyapunov function analysis [Derusso et al.] is based on finding a function $V(n)$ which has the following properties:

$$V(n) \geq 0 \qquad \text{for all } n \tag{4.63}$$

$$V(n) - V(n-1) < 0 \qquad \text{for all } n \tag{4.64}$$

Since $V(n)$ is always positive, it is sometimes termed a "pseudoenergy" function. The strict negativity of the first difference implies dissipation of that energy, making the process described by V passive. Under suitable regularity conditions, $V(n)$ tends to zero.

This concept is applied to convergence analysis by developing a positive function of all errors appearing in an adaptive processor. If the first difference of that function can be shown to be strictly negative, then, by Lyapunov's second theorem, the error must converge to zero and the adaptive algorithm is proven convergent.

At first glance this technique appears to satisfy many of the stated goals, since the evolution of $V(n)$ appears to provide some general indication of the speed, if not direction, of the convergence. Two points make it difficult to apply, however:

1. Finding a Lyapunov function $V(\cdot)$ for a given adaptive algorithm is often very difficult and, in fact, there is no guarantee that one exists.
2. The "error" used by the Lyapunov function frequently has "extraneous" terms in it which are mandatory to prove convergence but cloud the meaning of $V(\cdot)$.

No Lyapunov-type convergence proof has been discovered for any of the MMSE-based gradient algorithms discussed in Section 4.4 but [Johnson et al.

1981] have successfully built on earlier work in the control and identification literature to develop Lyapunov functions for the HARF and SHARF algorithms. This work concludes that SHARF's convergence can be guaranteed under reasonable circumstances (small μ and ρ, etc.) and that the output error and parameter error approach zero exponentially fast.

4.4.3 General Conclusions

The work on determining the convergence properties of IIR adaptive filters is currently in flux as new algorithms and variants are introduced. The convergence of MMSE-based algorithms has focused on the properties of the MMSE performance surface itself, and this work is inconclusive. The hyperstability-based algorithms have been examined in some detail however, principally in [Johnson et al. 1979(1)], with the following general conclusions:

1. If the adaptation constants in SHARF are chosen small enough (relative to the initial parameter and output estimate errors and driving input magnitude upper bound), then in a direct modeling format SHARF is asymptotically convergent (i.e., the output error decays to zero).
2. Given such sufficiently small adaptation constants, minimal-order direct modeling, and sufficiently rich excitation, both the output and parameter errors decay to zero exponentially fast.
3. SHARF does not follow a gradient descent strategy in the parameter (or combined parameter and output) error space.
4. An effective covariance (or sensitivity) matrix can be defined for SHARF, which is usually asymmetric. This asymmetry can lead to a mean convergence behavior described locally by complex eigenvalues and not the strictly real eigenvalues found in the analysis of adaptive FIR filters.
5. The choice of error smoothing coefficients in SHARF influences this effective covariance matrix and hence the convergence rate and behavior of SHARF. In particular, the choice of these coefficients can expand the allowable range of adaptation constant magnitudes, which translates into a possible improvement in convergence rate.

We note that the techniques described here for convergence analysis presently offer no final answers to the establishment of design rules, a key objective. Further development of these or other approaches is needed to provide guidelines for a priori error smoothing coefficient and adaptation constant selection. Clear at this stage should be the realization that the emerging class of hyperstable adaptive recursive filters exhibits behavior which is quite different from the typically gradient-descent-based adaptive FIR filters. These differences require acceptance of conceptual models which are new to adaptive filtering for proper interpretation of the behavior of adaptive IIR filters [Johnson 1984].

4.5 LIMITATIONS IN THE USE OF IIR ADAPTIVE FILTERS

While IIR adaptive filters offer the potential of relief from the high computational load imposed by FIR digital filtering, there are certain intrinsic limitations to IIR filtering as well. Some accrue from the basic feedback structure of an IIR filter and some hinge on the particular updating algorithm being employed. We examine both.

4.5.1 Coefficient Sensitivity

It is well known that time invariant IIR filters are generally more sensitive to coefficient errors than are FIR filters [Rabiner and Gold 1975]. In general, the frequency response of an IIR filter depends on the accurate placement of a few poles while that of an FIR filter depends on the somewhat less accurate placement of many zeros. This limitation carries over to adaptive IIR filters but with an additional problem. The iterative operation of all algorithms considered here leads to random coefficient noise which tends to dither the coefficients in the vicinity of the desired convergence point. Random movement of the coefficients will cause the filter's poles and zeros to shift. This movement is most extreme for high-Q, high-order, direct-form filters [Rabiner and Gold 1975]. Unfortunately, most of the MMSE-based and all the hyperstability-based IIR algorithms employ direct form filters, and high-Q filters are exactly those which offer the most computational improvement over FIR adaptive filters.

In the short term this problem can be relieved somewhat by employing smaller adaptation gains for the most sensitive coefficients. In the longer run, relief will come in the form of new adaptive algorithms that allow the use of the cascade, parallel, and lattice forms which have effectively replaced the direct form in non-adaptive IIR filters.

4.5.2 Inverse Modeling of Non-Minimum-Phase Filters

Consider again the signal path model shown in Figure 4.1 in which a transmitted signal is effectively filtered by its propagation path [e.g., specular reflection in a line-of-sight (LOS) radio link]. The received signal is then filtered adaptively to compensate for the "channel" filtering and to restore the original waveform. The compensated output is then processed (e.g., demodulated) in the conventional way as though no degradation had occurred. Such compensation can be achieved using either an FIR or IIR adaptive filter if the effective "channel" filter has a minimum phase characteristic; that is, all its zeros are inside the unit circle. Consider the diagram in Figure 4.1(a), however, and assume now that for some reason the reflected signal is stronger at the receiver than the direct transmission. When this occurs, a transmission zero moves outside the unit circle, and an IIR compensating filter must have a pole outside to neutralize the zero. This is unsatisfactory, of course, since keeping a pole outside the unit circle for an extended time will make the filter, and hence the IIR adaptive algorithm, go unstable. Note that in contrast an FIR equalizer is always stable and can compensate for both minimum and non-minimum-phase propagation channels, although sometimes at a great computational cost. This stability problem is not simply resolved and as a practical matter the use of IIR adaptive equalizers is limited to cases where the channel is known to remain minimum phase.

4.5.3 Order Matching

We have assumed here that the adaptive filter has at least as many numerator and denominator coefficients as does the underlying system being modeled. In certain practical cases, this is not true, either due to filter hardware limitations or to a sudden increase in the complexity of the signal to be processed. For FIR filtering the effects of such "reduced-order modeling" are well known but this is not the case for IIR algorithms. Further study is required to reliably predict the effects of reduced-order operation.

4.5.4 Conversion of Stability-Based Techniques to Inverse Modeling

Much of the development of IIR adaptive algorithms, and all of it done here for the hyperstability-based techniques, assumes the direct modeling formulation shown in Figure 4.2. A common input is supplied to both the system under test and to the IIR adaptive filter. The outputs are compared and the adaptive filter's coefficients are adjusted appropriately. While convenient for analysis and typical of some practical problems, most applications are more accurately modeled as shown in Figure 4.3, where the desired signal $d(n)$ is not necessarily a filtered version of the adaptive filter input $x(n)$. With some generality, $d(n)$ can be usually written as a sum of several components, a filtered version of $x(n)$, filtered versions of unobserved inputs, and additive broadband noise. Unfortunately, the performance of algorithms such as SHARF cannot yet be predicted for such inputs. Further analysis of the algorithms will be required to yield such an understanding.

4.6. CONCLUSION

Adaptive algorithms for IIR digital filters have yet to reach the level of maturity attained by their FIR counterparts. Even so, the computational cost of FIR filters and equalizers has served as steady encouragement for the development of simple, generally applicable IIR adaptive algorithms. The analytic effort surveyed in the chapter has led to several possible approaches, but considerable effort remains until the goal of a robust algorithm with well-understood convergence properties is attained.

5

RECURSIVE LEAST-SQUARES ESTIMATION AND LATTICE FILTERS

John M. Turner

5.1 INTRODUCTION

The lattice filter structure is an alternative means of realizing a digital filter transfer function. Although the lattice filter structure (also called the ladder structure) does not have the minimum number of multipliers and adders for a transfer function realization, it does have several advantageous properties. These include cascading of identical sections, coefficients with magnitudes less than 1, stability test by inspection, and good numerical round-off characteristics. Moreover, the lattice filter structure is particularly suited for adaptive filtering since the recursive solution of least-squares estimators naturally produces a lattice filter structure. Also, the lattice filter structure orthogonalizes the input signal on a stage-by-stage basis. This leads to very fast convergence and tracking capabilities of the lattice structure. Although many alternative techniques have been developed to estimate the reflection coefficients that parameterize the lattice structure, the recursive least-squares method updates the least squares estimate upon the observation of each data sample. This procedure leads to an optimal estimate and requires only slightly greater computational burden than alternative techniques.

This chapter presents the derivation of the recursive least-squares lattice filter using a recursive extension of the standard block data Levinson least-squares solution. The linear predictor filter presented here has been widely used for synthesis of speech waveforms (LPC), deconvolution of seismic data, high-resolution spectral estimation, adaptive line enhancement, adaptive noise canceling, and adaptive antenna array processing. The ideas embodied in this estimation technique are derived from the work of many persons since 1970. The approach presented here follows that of [Lee 1980(1)]. Recursive least-squares algorithms have been discussed in Chapter 3 for tapped delay line filters. The discussion presented here shows the natural use of lattice filters for recursive least-squares estimation.

Adaptive estimation techniques modify the estimation filter parameters according to the newly observed data sample. For every new data sample, recursive estimation using the lattice filter generates new reflection coefficients and prediction errors for every filter order. Changing every filter coefficient for each new data sample is important for applications where fast convergence or tracking of quickly time varying signals is required. However, for applications where the dynamics are slow, only the results after observing the signal for a certain time period are important. The recursive algorithms described here can also be used to accumulate signal properties over a particular time period. The procedure for converting between lattice filter coefficients and the more common equivalent tapped delay line filter coefficients is given in Section 5.2.

The mathematics of recursive least-squares estimation requires the updating of variables with time and order subscripts. The algorithms for this are often complicated. Therefore, an intuitive introduction in Section 5.3 presents the nature of the lattice filter structure, the stage-by-stage orthogonalizing property, and analogies with physical phenomena. Section 5.4 briefly presents approximation techniques for determining the reflection coefficients from observed data. The advantage of the lattice filter structure is that *exact least-squares solutions* that are *recursive in time* can be efficiently computed. The development of the recursive least-squares lattice estimation algorithm is presented in Sections 5.5 and 5.6. A square-root normalized least-squares lattice algorithm that has better numerical properties is presented in Section 5.7.

The computational complexity of these algorithms is discussed in Section 5.8. An efficient means of implementing the recursive least-squares algorithm using rotational arithmetic is presented. This rotational arithmetic, called CORDIC arithmetic, is not new, having been used for calculating trigonometric functions in hand-held calculators. The design of an integrated circuit chip to implement the least-squares lattice algorithm using CORDIC arithmetic is mentioned. An example of a hardware realization of the gradient lattice is presented later in Chapter 7.

To demonstrate the power of this adaptive estimator, simulation examples of convergence and tracking, examples using real speech data and electrophysiological data and adaptive equalizer examples are presented in Section 5.9. Since recursive least-squares estimation and lattice filter structures have been a very active area of research, Section 5.10 refers to extensions of the basic ideas.

Since the equations developed here are recursive in order and time, the following notation is used. A variable $x(t)$ is a general time-sampled data value while x_T is the specific data sample T samples after the beginning of the recursion (relative time T). Bold capital-letter variables represent matrices or vectors. When two subscripts are used, the first is the order and the second is the time parameter (i.e., $\mathbf{A}_{i,T}$ is the vector of *i*th-order predictor coefficients determined from data up to the specific time T).

5.2 GENERAL LATTICE DIGITAL FILTER STRUCTURE

The general lattice digital filter is a means of realizing a digital filter transfer function. The lattice structure itself is motivated by similar analog filter structures that have good properties. In this section, the digital lattice filter is introduced and related to the direct-form tapped delay line digital filter.

Since analog lattice and ladder filters have desirable characteristics, digital filters with similar structures were investigated. For example, the third-order *LC* Butterworth filter shown in Figure 5.1(a) is a simple analog lattice type of structure. It is noted for the relative insensitivity of its frequency response to slight perturbations of the circuit element values around their nominal values. This structure can be transformed into the general lattice network shown in Figure 5.1(b). The latter is the lattice structure of interest in this chapter. The analog lattice structure consists of a cascade of identical stages, each stage with a pair of input and output terminals. By developing a digital filter configuration that is similar to the analog lattice structure, the digital filter inherits many of the same properties. Since the structure of a digital filter realization influences its sensitivity to finite-word-length arithmetic, the digital lattice filter has good numerical properties.

The digital lattice filter realizations consist of cascaded stages with two input and two output ports, as in the analog structures. Possible digital lattice configura-

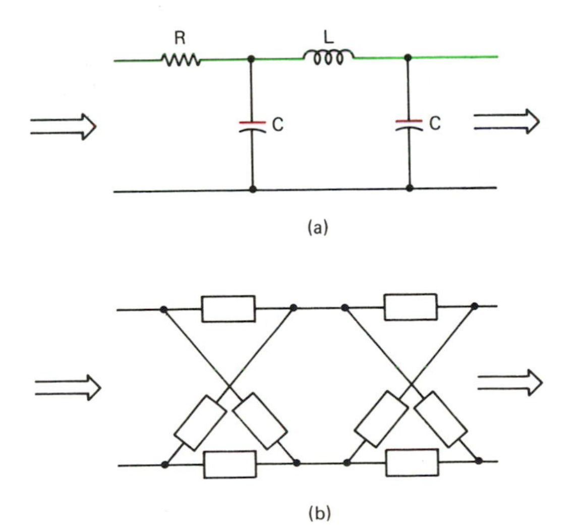

Figure 5.1 (a) Third-order *LC* analog ladder filter; (b) general analog impedance lattice filter.

tions for realizing a general digital transfer function include an asymmetric multiplier form (Figure 5.2) and a symmetric two-multiplier form (Figure 5.3). For the asymmetric multiplier lattice, the structure inside each stage realizes a single pole and zero equivalent transfer function. The algorithm for determining the asymmetric multiplier structure of Figure 5.2 can be found in [Mitra et al. 1977]. This form degenerates to a tapped delay line for either all-pole or all-zero transfer functions and thus is not of interest here. The symmetric two-multiplier form does not degenerate but it requires more multipliers than those required by an equivalent tapped delay line filter. This lattice filter can be modified to a one-multiplier form so as to have the minimum number of multipliers, but this requires extra adders [Figure 5.3(b)].

A cascade of lattice sections, forming a lattice filter, can implement a digital transfer function in a way that has advantages over the standard direct-form, parallel, or standard cascade-form realizations. The cascaded structure in the lattice filter propagates a forward signal $f_j(t)$ and a backward signal $b_j(t)$ at time t and section number j. The fundamental equation describing the lattice filter structure is (see Figure 5.4)

$$\begin{aligned} f_{j+1}(t) &= f_j(t) - k_{j+1} b_j(t-1) \\ b_{j+1}(t) &= b_j(t-1) - k_{j+1} f_j(t) \end{aligned} \tag{5.1}$$

The multipliers in the crossover portion of the lattice, k_i, are known as reflection coefficients or partial correlation (PARCOR) coefficients.

The implementation of digital filter transfer functions in lattice form have been examined [Gray and Markel 1973, 1975]. State-space canonical forms were also established [Morf 1974, Morf et al. 1977, Lee 1980(1)]. Algorithm 5.1 determines the reflection coefficients k_i and tap coefficients v_i for the lattice filter of Figure 5.3 that is equivalent to a (stable nonreducible) direct form transfer function with numerator coefficients b_i^P and denominator coefficients a_i^P (from [Gray and Markel 1973]). The one-multiplier form in Figure 5.3(b) uses coefficients v_i^*. Although the lattice coefficients and the direct-form coefficients are related in a nonlinear manner, this algorithm is invertible, so that a lattice structure can be converted uniquely to a direct-form filter, and vice versa (when all the roots are inside the unit circle).

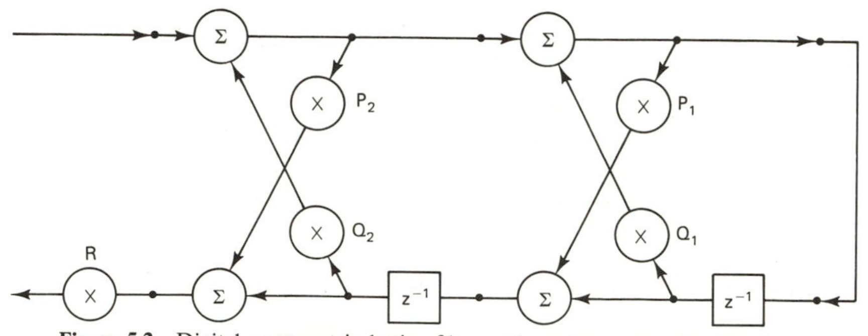

Figure 5.2 Digital asymmetric lattice filter, pole and zero transfer function.

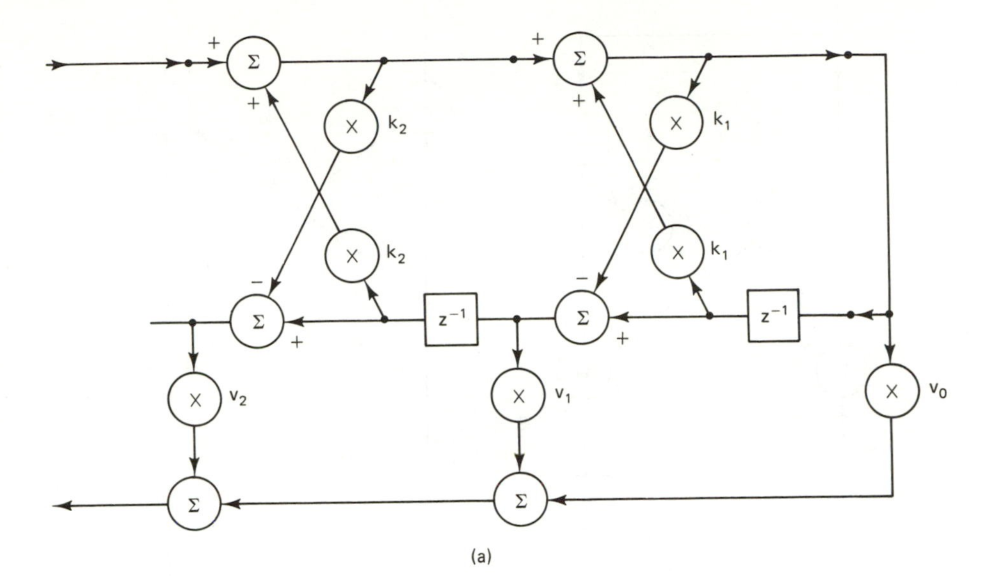

(a)

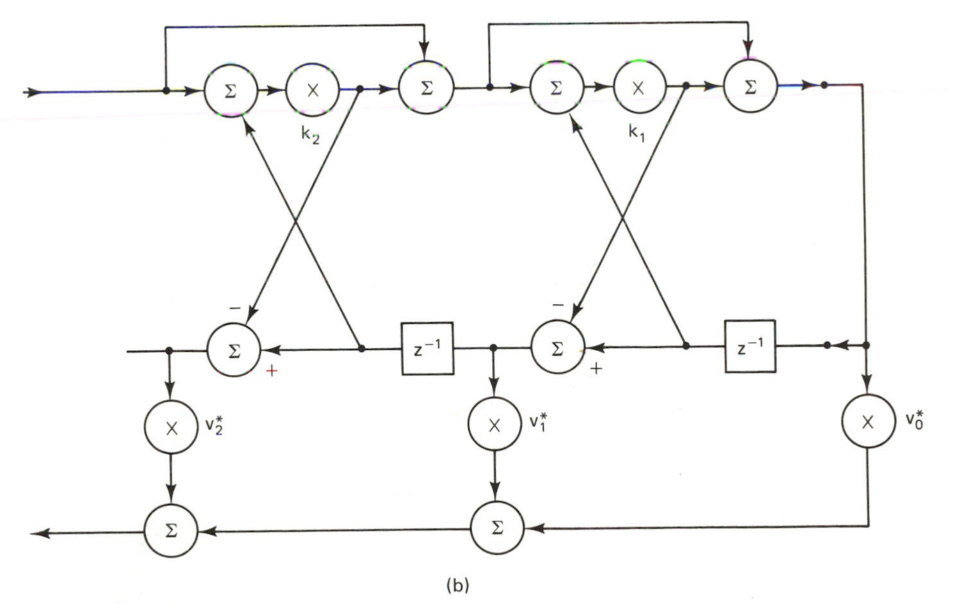

(b)

Figure 5.3 (a) Digital symmetric two multiplier lattice filter, pole and zero transfer function; (b) digital symmetric one-multiplier lattice filter, pole and zero transfer function.

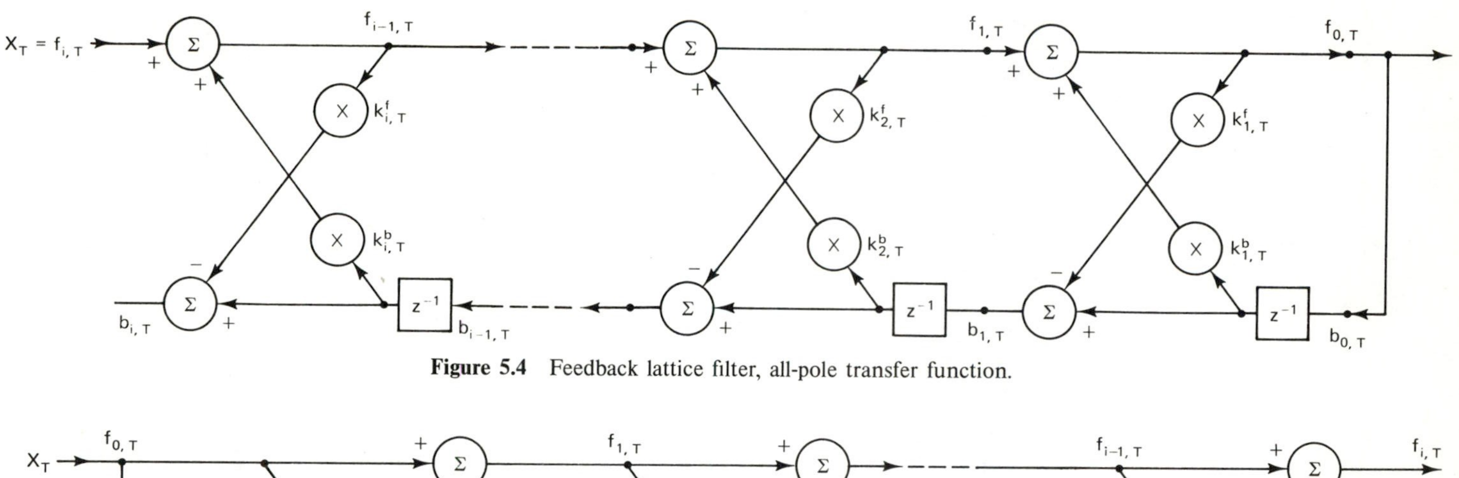

Figure 5.4 Feedback lattice filter, all-pole transfer function.

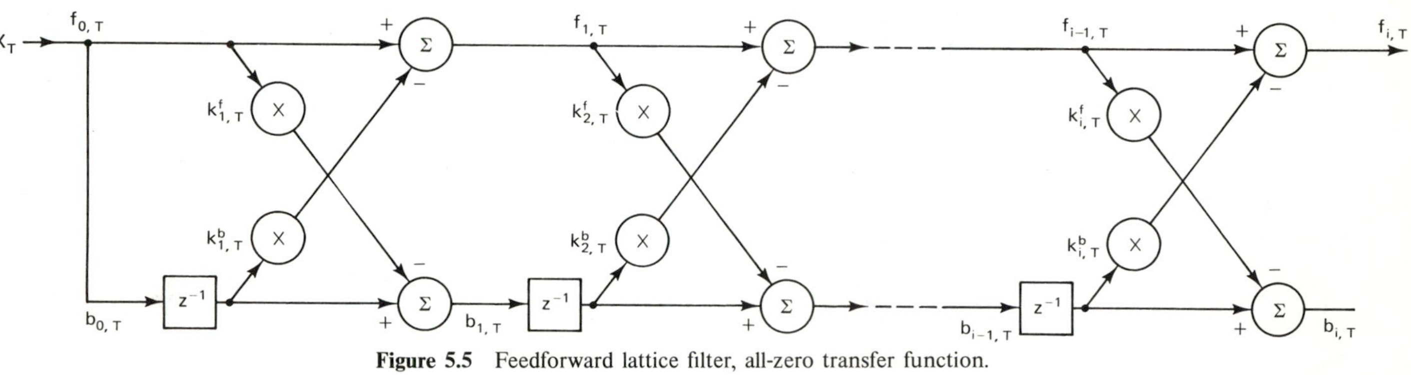

Figure 5.5 Feedforward lattice filter, all-zero transfer function.

Algorithm 5.1: General transfer function to lattice filter

$$H_P(z) = \frac{\sum_{j=0}^{P} b_j^P z^{-j}}{\sum_{j=0}^{P} a_j^P z^{-j}}, \qquad a_0^P = 1$$

$s_P = 1$
For $i = P$ to 1
  $k_i = -a_i^i$
  $v_i = b_i^i$
  $v_i^* = v_i / s_i$
  $s_{i-1} = s_i(1 + k_i)$
  For $j = 1$ to $i - 1$

$$a_j^{i-1} = \frac{a_j^i + k_i a_{i-j}^i}{1 - k_i^2}$$

$$b_j^{i-1} = b_j^i - v_i a_{i-j}^i$$

  continue
  $b_0^{i-1} = b_0^i + v_i k_i$
continue
$v_0 = b_0^0$
$v_0^* = v_o / s_i$
END

While Figure 5.3 and Algorithm 5.1 describe the lattice filter for a general transfer function with poles and zeros, the remainder of this chapter discusses all-pole transfer functions and their inverses, all-zero transfer functions. For an all-pole transfer function ($b_0^P = 1$, $b_j^P = 0$, $j > 1$), the lattice filter, called the feedback lattice filter, is shown in Figure 5.4. The inverse of the feedback lattice can be determined by applying Mason's rule to Figure 5.4. This finite impulse response filter, an all-zero transfer function, is the feedforward lattice (Figure 5.5). Thus a feedforward lattice and a feedback lattice with the same coefficients perform inverse operations on the input signal. If a signal is applied to a feedforward lattice filter and the result is applied to a feedback lattice filter, the original signal is returned. According to Mason's rule, the reflection coefficients parameterize both the feedback and feedforward lattice with the appropriate change in signal flow. Algorithm 5.2 gives the procedure for converting from reflection coefficients to tapped delay line coefficients where the signal flow specifies whether the all-pole or all-zero transfer function is implied.

Algorithm 5.2: Lattice coefficients to tapped delay line coefficients

$a_1^1 = -k_1$
For $i = 2$ to P
  $a_i^i = -k_i$
  For $j = 1$ to $i - 1$

$$a_j^i = a_j^{i-1} - k_i a_{i-j}^{i-1}$$

continue
continue
END

The coefficient sensitivity of the lattice implementation of a general digital transfer function has not been studied as thoroughly as have other common filter structures. Scaling conventions [Markel and Gray 1975(2)] and round-off noise characteristics [Markel and Gray 1975(1)] for finite-word-length arithmetic were developed for various lattice configurations. The one-multiplier lattice has round-off noise characteristics that are always better than those of the two-multiplier lattice. Both lattice filters are always better than the direct-form realization. The lattice filter structures have better round-off noise characteristics than other filter realizations, particularly when the width of the filter passband becomes small. A normalized lattice filter (requiring more multiplications) was developed that performs better than the other lattice structures or parallel-form realizations.

The implementation of a transfer function requires quantized coefficients. This can effect the stability of a filter and its inverse. The sensitivity of the roots of the transfer function to perturbations of the lattice filter and tapped delay line filter coefficients has been investigated [Chu and Messerschmitt 1980 and 1983]. The effect of varying the tapped delay line filter coefficients was the same for each coefficient. When the roots are close to the unit circle, quantization of the tapped delay line coefficients tends to move the roots perpendicular to the unit circle. For an all-pole transfer function, this quantization can cause the poles to move outside the unit circle and the transfer function to become unstable. For lattice filters, the effect on the root location of varying the low-order coefficients is much greater than for the higher-order coefficients. For reflection coefficients, the roots tend to move tangentially to the unit circle, thus changing the center frequency rather than the bandwidth of the roots. Low-order reflection coefficients, particularly those with magnitudes near unity, need to be more accurately quantized than do the higher-order reflection coefficients. No simple rule of thumb exists for tapped delay line filters. The effects of the lattice coefficient quantization have been studied most extensively for speech modeling applications. For typical prediction filters used in speech processing, substantially coarser quantization of the reflection coefficients than of the tapped delay line filter coefficients is possible while maintaining the subjectively perceived spectral response.

5.3 PROPERTIES OF THE LATTICE STRUCTURE

The lattice filter has a more complex structure and requires more numerical operations to implement a transfer function than does the straightforward direct-form realization. However, this increased complexity is offset by several advantageous properties of the lattice structure, including a stability test by inspection, a stage-by-stage orthogonalization of the input signal, and a physical interpretation as wave propagation in a stratified medium. The lattice filter structure naturally evolves from a prediction filter where orthogonality conditions are applied. The properties of the lattice estimation filter are presented in this section. An acoustical tube model of the human vocal tract is interpreted as a lattice filter in Section 5.3.2 to lend a physical interpretation to the reflection coefficients.

5.3.1 Orthogonalizing Properties

In early investigations of the lattice structure, a connection with orthogonal polynomials was noted [Itakura and Saito 1968, Burg 1975, Makhoul 1975, Markel and Gray 1976]. Lattice-form realizations are obtained by an orthogonalization of the state-space transfer function using Szego orthogonal polynomials. The theory of Szego polynomials and their applications in system theory (stability testing) and in stochastic problems (prediction theory and spectral analysis) has been discussed in [Grenander and Szego]. The Schur stability test uses the properties of orthogonal polynomials to determine whether the poles of a transfer function are inside the unit circle and hence a stable transfer function. The test is performed by using Algorithm 5.1 to compute the lattice filter, then checking that the magnitudes of all the reflection coefficients $\{k_i\}$ are less than 1.

For problems in estimation, the minimum mean-square-error estimation equations can be transformed into a stage-by-stage optimization by forming a recursion on the optimum filter order. The parameter in the recursion can be estimated stage by stage since it depends on quantities that are orthogonal between stages. This orthogonalization property is now developed for the prediction filter (an all-zero transfer function).

For a predictor of pth order, the data sample at time t, $x(t)$, is approximated as a linear combination of the previous p samples, $x(t-1), \ldots, x(t-p)$. The forward prediction error $f_p(t)$ must be orthogonal to the previous data samples to attain the minimum mean-square-error value. This determines the weighting factors $\{a_i\}$ on the previous data.

$$\begin{aligned} f_p(t) &= x(t) + a_1 x(t-1) + \cdots + a_p x(t-p) \\ E(f_p(t)x(t-j)) &= 0, \qquad 1 \le j \le p \end{aligned} \tag{5.2}$$

The operation $E(\cdot)$ represents the statistical expectation. This prediction error is also called a prediction residual or an innovation when the coefficients are chosen to attain the minimum mean-square error. A backward prediction error, $b_p(t-1)$, can similarly be defined to predict $x(t-p-1)$ from the same samples, $x(t-1), \ldots, x(t-p)$.

$$\begin{aligned} b_p(t-1) &= x(t-p-1) + c_1 x(t-p) + \cdots + c_p x(t-1) \\ E(b_p(t-1)x(t-j)) &= 0, \qquad 1 \le j \le p \end{aligned} \tag{5.3}$$

Here the $\{c_i\}$ are chosen to satisfy this condition. Notice that both prediction errors satisfy the same orthogonality conditions.

Increasing the prediction order to $p + 1$, $f_{p+1}(t)$ represents the component of $x(t)$ that is not predictable from $x(t-1), \ldots, x(t-p), x(t-p-1)$. The pth prediction error uses information up to $x(t-p)$, so now the information about $x(t)$ that can be predicted from $x(t-p-1)$ must be included. However, much of this information is already contained in $x(t-1), \ldots, x(t-p)$. The backward prediction error $b_p(t-1)$ represents the new information in the sample $x(t-p-1)$. The plausible recursion for $f_{p+1}(t)$ is (5.4), where the scalar k^f_{p+1} is determined so that $f_{p+1}(t)$ satisfies the new orthogonality conditions.

$$\begin{gathered} f_{p+1}(t) = f_p(t) - k^f_{p+1} b_p(t-1) \\ E(f_{p+1}(t)x(t-j)) = 0, \qquad 1 \leq j \leq p+1 \end{gathered} \tag{5.4}$$

The only constraint not immediately satisfied involves $x(t-p-1)$ and is given by (5.5). By substituting (5.3) and (5.4) in (5.5), the optimal k^f_{p+1} is determined (5.6).

$$\begin{aligned} E(f_{p+1}(t)x(t-p-1)) &= 0 \\ 0 = E(f_p(t)x(t-p-1)) &- k^f_{p+1} E(b_p(t-1)x(t-p-1)) \\ = E(f_p(t)b_p(t-1)) &- k^f_{p+1} E(b_p^2(t-1)) \end{aligned} \tag{5.5}$$

$$k^f_{p+1} = \frac{E(f_p(t)b_p(t-1))}{E(b_p^2(t-1))} \tag{5.6}$$

Similarly, the recursion for the backward predictor is obtained (5.7) and the optimal k^b_{p+1} is determined (5.8).

$$\begin{gathered} b_{p+1}(t) = b_p(t-1) - k^b_{p+1} f_p(t) \\ E(b_{p+1}(t-1)x(t)) = 0 \end{gathered} \tag{5.7}$$

$$\begin{gathered} 0 = E(f_p(t)b_p(t-1)) - k^b_{p+1} E(f_p^2(t)) \\ k^b_{p+1} = \frac{E(f_p(t)b_p(t-1))}{E(f_p^2(t))} \end{gathered} \tag{5.8}$$

Extending the prediction filter to the next higher order, $p + 2$, requires calculation of the new prediction errors, f_{p+1} and b_{p+1}, from (5.4) and (5.7). Thus a prediction filter can be constructed solely using the lattice structure by successively increasing the filter order. This is the stage-by-stage orthogonalization property of the lattice structure where each reflection coefficient is determined separately. This stage-by-stage computation of prediction coefficients does not hold for the tapped delay line filter (5.2). The coefficients $\{a_j\}$ are interdependent and they all change when the filter order increases.

Further insight into properties of the prediction errors is provided in [Makhoul 1978(2)]. The backward prediction error results from a Gram–Schmidt type of orthogonalization of delayed versions of the signal. This property of orthogonal variables makes the lattice structure advantageous for adaptive filtering.

Also, the decrease in signal energy after each prediction stage is easily determined. This feature can be used to scale the prediction errors to maintain good numerical properties. The most important properties are summarized here.

$$E(f_p(t)b_j(t-1)) = \begin{cases} \Delta, & j = p \\ 0, & 1 \le j < p \end{cases} \tag{5.9}$$

$$\begin{aligned} E(f_p^2(t)) &= E(f_p(t)x(t)) = \sigma_p^f \\ E(f_p(t)f_j(t)) &= \sigma_p^f, \qquad 1 \le j \le p \end{aligned} \tag{5.10}$$

$$\begin{aligned} E(b_p^2(t-1)) &= E(b_p(t-1)x(t-p-1)) = \sigma_p^b \\ E(b_p(t-1)b_j(t-1)) &= \begin{cases} \sigma_p^b, & j = p \\ 0, & 1 \le j < p \end{cases} \end{aligned} \tag{5.11}$$

$$\begin{aligned} \sigma_{p+1}^f &= \sigma_p^f(1 - k_p^f k_p^b) \\ \sigma_{p+1}^b &= \sigma_p^b(1 - k_p^f k_p^b) \end{aligned} \tag{5.12}$$

When the signal $x(\cdot)$ is stationary with known autocorrelation function, the forward and backward prediction error energies at each stage are identical ($\sigma_p^f = \sigma_p^b$). Then the two reflection coefficients are equal and the symmetric two-multiplier lattice structure computes these prediction error recursions. When the signal to be modeled is assumed to be stationary, a single reflection coefficient, k, is determined by combining sample data estimates of k^f and k^b. This lattice filter with constant coefficients is the feedforward lattice (5.1) of Section 5.2. For non-stationary signals, adaptive estimates are generated by making the reflection coefficients time varying.

The reflection coefficients are closely related to partial correlation factors, which have several interesting statistical properties. The correlation between $x(t)$ and $x(t-p-1)$, after their mutual linear dependence on the intervening samples $\{x(t-1), \ldots, x(t-p)\}$ has been removed, is $E(f_p(t)b_p(t))$. This relation arises from the orthogonalizing nature of the lattice. When this correlation is normalized by the variance of f_p and b_p, it is known as the pth-order partial correlation. The autocorrelation function of a stationary unit variance discrete-time process can be uniquely characterized by a sequence of reflection coefficients, having values less than or equal to 1 [Barndorff-Nielsen and Schou, Ramsey]. For any pth-order AR process, the partial correlation of higher order, lag $p + i$ $(i = 1, 2, \ldots)$, is zero. For a stationary AR process, the sample estimates of the partial correlations are asymptotically Gaussian and independent (see [Murthy and Narasimham] for more statistical properties).

In applications such as noise canceling or equalization, the orthogonalizing properties of the lattice are of primary interest to obtain fast tracking or convergence (see Section 5.6). The backward prediction errors, $b_j(t)$, are extensively used since they are a Gram–Schmidt orthogonalization of delayed versions of the input time series.

5.3.2 Physical Interpretation

The lattice structure and the reflection coefficients have a physical interpretation that for particular classes of signals lends understanding to the properties of the lattice structure. Modeling of wave propagation in a stratified medium leads to a cascade of lattice filters. This model has been applied in seismic signal processing by [Treitel and Robinson, Burg 1967] and others. The physical properties of scattering medium leads to inversion methods based on cascaded reflection elements: for example, the characterization of (electrical) transmission lines [Gopinath and Sondhi 1971] or the human vocal tract [Gopinath and Sondhi 1970]. Similarly, in the fields of acoustics and speech processing, an acoustic transmission line with step changes in impedance leads to a lattice cascade structure. The human vocal tract has been modeled as a cascade of acoustic tube sections with different impedances. This relationship between a physiological system and the lattice structure gives a physical meaning to the reflection coefficients and led to the development of speech synthesis systems using the lattice structure. The remainder of this section develops an acoustical tube model of the vocal tract into a lattice filter (see [Flanagan, Markel and Gray 1976, Rabiner and Schafer 1978]).

A lossless acoustical tube transmission line composed of cascaded cylinders of differing diameter but equal length was developed as a model of the vocal tract in [Kelly and Lochbaum]. This vocal tract model was studied to obtain a better understanding of the speech production mechanism and to synthesize speech by computer. Speech sounds result from pressure waves resonating in the vocal tract (acoustic tube). The significance of the model is that the cascaded cylinders become cascaded lattice stages. The cross-sectional areas of adjacent cylinders specify reflected and transmitted acoustic wave components, which translate into the lattice reflection coefficients.

Sound waves that propagate in a cylindrical section obey the conservation of momentum and mass equations (assuming standard conditions; see [Rabiner and Schafer 1978]). Since the cross-sectional area of the nth tube is constant, combining the conservation laws yields a one-dimensional wave equation. To satisfy this equation, the steady-state volume velocity $u_i(x, t)$ and the pressure wave $p_i(x, t)$ are composed of waves traveling in the forward, u^+, and backward, u^-, directions.

$$
\begin{aligned}
\frac{\partial p}{\partial x} &= -\frac{\rho}{A}\frac{\partial u}{\partial t} \\
\frac{\partial u}{\partial x} &= -\frac{A}{\rho c^2}\frac{\partial p}{\partial t} \\
u_i(x, t) &= u_i^+\left(t - \frac{x}{c}\right) - u_i^-\left(t + \frac{x}{c}\right) \qquad (5.13) \\
p_i(x, t) &= \frac{\rho c}{A_i}\left[u_i^+\left(t - \frac{x}{c}\right) + u_i^-\left(t + \frac{x}{c}\right)\right]
\end{aligned}
$$

The density of air is ρ, c is the speed of sound in air, and A_i is the cross-sectional area of the acoustic tube [see Figure 5.6(a)]. Assuming that all the tubes are of equal length, L, at the boundary between tubes i and $i + 1$, a continuous wave propagation is required.

$$u_i(L, t) = u_{i+1}(0, t)$$
$$p_i(L, t) = p_{i+1}(0, t)$$

Using the boundary conditions, the transmitted wave u_{i+1}^+ and the reflected wave u_i^- across the boundary are determined.

$$\begin{aligned} u_{i+1}^+(t) &= (1 + k_i)u_i^+(t - \tau) + k_i u_{i+1}^-(t) \\ u_i^-(t + \tau) &= -k_i u_i^+(t - \tau) + (1 - k_i)u_{i+1}^-(t) \end{aligned} \tag{5.14}$$

Here $\tau = L/c$ is the propagation time through the tube section and k_i is the wave reflection coefficient at the junction of A_i and A_{i+1}.

$$k_i = \frac{(A_{i+1} - A_i)}{(A_{i+1} + A_i)} \tag{5.15}$$

Since the cross-sectional areas are all positive, $-1 \leq k_i \leq 1$. The wave propagation due to the discontinuity in cross-sectional area is shown in Figure 5.6(a) and (b).

The lattice filter structure is obtained by normalizing variables and grouping time delays. By modifying (5.14), the waves in the ith physical section at the boundary with the $(i + 1)$th section can be written in terms of the $(i + 1)$th section.

$$\begin{aligned} u_i^+(t - \tau) &= \frac{u_{i+1}^+(t) - k_i u_{i+1}^-(t)}{1 + k_i} \\ u_i^-(t + \tau) &= \frac{u_{i+1}^-(t) - k_i u_{i+1}^+(t)}{1 + k_i} \end{aligned} \tag{5.16}$$

An absolute time reference is established at the output of the last tube section, which physically would be at the lips. Assuming that the vocal tract model has p tube sections, the time delay from the beginning of the ith tube to the lips is $t_i = (p + 1 - i)\tau$; so the time variable in the equation for the ith section is replaced by $t - t_i$. A scale factor c_i is introduced to combine the $(1 + k_j)$ factors from the ith tube to the last (pth) section (lips).

$$c_i = \prod_{j=i}^{p} (1 + k_j) = (1 + k_i)c_{i+1}$$

The lattice equations are obtained from (5.16) by using the absolute time reference and defining new variables [see Figure 5.6(c)].

$$\begin{aligned} f_i(t) &= c_i u_i^+(t - \tau - t_i) \\ b_i(t) &= c_i u_i^-(t + \tau - t_i) \\ f_i(t) &= f_{i+1}(t) - k_i b_{i+1}(t - 2\tau) \\ b_i(t) &= b_{i+1}(t - 2\tau) - k_i f_{i+1}(t) \end{aligned} \tag{5.17}$$

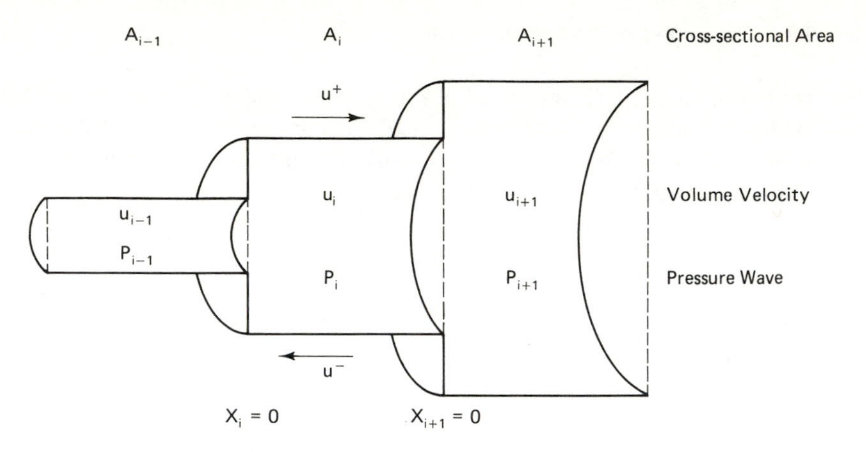

(a) Acoustic tube

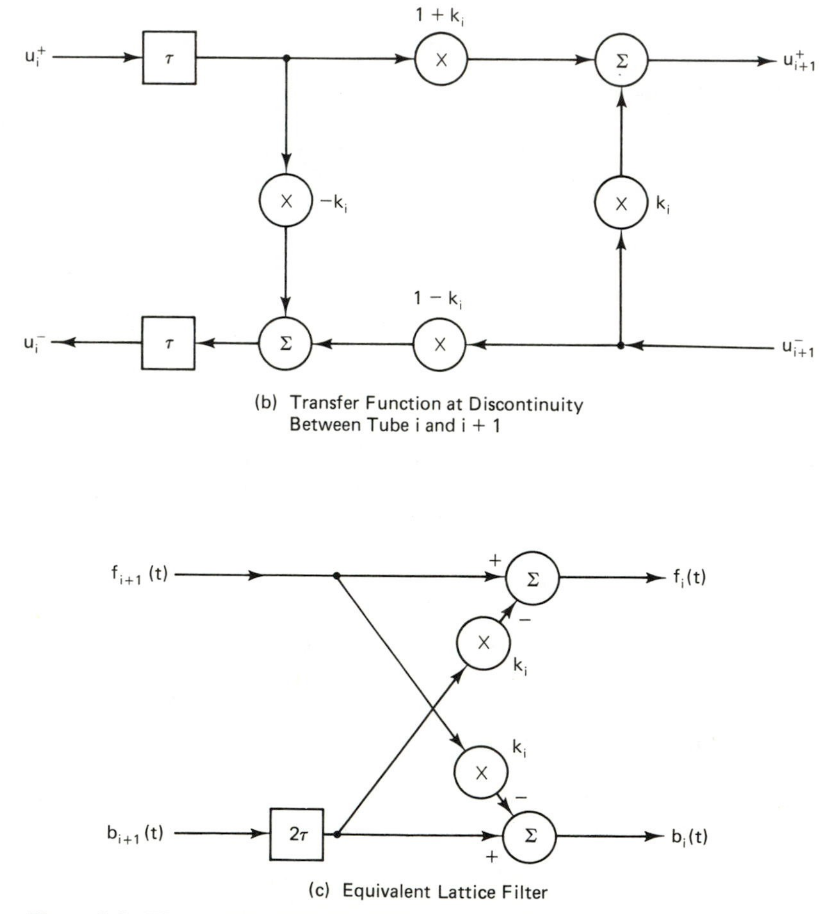

(b) Transfer Function at Discontinuity Between Tube i and i + 1

(c) Equivalent Lattice Filter

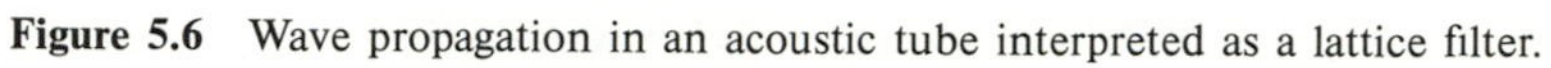

Figure 5.6 Wave propagation in an acoustic tube interpreted as a lattice filter.

This is the same equation as that developed earlier, (5.4) and (5.7), from orthogonality conditions except the unit delay is 2τ and the lattice sections are numbered in decreasing order.

Although modeling of the entire vocal tract includes other influences due to the vocal chords (glottis) and lip radiation, the wave propagation in the mouth ideally follows the lattice structure equations. Studies have indicated that every reasonable vocal tract shape could be generated by a lattice filter and that the reflection coefficients are directly related to the cross-sectional area of the vocal tract [Markel and Gray 1976].

For other types of signals, if they are generated by or can be modeled as wave propagation in a stratified medium, the lattice structure is intuitively motivated. For physically generated processes, the process is often nonstationary but there is a limit to the rate at which a process can change. The shape of the vocal tract (excluding the lips) can only change at a moderately slow rate determined primarily by the muscles in the tongue. Except when a sudden opening of the lips occurs, the cross-sectional area of the vocal tract changes slowly and hence the reflection coefficients also change slowly. This slow time evolution can be used advantageously in adaptive estimation or parameter quantization.

An intuitive understanding of the significance of reflection coefficient values is possible because the lattice filter structure can be thought of as wave propagation in the acoustical tube. The equivalent tapped delay line coefficients are not as easily interpreted. While the reflection coefficients are limited to $-1 \le k_i \le 1$, the equivalent tapped delay line coefficients are often a factor of 10 times larger. When the reflection coefficient is zero, the signal propagates without change since adjoining sections in the tube would have the same cross-sectional area. When the reflection coefficient is near -1, the signal is apt to have highly resonant or oscillator characteristics since if the next tube section is completely closed (i.e., zero cross section), the wave is totally reflected. Conversely, when the reflection coefficient is near $+1$, a decaying signal amplitude is usually found since if the cross-sectional area increases greatly across a boundary, there is full forward radiation. This connection between the physical properties of wave propagation in the acoustical tube and the analogous lattice filter structure greatly aids in an intuitive understanding of the effect of reflection coefficient values on signal characteristics.

5.4 SAMPLE DATA ESTIMATES OF REFLECTION COEFFICIENTS

The real advantage of the lattice filter is for adaptive filtering, where the characteristics of an unknown process are to be determined from the observed data samples. The remainder of this chapter presents the adaptive lattice filter where each new data sample is used to update the reflection coefficients. In this section, approximation techniques that estimate the reflection coefficients based on gradient approaches or sample (block) data estimates of statistical quantities are presented.

The development of a recursive exact solution for least-squares estimation, which naturally produces a lattice structure filter, begins in Section 5.5. A simpler set of recursive equations using normalized variables is presented in Section 5.7.

The earliest techniques for estimating reflection coefficients assumed that the signal was locally stationary. Therefore, sample data approximations were used for the statistical definition of the reflection coefficients, (5.6) and (5.8). When the process $x(\cdot)$ is stationary with known autocorrelation function, the forward and backward prediction error energies at each stage are identical ($\sigma_j^f = \sigma_j^b$). Thus the reflection coefficients for the forward and backward predictors are the same and the lattice filter stage requires a single parameter.

$$\begin{aligned} f_{j+1}(t) &= f_j(t) - k_{j+1} b_j(t-1) \\ b_{j+1}(t) &= b_j(t-1) - k_{j+1} f_j(t) \end{aligned} \tag{5.18}$$

The block data techniques use a time sequence of data and determine a single prediction filter for this entire block of data. A single reflection coefficient per lattice stage is calculated by combining sample data estimates of k^f and k^b. If the geometric mean of k^f and k^b is used, the reflection coefficient becomes the correlation coefficient between f_j and b_j. This parameter, k^I, was originally called a partial correlation (PARCOR) coefficient [Itakura and Saito 1968]. It is the normalized conditional correlation coefficient between $x(t)$ and $x(t-j-1)$ given the intervening data samples, $x(t-1), \ldots, x(t-j)$.

$$k_{j+1}^I = \frac{\sum_{t=1}^{T} f_j(t) b_j(t-1)}{\sqrt{\sum_{t=1}^{T} f_j^2(t) \sum_{t=1}^{T} b_j^2(t-1)}} \tag{5.19}$$

The expression for k_{j+1} that minimizes $E(f_{j+1}^2(t)) + E(b_{j+1}^2(t))$ is the harmonic mean of k^f and k^b. This estimate, k^B, is computationally simpler and is related to Burg's maximum entropy method [Burg 1975].

$$k_{j+1}^B = \frac{\sum_{t=1}^{T} f_j(t) b_j(t-1)}{\frac{1}{2} \sum_{t=1}^{T} [f_j^2(t) + b_j^2(t-1)]} \tag{5.20}$$

These two definitions for reflection coefficients use blocks of T data samples and thus require many computations. For k_{j+1}, the T samples of f_j and b_j are required. To calculate k_{j+2}, the lattice filtering at stage $j+1$ must be performed to obtain T new values for f_{j+1} and b_{j+1}. Clearly, this requires P filtering steps of $2T$ samples to determine P reflection coefficients.

5.4.1 Gradient Estimates of Reflection Coefficients

Adaptive gradient algorithms for determining the reflection coefficients greatly reduce the computational complexity of the estimation technique. Only the prediction errors at the preceding time instant are needed for the gradient methods. The block data approach required all the past error values. Several techniques have been proposed for adapting the reflection coefficients for every newly observed data sample. These techniques do not minimize any criterion but try to change the reflection coefficient in the direction of decreasing prediction error energy (gradient descent). Two classes of gradient techniques either approximate the reflection definition of (5.20) as the current reflection coefficient plus a correction or approximate the numerator and denominator separately. The simplest update of the reflection coefficient [Griffiths 1977] uses the forward and backward prediction errors weighted by the constant α.

$$k_j(t+1) = k_j(t) + \alpha\{f_j(t)b_{j-1}(t-1) + f_{j-1}(t)b_j(t)\} \qquad (5.21)$$

This estimate can be improved by replacing the weighting factor by an energy-normalized term, $1/\sigma_j(t)$, where $\sigma_j(t)$ is the accumulated average of $f_{j-1}^2(t)$ and $b_{j-1}^2(t-1)$ [Griffiths 1978, Makhoul 1978(2)].

$$\begin{aligned} k_j(t+1) &= k_j(t) + \frac{1}{\sigma_j}\{f_j(t)b_{j-1}(t-1) + f_{j-1}(t)b_j(t)\} \\ \sigma_j(t) &= (1-\beta)\sigma_j(t-1) + (f_{j-1}^2(t) + b_{j-1}^2(t-1))\beta \end{aligned} \qquad (5.22)$$

Another adaptive estimate [Makhoul and Viswanathan 1978(1), Makhoul 1978(2)] approximates the numerator and denominator of (5.20) separately. The same weighting factor α is used for both terms. The ratio of these two terms becomes the estimate of the reflection coefficient.

$$\begin{aligned} c_j(t+1) &= (1-\alpha)c_j(t) + 2f_{j-1}(t)b_{j-1}(t-1) \\ d_j(t+1) &= (1-\alpha)d_j(t) + f_{j-1}^2(t) + b_{j-1}^2(t-1) \\ k_j(t+1) &= \frac{c_j(t+1)}{d_j(t+1)} \end{aligned} \qquad (5.23)$$

This ratio is a biased estimator since in general $E(x/y) \neq E(x)/E(y)$ but simulations indicate that this bias is generally very small [Honig and Messerschmitt 1981].

The convergence of the lattice filter is much faster than that of the adaptive tapped delay line filter [Satorius and Alexander 1979(2), Horvath]. This is because the lattice filter tries to orthogonalize the input signal so that the coefficient estimates are decoupled. In fact, the convergence time is almost independent of the eigenvalue spread of the signal (i.e., independent of the signal's spectral dynamic range) [Griffiths 1977]. Quantitative characterizations of the convergence properties of the gradient reflection coefficient estimators (5.22) and (5.23) have been studied [Honig and Messerschmitt 1981]. A two-stage gradient lattice algorithm was compared with a two-stage LMS gradient tapped delay line filter to show that it is possible but unlikely for the tapped delay line filter to converge faster than the lattice filter. A comparison of lattice estimation techniques using the gradient and block data reflection coefficient definitions (5.21), (5.22), and (5.23) has been presented in [Gibson and Haykin].

This orthogonalizing and decoupling property of the lattice is asymptotically obtained only by using the gradient estimation techniques of this section. The recursive least-squares lattice, developed in the next section, exactly solves the orthogonalization for every new data sample. The optimal solution is similar to the energy-normalized gradient lattice (5.22) except that the optimum weighting factor is computed (instead of the constant β). This least-squares lattice estimation technique has even faster convergence than the gradient lattice methods above. However, as the number of data samples (from a stationary process) gets large, the results from the gradient lattice and the least-squares lattice become similar.

5.5 RECURSIVE LEAST-SQUARES LATTICE ALGORITHM

The recursive least-squares lattice (LSL) algorithm allows the exact solution to the least-squares problem to be updated for every newly observed data sample. This adaptive estimation technique uses the properties of the lattice filter to efficiently implement the adaptation. The LSL algorithm looks similar to the gradient techniques of the preceding section except that optimal weighting factors are calculated. The LSL algorithm is developed in this section in the context of an extension to the Levinson algorithm for solving the normal equation.

The least-squares solution to a linear modeling problem can be reduced to a simple set of linear equations called the normal or Yule–Walker equations (see Section 3.2). These equations, which involve the inversion of a covariance matrix, has been widely studied to reduce the computational burden, guarantee stable models, and handle nonstationary processes. The linear predictor form of the linear modeling problem is presented here.

The linear prediction model assumes that a data sample at time T, x_T, can be approximated as $\hat{x}_T$, a weighted sum of previous data samples. For a pth-order linear predictor with coefficients $(a_1, \ldots, a_p)$,

$$\hat{x}_{p,T} = -a_1 x_{T-1} - \cdots - a_p x_{T-p} \tag{5.24}$$

The coefficients are to be chosen so as to minimize the mean-square error between x_T and the estimate, $\hat{x}_{p,T}$. The pth-order covariance matrix of the process $x(\cdot)$ is $\mathbf{R}_p$ and is composed of elements $r_{i,j}$.

$$\begin{aligned} \mathbf{R}_p &= E[\mathbf{x}_{|T:T-p|}\mathbf{x}^{\mathrm{T}}_{|T:T-p|}] \\ \mathbf{x}_{|T:T-p|} &= [x_T \quad x_{T-1} \quad \cdots \quad x_{T-p}]^{\mathrm{T}} \\ r_{i,j} &= E[x_{T-i}x_{T-j}], \qquad 0 \le i, j \le p \end{aligned} \tag{5.25}$$

Minimizing the square of the prediction error with respect to the predictor coefficients $\{a_i\}$ requires that the predictor coefficients satisfy (5.26), called the normal equation, where σ_p is the minimum error.

$$\min_{\{a_i\}} E\{(x_T - \hat{x}_{p,T})^2\} \longrightarrow \mathbf{R}_p \begin{bmatrix} 1 \\ a_1 \\ \cdot \\ \cdot \\ \cdot \\ a_p \end{bmatrix} = \begin{bmatrix} \sigma_p \\ 0 \\ \cdot \\ \cdot \\ \cdot \\ 0 \end{bmatrix} \tag{5.26}$$

Unless the process is a deterministic one, a unique solution to (5.26) exists.

In general, for a pth-order linear model, the solution of the normal equation involves the inversion of the $p \times p$ covariance matrix. Standard matrix inversion methods such as Gaussian elimination require $O(p^3)$ computations (multiplications). However, for stationary random processes, the covariance matrix is a Toeplitz matrix.

$$r_{i,j} = r_{|i-j|}, \qquad 0 \leq i, \quad j \leq p$$

Using the Levinson algorithm, the normal equation in Toeplitz form can be solved in $O(p^2)$ computations. The Levinson algorithm is an order-recursive technique that uses the solution for an ith-order predictor to generate the solution for the $(i+1)$th-order predictor. This algorithm performs an orthogonalization as discussed is Section 5.3. The reflection coefficients are related to the predictor coefficients and are generated as a by-product of this algorithm. In the Toeplitz case, the reflection coefficients can be determined directly without using predictor coefficients [Le Roux and Gueguen]. This is an application of the Schur algorithm. If the covariance sequence $(r_0, r_1, \ldots, r_p)$ is fed into a growing order lattice filter, the reflection coefficient at each stage can be determined by dividing the forward error by backward error at the input of that stage [Morf et al. 1977].

5.5.1 Formulation of Recursive Estimates

In the development of recursive least-squares lattice algorithms, two aspects of the solution of the normal equation are important. The first aspect is the efficient inversion of the covariance matrix, Toeplitz and non-Toeplitz, that gives rise to the order-update recursions. Second, the time-update structure allows exact least-squares solutions to be computed in a recursive manner for each new data sample. This enables the lattice algorithms to achieve extremely fast convergence and excellent tracking capability. The first derivation of the least-squares lattice came from extending the Levinson approach and was based on the predictor coefficients. Once the recursions were obtained, it was noted that they could be written compactly using just the lattice filter parameters. A subsequent direct derivation of the recursive equation, using a geometric approach solely in terms of reflection coefficients and lattice parameters, was reported in [Lee 1980(1), Lee et al. 1981, Shensa 1981]. An overview of the first derivation is presented here since it gives insight into the nature of the recursions. The development of the order- and time-update recursions for the lattice algorithms based on the special structures of the normal equation follows the approach in [Lee 1980(1)].

First the structural properties of the sample covariance matrix must be exploited as the order of the predictor changes and as new time samples are added. The covariance matrix of order p for data samples $\{x_j,\ j = 0, T\}$ used here is the prewindowed form.

$$\mathbf{R}_{i,T} = \mathbf{X}_{i,T}\mathbf{X}_{i,T}^{\mathrm{T}} \tag{5.27}$$

$$\mathbf{R}_{i,T} = \begin{bmatrix} x_0 & \cdots & x_i & \cdots & x_T \\ & \ddots & \vdots & & \vdots \\ 0 & & x_0 & \cdots & x_{T-i} \end{bmatrix} \tag{5.28}$$

The covariance matrix $\mathbf{R}_{i,T}$ has the following structural properties. As a new data sample x_{T+1} is included, the new covariance matrix is composed of the preceding covariance matrix plus a matrix of special form. The time-update matrix equation is

$$\mathbf{R}_{i,T+1} = \mathbf{R}_{i,T} + \begin{bmatrix} x_{T+1} \\ \vdots \\ x_{T-i+1} \end{bmatrix} [x_{T+1} \quad \cdots \quad x_{T-i+1}] \tag{5.29}$$

Also, the covariance matrix of order $i + 1$ contains the covariance matrix of order i. The order-update matrix equations for the covariance are given in (5.30), where $*$ denotes unspecified elements along the outer row and column.

$$\mathbf{R}_{i+1,T} = \begin{bmatrix} * & * \\ * & \mathbf{R}_{i,T-1} \end{bmatrix} = \begin{bmatrix} \mathbf{R}_{i,T} & * \\ * & * \end{bmatrix} \tag{5.30}$$

The development of a recursive solution requires that the normal equation, (5.26), be extended with auxiliary vectors. The forward and backward predictors vectors are $\mathbf{A}_{i,T}$ and $\mathbf{C}_{i,T}$, respectively. A vector $\mathbf{D}_{i,T}$ is defined to account for new data samples. The matrix equation (5.31), an extension of (5.26), defines these vectors.

$$\mathbf{R}_{i,T}[\mathbf{A}_{i,T} \quad \mathbf{C}_{i,T} \quad \mathbf{D}_{i,T}] = \left[\begin{array}{c|c|c} \sigma_{i,T}^{f} & 0 & x_T \\ 0 & 0 & x_{T-1} \\ \vdots & \vdots & \vdots \\ 0 & 0 & x_{T-i+1} \\ 0 & \sigma_{i,T}^{b} & x_{T-i} \end{array}\right] \tag{5.31}$$

$$\mathbf{A}_{i,T} = [1 \quad a_{1,T}^{i} \quad \cdots \quad a_{i,T}^{i}]^{\mathrm{T}}$$

$$\mathbf{C}_{i,T} = [c_{i,T}^{i} \quad \cdots \quad c_{1,T}^{i} \quad 1]^{\mathrm{T}}$$

The forward prediction error $f_{i,T}$ and the backward prediction error $b_{i,T}$ are defined as in (5.2) and (5.3) with $\mathbf{x}_{|T:T-i|}$ defined as in (5.25).

$$\begin{aligned} f_{i,T} &= \mathbf{A}_{i,T}^{\mathrm{T}}\mathbf{x}_{|T:T-i|} \\ b_{i,T} &= \mathbf{C}_{i,T}^{\mathrm{T}}\mathbf{x}_{|T:T-i|} \end{aligned} \tag{5.32}$$

To account for the end of the sample data set, an auxiliary vector $\mathbf{D}_{i,T}$ and related scalar $\gamma_{i,T}$ are introduced.

$$\mathbf{D}_{i,T} = \mathbf{R}_{i,T}^{-1}\mathbf{x}_{|T:T-i|} \tag{5.33}$$

$$\gamma_{i,T} = \mathbf{D}_{i,T}^{\mathrm{T}}\mathbf{x}_{|T:T-i|} = \mathbf{x}_{|T:T-i|}^{\mathrm{T}}\mathbf{R}_{i,T}^{-1}\mathbf{x}_{|T:T-i|} \tag{5.34}$$

This parameter, $\gamma_{i,T}$, can be interpreted as a likelihood variable and is limited to the range $0 \leq \gamma_{i,T} \leq 1$ (see Section 5.5.5).

5.5.2 Order-Update Equations

As in the Levinson solution, an efficient means of determining the nth-order solution is to develop recursive equations for updating the predictor order (at a fixed time). Following the usual development of recursive equations, assuming that the predictor vectors $\mathbf{A}_{i,T}$ and $\mathbf{C}_{i,T}$ are known, the predictors of order $i+1$ are to be determined. Thus the vectors $\mathbf{A}_{i+1,T}$ and $\mathbf{C}_{i+1,T}$ must satisfy the following normal equation:

$$\mathbf{R}_{i+1,T}[\mathbf{A}_{i+1,T} \quad \mathbf{C}_{i+1,T}] = \begin{bmatrix} \sigma_{i+1,T}^{f} & 0 \\ 0 & 0 \\ \vdots & \vdots \\ 0 & 0 \\ 0 & \sigma_{i+1,T}^{b} \end{bmatrix}$$

Since $\mathbf{A}_{i+1,T}$ is to be assembled from $\mathbf{A}_{i,T}$ and $\mathbf{C}_{i,T-1}$, these ith-order predictors are augmented with a final (or initial) zero to make an $i+1$ vector. From (5.30), the augmented $\mathbf{A}_{i,T}$ satisfies the new normal equation except for the last entry.

$$\mathbf{R}_{i+1,T}\begin{bmatrix}\mathbf{A}_{i,T} \\ 0\end{bmatrix} = \begin{bmatrix}\mathbf{R}_{i,T} & * \\ * & *\end{bmatrix}\begin{bmatrix}\mathbf{A}_{i,T} \\ 0\end{bmatrix} = [\sigma_{i,T}^{f} \quad 0 \quad \cdots \quad 0 \quad \Delta_{i+1,T}]^{\mathrm{T}} \tag{5.35}$$

$$\Delta_{i+1,T} = [\text{last row of } \mathbf{R}_{i+1,T}]\begin{bmatrix}\mathbf{A}_{i,T} \\ 0\end{bmatrix} \tag{5.36}$$

Similarly, augmenting $\mathbf{C}_{i,T-1}$ with a leading zero satisfies the normal equations except for the first entry.

$$\begin{aligned}\mathbf{R}_{i+1,T}\begin{bmatrix}0 \\ \mathbf{C}_{i,T-1}\end{bmatrix} &= \begin{bmatrix}* & * \\ * & \mathbf{R}_{i,T-1}\end{bmatrix}\begin{bmatrix}0 \\ \mathbf{C}_{i,T-1}\end{bmatrix} \\ &= [\Gamma_{i+1,T} \quad 0 \quad \cdots \quad 0 \quad \sigma_{i,T-1}^{b}]^{\mathrm{T}} \end{aligned} \tag{5.37}$$

$$\Gamma_{i+1,T} = [\text{first row of } \mathbf{R}_{i+1,T}]\begin{bmatrix}0 \\ \mathbf{C}_{i,T-1}\end{bmatrix}$$

Since $\mathbf{R}_{i+1,T}$ is symmetric, it can be seen that $\Delta_{i+1,T} = \Gamma_{i+1,T}$ [Burg 1975].

By appropriately combining these two equations to cancel the leading or final term in the normal equation, the order-update equations are obtained. Multiplying (5.37) by $\boldsymbol{\Delta}_{i+1,T}/\sigma^b_{i,T-1}$ and then subtracting the result from (5.35), the order-update recursion for $\mathbf{A}_{i,T}$ and $\sigma^f_{i,T}$ are obtained.

$$\mathbf{A}_{i+1,T} = \begin{bmatrix} \mathbf{A}_{i,T} \\ 0 \end{bmatrix} - \begin{bmatrix} 0 \\ \mathbf{C}_{i,T-1} \end{bmatrix} \frac{\boldsymbol{\Delta}_{i+1,T}}{\sigma^b_{i,T-1}} \tag{5.38}$$

$$\sigma^f_{i+1,T} = \sigma^f_{i,T} - \frac{\boldsymbol{\Delta}^2_{i+1,T}}{\sigma^b_{i,T-1}} \tag{5.39}$$

Similarly, the order-update recursions for $\mathbf{C}_{i,T}$ and $\sigma^b_{i,T}$ are obtained.

$$\mathbf{C}_{i+1,T} = \begin{bmatrix} 0 \\ \mathbf{C}_{i,T-1} \end{bmatrix} - \begin{bmatrix} \mathbf{A}_{i,T} \\ 0 \end{bmatrix} \frac{\boldsymbol{\Delta}_{i+1,T}}{\sigma^f_{i,T}} \tag{5.40}$$

$$\sigma^b_{i+1,T} = \sigma^b_{i,T-1} - \frac{\boldsymbol{\Delta}^2_{i+1,T}}{\sigma^f_{i,T}} \tag{5.41}$$

When the order-update equations of the predictors are premultiplied by $[x_T \quad \cdots \quad x_{T-i-1}]$, the lattice equations result. The reflection coefficients are $k^f_{i+1,T}$ and $k^b_{i+1,T}$.

$$f_{i+1,T} = f_{i,T} - k^b_{i+1,T} b_{i,T-1} \tag{5.42}$$

$$b_{i+1,T} = b_{i,T-1} - k^f_{i+1,T} f_{i,T} \tag{5.43}$$

$$k^f_{i+1,T} = \frac{\boldsymbol{\Delta}_{i+1,T}}{\sigma^f_{i,T}} \tag{5.44}$$

$$k^b_{i+1,T} = \frac{\boldsymbol{\Delta}_{i+1,T}}{\sigma^b_{i,T-1}} \tag{5.45}$$

So far the development has been parallel to the nonadaptive approach in Section 5.3. To develop adaptive solutions, time-update equations for the predictors must be developed. Before proceeding to time updates, the order-update recursion for $\mathbf{D}_{i,T}$ is developed in a similar fashion.

$$\mathbf{R}_{i+1,T} \begin{bmatrix} \mathbf{D}_{i,T} \\ 0 \end{bmatrix} = \begin{bmatrix} \mathbf{R}_{i,T} & * \\ * & * \end{bmatrix} \begin{bmatrix} \mathbf{D}_{i,T} \\ 0 \end{bmatrix} = [x_T \quad x_{T-1} \quad \cdots \quad x_{T-i} \quad *]^{\mathrm{T}}$$

The last element of $\mathbf{D}_{i+1,T}$ can be found from the last row of $\mathbf{R}^{-1}_{i+1,T}$, by (5.32) and (5.33).

$$\text{last element } \mathbf{D}_{i+1,T} = \text{last element } (\mathbf{R}^{-1}_{i+1,T})\mathbf{x}_{|T:T-i-1|} = \frac{b_{i+1,T}}{\sigma^b_{i+1,T}}$$

Since the last element of $\mathbf{C}_{i+1,T}$ is 1 and the last element of $\mathbf{D}_{i+1,T}$ has been determined, the order-update equation for $\mathbf{D}_{i+1,T}$ becomes (5.46).

$$\mathbf{D}_{i+1,T} = \begin{bmatrix} \mathbf{D}_{i,T} \\ 0 \end{bmatrix} + \mathbf{C}_{i+1,T} \frac{b_{i+1,T}}{\sigma^b_{i+1,T}} \tag{5.46}$$

The order update for $\gamma_{i,T}$ is determined by premultiplying (5.46) by $[x_T \quad \cdots \quad x_{T-i-1}]$.

$$\gamma_{i+1,T} = \gamma_{i,T} + \frac{b_{i+1,T}^2}{\sigma_{i+1,T}^b} \tag{5.47}$$

5.5.3 Time-Update Equations

Next the time update of the covariance matrix is used to determine time updates for the predictor vectors. From the time update of the covariance matrix (5.29) and the definition of the forward prediction error (5.32);

$$\mathbf{R}_{i,T+1}\mathbf{A}_{i,T} = \begin{bmatrix} \sigma_{i,T}^f \\ 0 \\ \vdots \\ 0 \\ 0 \end{bmatrix} + \begin{bmatrix} x_{T+1} \\ x_T \\ \vdots \\ x_{T-i} \\ x_{T-i+1} \end{bmatrix} [x_T \quad \cdots \quad x_{T-i}]^{\mathrm{T}}\mathbf{A}_{i,T} \tag{5.48}$$

The auxiliary vector $\mathbf{D}_{i,T}$ is used to account for the new data samples.

$$\mathbf{R}_{i,T+1}\begin{bmatrix} 0 \\ \mathbf{D}_{i-1,T} \end{bmatrix} = \begin{bmatrix} * & * \\ * & \mathbf{R}_{i-1,T} \end{bmatrix}\begin{bmatrix} 0 \\ \mathbf{D}_{i-1,T} \end{bmatrix} = \begin{bmatrix} * \\ x_T \\ \vdots \\ x_{T-i} \\ x_{T-i+1} \end{bmatrix} \tag{5.49}$$

The time-update recursion for $\mathbf{A}_{i,T}$ is obtained from (5.48) and (5.49).

$$\mathbf{A}_{i,T+1} = \mathbf{A}_{i,T} - \begin{bmatrix} 0 \\ \mathbf{D}_{i-1,T} \end{bmatrix} [x_T \quad \cdots \quad x_{T-i}]^{\mathrm{T}}\mathbf{A}_{i,T} \tag{5.50}$$

By premultiplying (5.50) by $[x_{T+1} \quad \cdots \quad x_{T-i+1}]$ and using the definition of $\gamma_{i-1,T}$, (5.34), the expression can be simplified.

$$[x_T \quad \cdots \quad x_{T-i}]^{\mathrm{T}}\mathbf{A}_{i,T} = \frac{f_{i,T+1}}{1-\gamma_{i-1,T}} \tag{5.51}$$

The time update for $\mathbf{A}_{i,T}$ becomes a simple expression.

$$\mathbf{A}_{i,T+1} = \mathbf{A}_{i,T} - \begin{bmatrix} 0 \\ \mathbf{D}_{i-1,T} \end{bmatrix} \frac{f_{i,T+1}}{1-\gamma_{i-1,T}} \tag{5.52}$$

From the preceding relation, the time update for $\sigma_{i,T}^f$ can be determined using (5.52) and (5.29).

$$\sigma_{i,T+1}^f = \mathbf{A}_{i,T+1}^{\mathrm{T}}\mathbf{R}_{i,T+1}\mathbf{A}_{i,T+1}$$

$$\sigma_{i,T+1}^f = \sigma_{i,T}^f + \frac{f_{i,T+1}^2}{1-\gamma_{i-1,T}} \tag{5.53}$$

By applying similar techniques, the time-update recursions for $\mathbf{C}_{i,T}$ and $\sigma^b_{i,T}$ are determined.

$$\mathbf{C}_{i,T+1} = \mathbf{C}_{i,T} - \begin{bmatrix} \mathbf{D}_{i-1,T+1} \\ 0 \end{bmatrix} \frac{b_{i,T+1}}{1 - \gamma_{i-1,T+1}} \tag{5.54}$$

$$\sigma^b_{i,T+1} = \sigma^b_{i,T} + \frac{b^2_{i,T+1}}{1 - \gamma_{i-1,T+1}} \tag{5.55}$$

To update the reflection coefficients recursively, the time-update equation for Δ is needed.

$$[0 \quad \mathbf{C}^{\mathrm{T}}_{i,T}]\mathbf{R}_{i+1,T+1} \begin{bmatrix} \mathbf{A}_{i,T+1} \\ 0 \end{bmatrix} = \Delta_{i+1,T+1} \tag{5.56}$$

By using the covariance time update (5.29), the time update for $\mathbf{C}_{i,T}$, and the forward and backward predictors, the time-update equation is obtained.

$$\Delta_{i+1,T+1} = \Delta_{i+1,T} + \frac{b_{i,T} f_{i,T+1}}{1 - \gamma_{i-1,T}} \tag{5.57}$$

The exact time update of $\Delta_{i+1,T}$ is a time average of the cross-correlations between $b_{i,T}$ and $f_{i,T+1}$, with the special gain factor $1/(1 - \gamma_{i-1,T})$. This relates $\Delta_{i+1,T}$ to the partial correlations discussed in the preceding section.

The development of the time-update equations allows the influence of all prior data to be accumulated into the current parameter estimates. If the process that is being approximated contains time-varying attributes, it is desirable to weight more recent observations more heavily. An exponential weighting factor λ on the accumulated covariances can be included in the development of this algorithm [see (3.16)]. Typical values of λ are from 0.98 to 1.00 (corresponding to full weighting of past samples). The algorithms presented in subsequent sections include this exponential weighting factor.

5.5.4 Exact Least-Squares Lattice Recursions

The recursions so far have focused on the predictor vectors, $\mathbf{A}_{\mathrm{i,\,T}}$, $\mathbf{C}_{\mathrm{i,\,T}}$, and $\mathbf{D}_{\mathrm{i,\,T}}$. For a Pth-order prediction filter, these recursions require $O(P^2)$ operations per time sample since all the predictor coefficients change in an order update. However, for the lattice structure, due to its orthogonalizing nature, only the ith reflection coefficient changes in the ith order update. The exact least-squares recursion can be written directly in terms of lattice filter variables which require only $O(P)$ operations to update per time sample.

The order-update recursions for the lattice filter variables $f_{i,T}$ and $b_{i,T}$ were developed in (5.42)–(5.45). The order updates for $\sigma^f_{i,T}$ and $\sigma^b_{i,T}$ are given in (5.39) and (5.41) and the time updates are given in (5.53) and (5.55). The reflection coefficients defined above depend on $\Delta_{i+1,T+1}$, which is related to partial correlations. Here, the time update for $\Delta_{i+1,T+1}$ (5.57) is required to augment the correlation for the new data sample. These updates also require the order update of $\gamma_{i,T}$, determined in (5.47).

These recursions are seen to compute the sample cross-covariance of the forward and backward prediction errors, using the optimal weighting $1/(1-\gamma_{i-1,T})$. From Section 5.4, the gradient lattice equations have the same form as above except that the exact recursive least-squares solution calculates and uses the optimal weighting factor for the new data sample.

The complete set of order-update and time-update recursions to obtain the exact least-squares lattice predictor (LSL) is presented in Algorithm 5.3. When starting the lattice filter, only the stages that receive data are executed, until P data samples have been observed [i.e., min (T, P) filter stages are used, where T is the data sample number].

Algorithm 5.3: Recursive Least-Squares Lattice (LSL)—Scalar case for exponentially weighted data

Input parameters:

$$P = \text{maximum order of lattice}$$
$$\lambda = \text{exponential weighting factor (usually 0.98 to 1.0)}$$
$$x_T = \text{data sample at time } T$$

Variables:

$$\Delta_{i,T} = \text{partial autocorrelation coefficients}$$
$$\gamma_{i,T} = \text{likelihood variable}$$
$$f_{i,T},\, b_{i,T} = \text{forward (backward) prediction errors}$$
$$\sigma^f_{i,T},\, \sigma^b_{i,T} = \text{forward (backward) prediction error covariances}$$
$$k^f_{i,T},\, k^b_{i,T} = \text{forward (backward) reflection coefficient}$$

Initialization:

$$f_{0,0} = b_{0,0} = x_0, \quad \sigma^f_{0,0} = \sigma^b_{0,0} = x_0^2$$
$$\Delta_{i,i} = 0, \quad \gamma_{-1,i} = 0, \quad 1 \le i \le P$$

Iteration for every new data sample:

For data sample x_T and previous results Δ, γ, b, σ^f, σ^b:

$$f_{0,T} = b_{0,T} = x_T$$
$$\sigma^f_{0,T} = \sigma^b_{0,T} = \lambda\sigma^f_{0,T-1} + x_T^2$$

For each stage of the lattice, $i = 0, \min(T, P) - 1$

$$\Delta_{i+1,T} = \lambda\Delta_{i+1,T-1} + \frac{b_{i,T-1}f_{i,T}}{1-\gamma_{i-1,T-1}}$$

$$\gamma_{i,T} = \gamma_{i-1,T} + \frac{b^2_{i,T}}{\sigma^b_{i,T}}$$

$$k^f_{i+1,T} = \frac{\Delta_{i+1,T}}{\sigma^f_{i,T}}$$

$$k^b_{i+1,T} = \frac{\Delta_{i+1,T}}{\sigma^b_{i,T-1}}$$

$$f_{i+1,T} = f_{i,T} - k^b_{i+1,T} b_{i,T-1}$$

$$b_{i+1,T} = b_{i,T-1} - k^f_{i+1,T} f_{i,T}$$

$$\textit{when } T \le P \quad \sigma^f_{i+1,T} = \sigma^f_{i,T} - k^b_{i+1,T}\Delta_{i+1,T}$$

$$\sigma^b_{i+1,T} = \sigma^b_{i,T-1} - k^f_{i+1,T}\Delta_{i+1,T}$$

$$\textit{else} \quad \sigma^f_{i+1,T} = \lambda\sigma^f_{i+1,T-1} + \frac{f^2_{i+1,T}}{1-\gamma_{i,T-1}}$$

$$\sigma^b_{i+1,T} = \lambda\sigma^b_{i+1,T-1} + \frac{b^2_{i+1,T}}{1-\gamma_{i,T}}$$

5.5.5 Likelihood Variable

Information about changes in the nature of the observed process can be determined from $\gamma_{P,T}$. This variable can be interpreted as a sample data approximation of a statistical likelihood variable. For a zero-mean Gaussian random process, x, the joint distribution for $\{x_T \quad x_{T-1} \quad \cdots \quad x_{T-P}\}$ is given by (5.58) with the covariance matrix as defined in (5.25).

$$p(x_T \quad \cdots \quad x_{T-P}) = |2\pi\mathbf{R}_P|^{-1/2} \exp\{-\tfrac{1}{2}\mathbf{x}^{\mathrm{T}}_{|T:T-P|}\mathbf{R}_P^{-1}\mathbf{x}_{|T:T-P|}\} \tag{5.58}$$

The determinant of the covariance matrix $|\mathbf{R}_P|$ is related to the reflection coefficients and the process variance R_0 [Markel and Gray 1976].

$$|\mathbf{R}_P| = R_0 \prod_{i=1}^{P} (1 - K_i^2) \tag{5.59}$$

The logarithm of (5.58) becomes a log-likelihood function composed of two parts. The first two terms depend on the covariance of the process and the third term relies on the data samples observed.

$$\text{log likelihood} = \left\{\ln R_0 + \sum_{i=1}^{P} \ln(1 - K_i^2)\right\} + \mathbf{x}^{\mathrm{T}}_{|T:T-P|}\mathbf{R}_P^{-1}\mathbf{x}_{|T:T-P|} \tag{5.60}$$

The variable $\gamma_{P,T}$ obtained in the exact least-squares recursions can be interpreted as the sample estimate of the third term in (5.60). The definition of $\gamma_{P,T}$ uses the sample estimate of the covariance matrix, $R_{P,T}$, instead of the known covariance matrix, R_P. Thus $\gamma_{P,T}$ is a measure of the likelihood that the P most recent data samples, $\{x_T, \ldots, x_{T-P}\}$ come from a Gaussian process with sample covariance $R_{P,T}$ determined from all the past observations $\{x_j, 0 \le j \le T\}$. Since $0 \le \gamma_{P,T} \le 1$, a small value of $\gamma_{P,T}$ indicates that the recent data samples are likely observations from a Gaussian process with covariance $R_{P,T}$. However, a value of $\gamma_{P,T}$ near 1 implied that given the current Gaussian process assumption, the observations are unexpected; either the new observations come from a different Gaussian process due to a time-varying nature of the physical process or there is a non-Gaussian component in the observations. So $\gamma_{P,T}$ can be used as a detection statistic for changes in the process characteristics or for unexpected (non-Gaussian) components in the observations. Simulations indeed demonstrated that $\gamma_{P,T}$ does take values close to 1 when sudden changes in the observations occurred. In the LSL algorithm, $\gamma_{P,T}$ acts as an optimal gain control since the new observation influences the accumulated estimate by a factor of $(1 - \gamma_{P,T})^{-1}$. With this gain factor, changes in the process statistics can instantaneously influence the estimates more than just being averaged with past observations. Simulation results that demonstrate this behavior on synthetic signals and speech signals are shown in Section 5.9.

5.6 JOINT-PROCESS LATTICE FILTER

Many practical problems require the joint interaction of two or more processes rather than the prediction of a process based on its own past observations. For example, a channel equalizer in its adaptation phase uses the received distorted signal from the channel and the actual transmitted channel symbols to determine the channel distortion characteristics. In a noise canceler, the information signal plus noise and a reference or noise signal are used to extract the information. This joint-process recursive least-squares technique provides very fast tracking or adaptation for channel equalization or noise canceling. The estimation of autoregressive moving-average (ARMA) processes with known input involves the joint interaction of two processes. In general, a multichannel problem can be formulated as a single vector process problem which can be solved by extending the previous least-squares solution to the vector case.

When one process y is to be estimated from observations of a related process x, it is possible to combine them into a joint process (x, y), that can be solved as a joint-process lattice filter. The exact least-squares solution for joint-process estimation is an extension of the development in Section 5.5. A new prediction error is defined that includes samples from both processes. The Pth-order joint prediction error $j_{P,T}$ is the error in estimating y_T from $\{x_T, x_{T-1}, \ldots, x_{T-P}\}$, where $\{g_i^P\}$ are the prediction coefficients obtained by minimizing the sum of the squared errors.

$$j_{P,T} = y_T + \sum_{i=0}^{P} g_i^P x_{T-i} \tag{5.61}$$

The solution of (5.61) can be formulated in terms of the lattice structure just as the single process predictor (5.32) was translated into (5.42). A prediction lattice filter (LSL) for x_T performs a Gram-Schmidt orthogonalization of $\{x_{T-i}\}$ into the mutually orthogonal backward prediction errors $\{b_{T-i}\}$. The advantage of using the orthogonal $\{b_{T-i}\}$ instead of $\{x_{T-i}\}$ in (5.61) is that the joint predictor coefficients $\{g_i^P\}$ become decoupled from one another, so that faster convergence is possible.

The joint-process lattice solution involves the LSL for the x process and a similar lattice recursion of the joint prediction error, $j_{i,T}$. From the LSL for the x

process, at the ith lattice stage, $b_{i,T}$ is the backward prediction error, $\sigma^b_{i,T}$ is its variance, and the likelihood variable is $\gamma_{i-1,T}$. A new cross-correlation term, $\Delta^j_{i,T}$, similar to (5.57), can be defined between signals available after the ith lattice stage, $j_{i-1,T}$ and $b_{i,T}$. This new term can be recursively updated.

$$\Delta^j_{i,T} = \lambda \Delta^j_{i,T-1} + \frac{j_{i-1,T} b_{i,T}}{1 - \gamma_{i-1,T}} \tag{5.62}$$

The recursion for the joint prediction error $j_{i,T}$ is similar to (5.42) except that quantities after the ith lattice stage are used. The initial condition is $j_{-1,T} = y_T$ and the output is $j_{P,T}$.

$$j_{i,T} = j_{i-1,T} - \frac{\Delta^j_{i,T}}{\sigma^b_{i,T}} b_{i,T} \tag{5.63}$$

The previous single-channel LSL equations augmented with (5.62) and (5.63) form the complete solution to the joint estimation problem and lead to the joint-process lattice filter (Figure 5.7).

For noise-canceling problems, noisy data containing the signal of interest, $\{y_T\}$, are observed together with the noise estimate or reference signal, $\{x_T\}$, (see [Satorius et al. 1979(1), Griffiths 1979(2)]). For channel equalization problems, $\{y_T\}$ is a known training sequence sent through the channel and $\{x_T\}$ is the distorted channel output. Applications of the joint-process lattice estimation algorithm to adaptive data equalization have been investigated [Satorius and Alexander 1979(2), Satorius and Pack 1981].

The ARMA estimation problem with known input and with bootstrap estimated input was formulated as a two-channel lattice filter in [Lee et al. 1982]. For an input process y, the output ARMA process is x generated in the following manner:

$$x_T + \sum_{i=1}^{P} a_i x_{T-i} = y_T + \sum_{i=1}^{P} b_i y_{T-i}$$

A prediction equation can be written for the process x if the input y is considered known. This predictor follows from Section 5.5 but now includes a weighted combination of past inputs y.

$$\hat{x}_T = -\sum_{i=1}^{P} a_i x_{T-i} + \sum_{i=1}^{P} b_i y_{T-i}$$

Similarly, a prediction equation for the y process with x known is generated by extending (5.61) to include a weighted combination of past inputs y.

$$\hat{y}_T = -\sum_{i=1}^{P} c_i y_{T-i} + \sum_{i=1}^{P} d_i x_{T-i}$$

The vector process $[x_T, y_T]$ has a structure similar to the scalar AR processes discussed earlier. Prediction errors are now vectors and covariances are matrices. The ARMA lattice estimation algorithm follows the scalar LSL or SQNLSL algorithm, but the quantities are now vectors and matrices. The reflection coefficient has become a 2×2 matrix. Further details are found in [Lee et al. 1982].

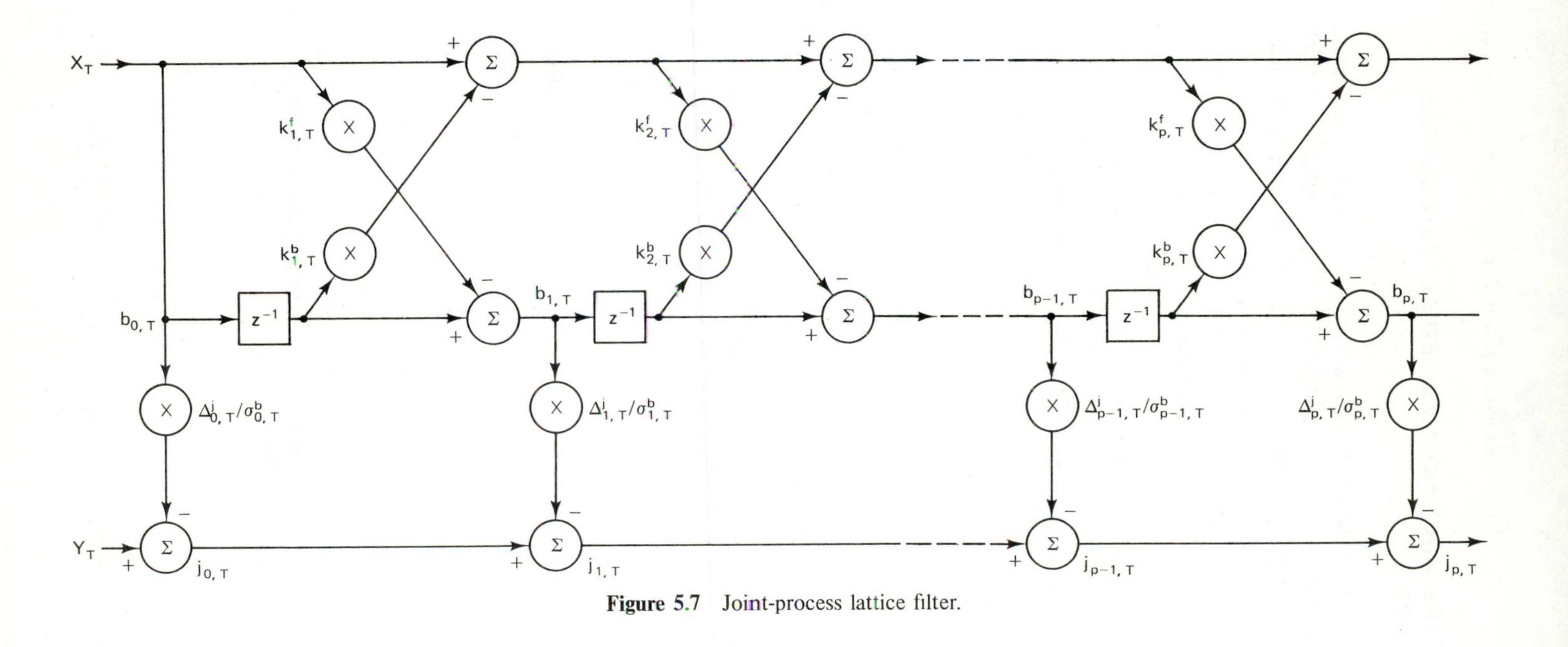

Figure 5.7 Joint-process lattice filter.

5.7 SQUARE-ROOT-NORMALIZED LEAST-SQUARES LATTICE

The complexity of the recursions can be reduced and the numerical properties of the variables improved by rewriting the least-squares lattice algorithm in terms of normalized variables. The square-root-normalized least-squares lattice (SQNLSL) developed in [Lee 1980(1)] has only three recursions per order for each time sample where all three variables have unit variance. The reduction of the LSL into the SQNLSL requires two types of normalizations. A variance normalization scales the variables by their respective variances. A normalization by the optimal weighting factor using the likelihood variable γ is also necessary. A brief development of the SQNLSL is presented here (see [Lee et al. 1981] for more details).

The forward and backward prediction errors when normalized become $\nu_{i,T}$ and $\eta_{i,T}$, respectively. The normalizing factors are the square roots of the variances, $\sigma^f_{i,T}$ and $\sigma^b_{i,T}$, and the square root of the optimal weighting factor $(1 - \gamma_{i-1,T-1})$.

$$
\begin{aligned}
\nu_{i,T} &= f_{i,T}(\sigma^f_{i,T})^{-1/2}(1 - \gamma_{i-1,T-1})^{-1/2} \\
\eta_{i,T-1} &= b_{i,T-1}(\sigma^b_{i,T-1})^{-1/2}(1 - \gamma_{i-1,T-1})^{-1/2}
\end{aligned}
\tag{5.64}
$$

By combining the two reflection coefficients from the LSL, the normalized partial correlation, $\rho_{i,T}$ is defined like a correlation coefficient. This single parameter is the new reflection coefficient.

$$
\rho_{i+1,T} = (\sigma^f_{i,T})^{-1/2}\Delta_{i+1,T}(\sigma^b_{i,T-1})^{-1/2} \tag{5.65}
$$

First the recursion for the partial correlation (5.65) will be developed from the LSL algorithm. The variances of the prediction errors (with exponential weighting λ) has a time-update recursion given by (5.53).

$$
\lambda\sigma^f_{i,T-1} = \sigma^f_{i,T} - \frac{f^2_{i,T}}{1 - \gamma_{i-1,T-1}}
$$

By dividing by $\sigma^f_{i,T}$ and using the definition for $\nu_{i,T}$, the time update for the variance can be related to the new variables (5.66). A similar relation (5.67) exists for the backward prediction error variance.

$$
\frac{\lambda\sigma^f_{i,T-1}}{\sigma^f_{i,T}} = 1 - \nu^2_{i,T} \tag{5.66}
$$

$$
\frac{\lambda\sigma^b_{i,T-1}}{\sigma^b_{i,T}} = 1 - \eta^2_{i,T} \tag{5.67}
$$

The time-update recursion for the normalized partial correlations are obtained by substituting the expression for $\Delta_{i+1,T}$ (5.57) (including the exponential weighting λ) in (5.65).

$$
\begin{aligned}
\Delta_{i+1,T} &= \Delta_{i+1,T-1} + \frac{b_{i,T-1}f_{i,T}}{1 - \gamma_{i-1,T}} \\
\rho_{i+1,T} &= (\sigma^f_{i,T})^{-1/2}\left(\lambda\Delta_{i+1,T-1} + \frac{b_{i,T-1}f_{i,T}}{1 - \gamma_{i-1,T-1}}\right)(\sigma^b_{i,T-1})^{-1/2}
\end{aligned}
\tag{5.68}
$$

The first term is replaced by the definition of $\rho_{i+1,T-1}$ and the second term is $\nu_{i,T}\eta_{i,T-1}$.

$$\rho_{i+1,T} = \lambda(\sigma^f_{i,T})^{-1/2}(\sigma^f_{i,T-1})^{1/2}\rho_{i+1,T-1}(\sigma^b_{i,T-2})^{-1/2}(\sigma^b_{i,T-1})^{1/2} + \nu_{i,T}\eta_{i,T-1}$$

Using (5.66) and (5.67), the new time-update equation for the normalized partial correlation simplifies to (5.69).

$$\rho_{i+1,T} = (1 - \nu^2_{i,T})^{1/2}\rho_{i+1,T-1}(1 - \eta^2_{i,T-1})^{1/2} + \nu_{i,T}\eta_{i,T-1} \tag{5.69}$$

Now the lattice recursions can be written in terms of these new variables. The order-update recursions for the forward prediction errors (5.42) can be written using the normalized partial correlation.

$$\begin{aligned}(1 - \gamma_{i,T-1})^{1/2}\nu_{i+1,T}(\sigma^f_{i+1,T})^{1/2} \\ = (\sigma^f_{i,T})^{1/2}(\nu_{i,T} - \rho_{i+1,T}\eta_{i,T-1})(1 - \gamma_{i-1,T-1})^{1/2}\end{aligned} \tag{5.70}$$

To simplify this expression, two order-update equations from the development of LSL are needed: for the likelihood variable (5.47) and for the prediction error variances.

$$1 - \gamma_{i,T} = (1 - \gamma_{i-1,T})(1 - \eta^2_{i,T})$$

$$\frac{\sigma^f_{i+1,T}}{\sigma^f_{i,T}} = 1 - \rho^2_{i+1,T}$$

Using these relations, (5.70) can be reduced to a simple expression for the normalized forward prediction errors (5.71). A similar development for the backward prediction error leads to (5.72).

$$\nu_{i+1,T} = (1 - \rho^2_{i+1,T})^{-1/2}(\nu_{i,T} - \rho_{i+1,T}\eta_{i,T-1})(1 - \eta^2_{i,T-1})^{-1/2} \tag{5.71}$$

$$\eta_{i+1,T} = (1 - \rho^2_{i+1,T})^{-1/2}(\eta_{i,T-1} - \rho_{i+1,T}\nu_{i,T})(1 - \nu^2_{i,T})^{-1/2} \tag{5.72}$$

The lattice recursions have now become three equations, (5.69), (5.71), and (5.72), that compute the normalized prediction errors, $\{\nu\}$ and $\{\eta\}$, and the reflection coefficients, $\{\rho\}$, for each lattice stage and for every data sample. Proper initialization is required to start the recursions with unit variance quantities.

The reflection coefficients in the SQNLSL still have magnitudes bounded by 1, but now the prediction errors are also bounded. The complexity of the lattice recursions has been reduced from six recursions to only three recursions per order and time update. Square-root operations are required. They can be efficiently computed by bit recursive algorithms such as the CORDIC technique (discussed in the next section). Simple recursions and their potentially better numerical behavior makes the SQNLSL algorithm preferable over the unnormalized LSL version.

The SQNLSL just developed applies for exponentially weighted data. However, the exponential weight λ in (5.68) is not evident in these recursions. By combining the three time-update recursions for $\sigma^f_{i,T}$, $\sigma^b_{i,T}$, and $\Delta_{i+1,T}$ into one recursion for $\rho_{i+1,T}$, the effect of λ is carried through unseen. When a new data sample is used, the exponential weighting is applied to the sample variance estimate. Algorithm 5.4 summarizes the square-root-normalized least-squares lattice

(SQNLSL) estimation method. The sample variance R_T is initialized to some value σ to avoid dividing by zero.

Although SQNLSL is a very powerful and compact algorithm, the necessity of computing square roots can lead to problems. The fixed-point error analysis of this algorithm [Samson and Reddy] indicated that finite-word-length arithmetic computation of the square roots led to small biases in the reflection coefficients. This bias was more predominant than the variance of the error in the estimate and generally quite small. The bias increased as the word length became shorter or the exponential weighting factor λ approached 1.

Algorithm 5.4: Square-root-normalized least-squares lattice (SQNLSL)—exponentially weighted scalar data

Input parameters:

$$P = \text{maximum order of lattice filter}$$

$$\lambda = \text{exponential weighting factor (usually 0.98 to 1.0)}$$

$$\sigma = \text{prior variance}$$

$$x_T = \text{data sample at time } T$$

Variables:

$$R_T = \text{estimated variance of } x$$

$$\rho_{i,T} = \text{reflection coefficients}$$

$$\nu_{i,T} = \text{normalized forward prediction error}$$

$$\eta_{i,T} = \text{normalized backward prediction error}$$

Initialization:

$$R_o = \sigma + x_o^2$$

$$\nu_{o,o} = \eta_{o,o} = \frac{x_o}{\sqrt{R_o}}$$

$$\rho_{i,o} = 0, \qquad 1 \le i \le P$$

Iteration for every new data sample:

New data sample x_T and previous results $\eta_{i,T-1}$, $\rho_{i+1,T-1}$, R_{T-1}:

$$R_T = \lambda R_{T-1} + x_T^2$$

$$\nu_{o,T} = \eta_{o,T} = \frac{x_T}{\sqrt{R_T}}$$

For each stage of the lattice, $i = 0$ to $\min(T, P) - 1$

$$\rho_{i+1,T} = \sqrt{1 - \nu_{i,T}^2}\sqrt{1 - \eta_{i,T-1}^2}\,\rho_{i+1,T-1} + \nu_{i,T}\eta_{i,T-1}$$

$$\nu_{i+1,T} = \frac{\nu_{i,T} - \rho_{i+1,T}\eta_{i,T-1}}{\sqrt{1 - \rho_{i+1,T}^2}\sqrt{1 - \eta_{i,T-1}^2}}$$

$$\eta_{i+1,T} = \frac{\eta_{i,T-1} - \rho_{i+1,T}\nu_{i,T}}{\sqrt{1 - \rho_{i+1,T}^2}\sqrt{1 - \nu_{i,T}^2}}$$

5.8 COMPUTATIONAL COMPLEXITY AND CORDIC ARITHMETIC

The complexity of the lattice filter is greater than the equivalent tapped delay line filter. However, the lattice filter has many advantageous properties not shared by the simpler filter. Similarly, the adaptive lattice filter has a few more operations than does an adaptive tapped delay line filter. Table 5.1 compares the computational complexity of several adaptive estimation algorithms. The lattice methods require three to six times as many computations as the simplest adaptive tapped delay line filter (LMS). However, this increase in computational complexity provides for substantially faster convergence, better numerical properties of the coefficients, and an assurance of a stable filter. The complexity of several adaptive algorithms is presented in the following table. The scaling by a constant weighting factor (e.g., λ or β is usually approximated as a shifting by a power of 2). Thus this fixed scaling is not included in the count of operations. The LMS algorithm is the tapped delay line gradient least-mean-squares technique. The gradient lattice algorithms are (5.22) and (5.18). Algorithm 5.3 is denoted LSL and Algorithm 5.4 is called SQNLSL. The number of operations are counted for executing a single filter stage on a single data sample. To process T data samples in an Nth-order filter would require NT times as many computations.

TABLE 5.1 Computational Complexity

Algorithm	$\times$	$\div$	$\sqrt{}$	$\pm$
LMS	2	0	0	2
Gradient lattice	6	1	0	6
LSL	6	6	0	7
SQNLSL	10	2	3	6

The SQNLSL algorithm is the most complex of the algorithms, requiring three square roots, 10 multiplications, and two divisions to execute an update for each stage in the lattice for every new data sample. However, the SQNLSL has a very compact form with only three equations involving variables that have constrained magnitudes. The implementation of the SQNLSL equations in hardware would require special multiplier hardware for fast execution or would require considerable execution time on general-purpose microprocessors. Even using shift-and-add instead of multiplication and Newton's method for square-root operations would require considerable execution time.

By interpreting the lattice equations as rotations, an efficient realization of the square-root-normalized lattice algorithm using CORDIC arithmetic was developed in [Ahmed 1981(1)]. The Coordinate Rotation Digital Computer (CORDIC) developed in [Volder] is an iterative algorithm for computing trigonometric functions, multiplications, divisions, and square roots. The CORDIC algorithm interprets the foregoing functions as rotations of a vector in different coordinate systems. The rotation is implemented by a sequence of shift-and-add operations. This type of arithmetic is not new; it has been used to compute trigonometric functions and their inverses in hand-held calculators. Many other signal processing algorithms, such as DFT and matrix inversion, can also be implemented in arrays of CORDIC processors [Ahmed et al. 1982].

5.8.1 CORDIC Arithmetic

The well-known equation for rotating a vector $[x_i, y_i]^T$ to a new vector $[x_{i+1}, y_{i+1}]^T$ uses the *sine* and *cosine* of the rotation angle θ.

$$\begin{bmatrix} x_{i+1} \\ y_{i+1} \end{bmatrix} = \begin{bmatrix} \cos\theta & \sin\theta \\ -\sin\theta & \cos\theta \end{bmatrix} \begin{bmatrix} x_i \\ y_i \end{bmatrix}$$

Four multiplications by two trigonometric quantities are required. This operation can be made more amenable to fast computer implementation by modifying this equation into a sequence of small rotations of a specific form, each implemented using only additions and shifts.

CORDIC arithmetic was unified into a single equation [Walther] that allows rotations on either a circle $(m = 1)$, along a line $(m = 0)$, or along a hyperbola $(m = -1)$. The incremental unit of rotation at the ith iteration is the predetermined sequence $\{\delta_i\}$ and $\mu_i = \pm 1$ that determines the direction of rotation.

$$\begin{bmatrix} x_{i+1} \\ y_{i+1} \end{bmatrix} = \begin{bmatrix} 1 & m\mu_i\delta_i \\ -\mu_i\delta_i & 1 \end{bmatrix} \begin{bmatrix} x_i \\ y_i \end{bmatrix} \tag{5.73}$$

The $\{\delta_i\}$ must be chosen to satisfy certain constraints that assure convergence of the iterations. To obtain computational efficiency on current computer hardware using binary representations, a negative integer power of 2 is chosen for δ_i. This allows the multiplication by δ_i to be performed as a shift.

The effect of (5.73) is a rotation and a change of scale interpreted in the appropriate coordinate space. The vector $[x_0, y_0]^T$ can be represented as a generalized radial component R_0 and a generalized angular component, Φ_0.

$$R_0 = \sqrt{x_0^2 + my_0^2}$$

$$\Phi_0 = \sqrt{m}\tan^{-1}\frac{y_0\sqrt{m}}{x_0}$$

For the circular rotation, this is a true polar coordinate representation. Performing the operation in (5.73) scales the radial component by $r_i = \sqrt{1 - m\delta_i^2}$ and changes the angular component by $\phi_i = m^{-1/2}\tan^{-1}(\delta_i\sqrt{m})$. After p iterations, the new radial and angular components are R_p and Φ_p.

$$\begin{aligned} R_p &= R_0\prod_{i=1}^{p} r_i = R_0\prod_{i=1}^{p}\sqrt{1 + m\delta_i^2} \\ \Phi_p &= \Phi_0 - \sum_{i=1}^{p}\mu_i\phi_i = \Phi_0 - \sum_{i=1}^{p}\mu_i m^{-1/2}\tan^{-1}(\delta_i\sqrt{m}) \end{aligned} \tag{5.74}$$

The convergence of the iterations and the efficient implementation of the rotations depend critically on the predetermined choice of δ_i. Each type of rotation, $(m = -1, 0, +1)$ has a different predetermined set of positive increments (δ_i) which specify fixed radial and angular increments (r_i, ϕ_i) from (5.74). Within the domain of convergence (limited by the total possible rotation) constraints were developed on the sequence $\{\phi_i\}$ such that any angle Φ could be rotated to within ϕ_{p-1} of zero in p steps [Walther]. This guarantees that the granularity of the calculation (the angular resolution) is ϕ_{p-1} in p steps. With the proper choice of the set of increments $\{\delta_i\}$, each successive iteration yields approximately one more bit of accuracy in the final result.

The CORDIC equation, (5.73), is augmented by an additional variable z_p that accumulates the angular component of the rotation.

$$z_p = z_0 - \sum_{i=1}^{p} \mu_i \phi_i \tag{5.75}$$

The three parameters (x, y, z) are manipulated by successive application of the CORDIC equation. The function to be computed is obtained by forcing either y or z to zero. Often the initial value of the other variable, z or y, is zero or 1. The sign of the rotation, μ_i is chosen at each iteration to move the desired parameter toward zero. Rotation operations are obtained by forcing z_p to zero, and vectoring operations result if y_p is forced to zero.

The interpretation of (5.73) as a rotation on a circle, line, and hyperbola can be seen from a graphical perspective. For example, a circular rotation $(m = 1)$ of a vector $[x_o, y_o]^T$ into the vector $[x_p, 0]^T$ computes $\tan^{-1}(y_o/x_o)$. The direction of rotation μ_i is chosen at each iteration to force y_i closer to zero. The sequence of small angular step $\{\phi_i\}$, predetermined by $\{\delta_i\}$, is accumulated with appropriate sign in z_p, giving the answer. Figure 5.8(a) indicates how this rotation proceeds. The radius of the circle increases a predetermined amount r_i with each rotation step. It is not necessary to account for this change in radius when computing $\tan^{-1}(y_o/x_o)$. However, the value of x_p has become a scaled square root where the scale factor is known in advance.

$$x_p = \sqrt{x_0^2 + y_0^2} \prod_{i=1}^{p} r_i$$

The three-input/output box of Figure 5.8(a) is used to describe the function evaluated by the CORDIC rotation. With a nonzero initial value of z_0, forcing z_i to zero generates $\sin z_0$ and $\cos z_0$ [see Figure 5.8(a)].

For a rotation on a line, the radial component is always 1 and the angle component is interpreted as the y_i value. The increments become $r_i = 1$ and $\phi_i = \delta_i$. The result of applying the CORDIC operation is shown in Figure 5.8(b). Multiplication and division can be calculated this way.

The hyperbolic rotation computes sinh, cosh, arctanh, and square roots [see Figure 5.8(c)]. The surface of rotation is the set of points that is a constant distance $\sqrt{x^2 - y^2}$ from the origin. This hyperbola moves a fixed amount, $\sqrt{1 - d_i^2}$, closer to the origin after each CORDIC operation.

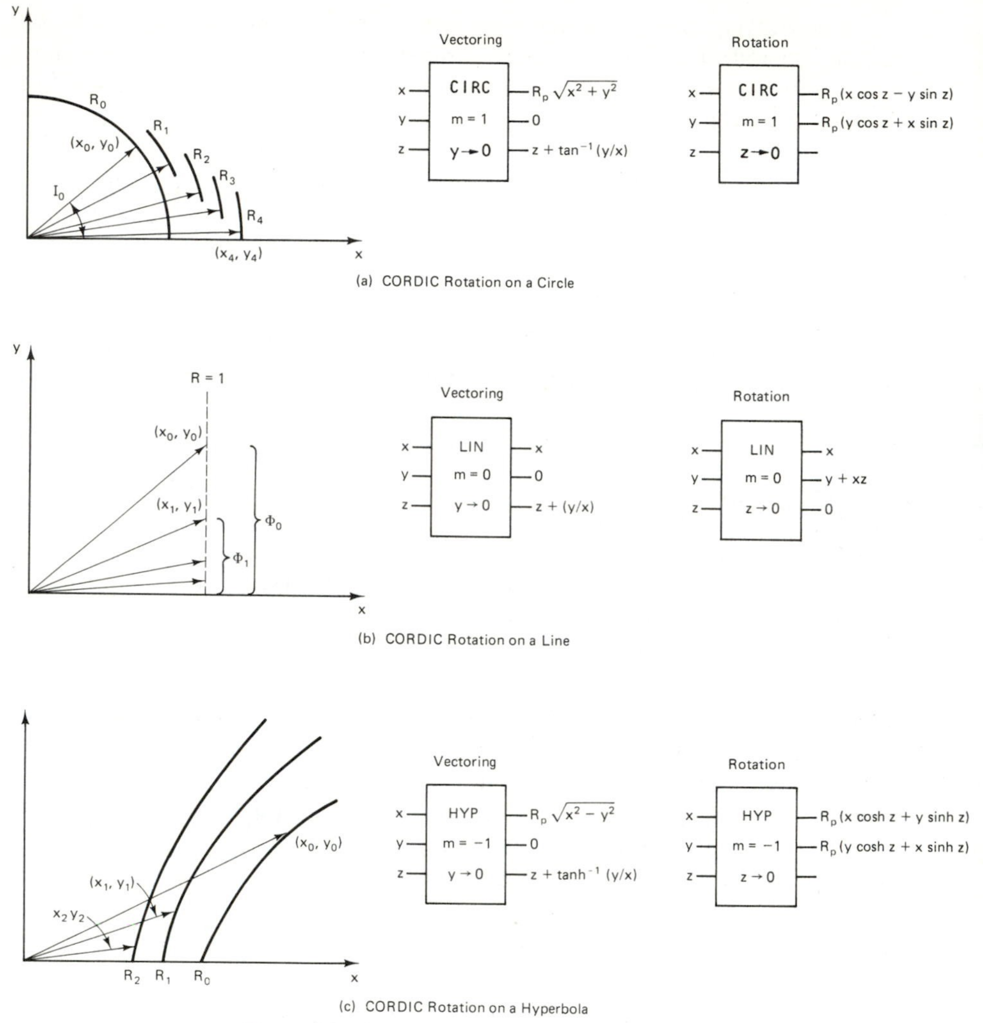

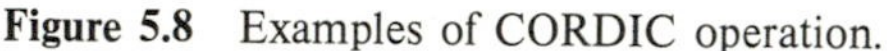
Figure 5.8 Examples of CORDIC operation.

5.8.2 Lattice Filtering by Rotations

The square-root-normalized lattice equations have a natural interpretation as rotations. The three recursive equations can be efficiently realized using CORDIC arithmetic [Ahmed et al. 1981(1) and 1981(2)]. The structure of the SQNLSL algorithm suggests an implementation via rotations. Since ν, ρ, and η always have magnitudes less than 1, they can be interpreted as cosines of angles. Furthermore, if $x = \cos \theta_x$, the complement of x is $x^c = \sqrt{1 - x^2} = \sin \theta_x$. The SQNLSL equations can be written using a compact notation.

$$\rho_{i+1,T} = \sqrt{1 - \nu_{i,T}^2}\sqrt{1 - \eta_{i,T-1}^2}\,\rho_{i+1,T-1} + \nu_{i,T}\eta_{i,T-1} \longrightarrow \rho_+ = \nu^c\eta^c\rho + \nu\eta$$

$$\nu_{i+1,T} = \frac{\nu_{i,T} - \rho_{i+1,T}\eta_{i,T-1}}{\sqrt{1 - \rho_{i+1,T}^2}\sqrt{1 - \eta_{i,T-1}^2}} \longrightarrow \nu_+ = \frac{\nu - \rho_+\eta}{\rho_+^c\eta^c}$$

$$\eta_{i+1,T} = \frac{\eta_{i,T-1} - \rho_{i+1,T}\nu_{i,T}}{\sqrt{1 - \rho_{i+1,T}^2}\sqrt{1 - \nu_{i,T}^2}} \longrightarrow \eta_+ = \frac{\eta - \rho_+\nu}{\rho_+^c\nu^c}$$

For notational convenience, the following abbreviations were used:

$$\rho = \rho_{i+1,T-1} \qquad \rho_+ = \rho_{i+1,T}$$

$$\nu = \nu_{i,T} \qquad \nu_+ = \nu_{i+1,T}$$

$$\eta = \eta_{i,T-1} \qquad \eta_+ = \eta_{i+1,T}$$

The SQNLSL update equations can be written almost entirely in a single matrix equation (5.76) using the rotation matrices for ν and η.

$$\begin{bmatrix} \nu^c & \nu \\ -\nu & \nu^c \end{bmatrix}\begin{bmatrix} \rho & 0 \\ 0 & 1 \end{bmatrix}\begin{bmatrix} \eta^c & -\eta \\ \eta & \eta^c \end{bmatrix} = \begin{bmatrix} \nu^c\eta^c\rho + \nu\eta & \nu\eta^c - \rho\eta\nu^c \\ \nu^c\eta - \rho\nu\eta^c & \nu^c\eta^c + \rho\nu\eta \end{bmatrix} = \begin{bmatrix} \rho_+ & \rho_+^c\nu_+ \\ \rho_+^c\eta_+ & * \end{bmatrix} \tag{5.76}$$

On the left-hand side of (5.76), the first matrix performs a rotation by $\theta_\nu = \cos^{-1}\nu$ and the third matrix rotates by $\theta_\eta = \cos^{-1}\eta$. The result is the complete update for ρ and partial updates for ν and η and a term $(*)$ of no interest. The updates for ν and η are completed by dividing by ρ_+^c. This matrix equation (5.76) is directly realizable using CORDIC operations.

The implementation of the SQNLSL algorithm in an integrated circuit proposes using two processors in parallel, each executing sequentially five functions [Ahmed et al. 1981(1) and 1981(2)]. The computation proceeds from (5.76) in three steps: an angular rotation by θ_ν, the multiplication of ρ matrix by the η rotation matrix, and the divisions by ρ_+^c. The sequence of CORDIC operations shown in Figure 5.10 begins with ρ, η, and ν and computes ρ_+, η_+, and ν_+. The functional elements use the notation of Figure 5.9. The rotation angle θ_ν is calculated as $\tan^{-1}(\nu/\nu^c)$ using a circular CORDIC operation by processor 2 in time slots 1 and 2. Rotating the ρ matrix by θ_η is computed as two multiplications (linear CORDIC): $\rho\eta^c$ by processor 1 during time slots 1 and 2, $\rho\eta$ by processor 2 in time slot 3. These quantities are rotated by θ_ν using circular CORDIC operations in processor 1 (time slot 3) and processor 2 (time slot 4). This generates the ρ update and partial updates for ν and η. In time slot 5, the processors generate the updates for ν and η by dividing the earlier results by ρ_+^c. The signals that flow farther than adjacent time slots must be held in temporary buffers. Each CORDIC operation uses 16 iterations and results in almost 16 bits of accuracy. The integrated circuit could perform the SQNLSL algorithm of tenth order on an 8-kHz sampled signal in real time. This assumes standard integrated circuit design rules to generate a moderate-size chip running at a 20-MHz clock rate.

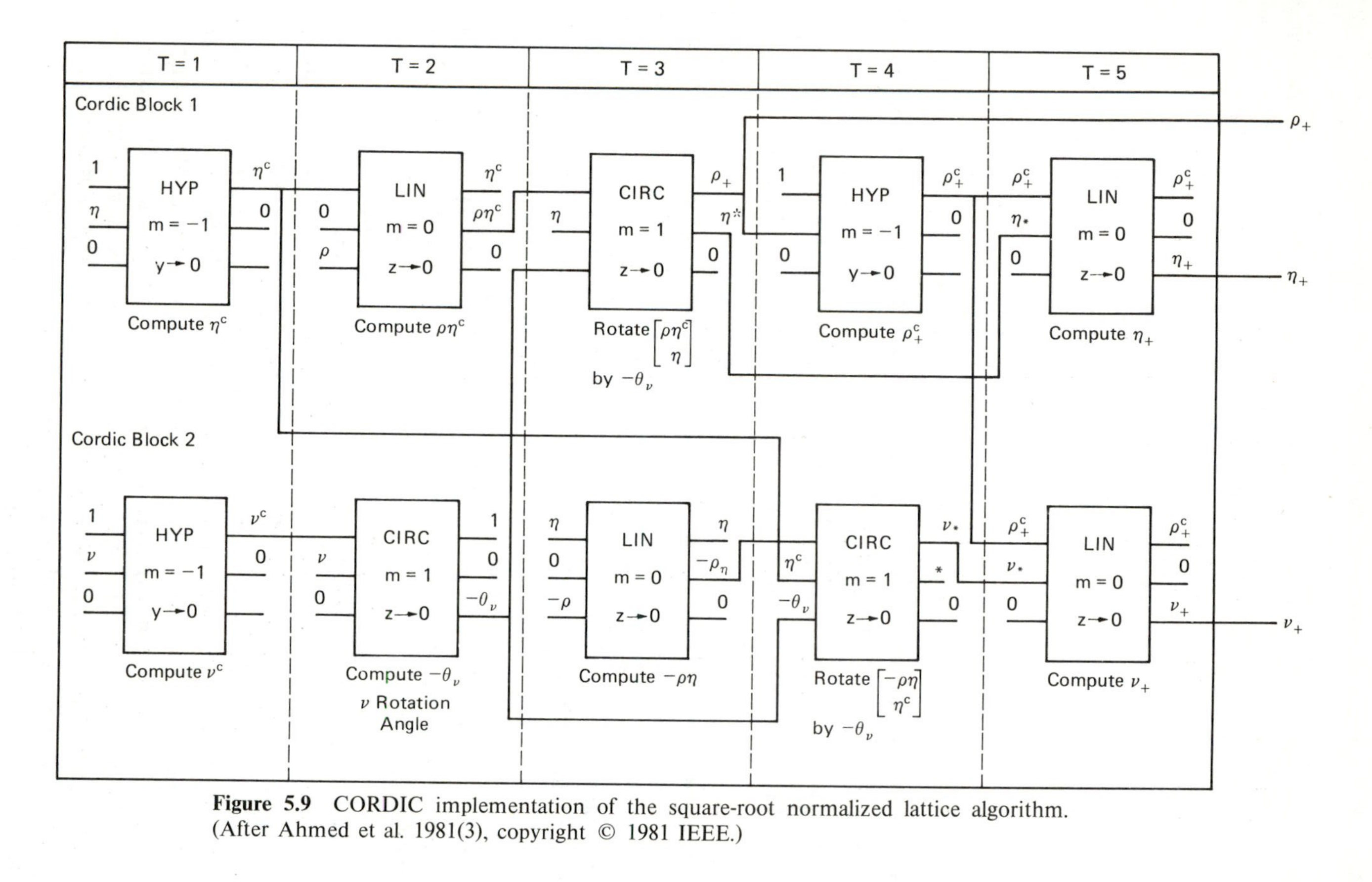

Figure 5.9 CORDIC implementation of the square-root normalized lattice algorithm. (After Ahmed et al. 1981(3), copyright © 1981 IEEE.)

5.9 SIMULATIONS AND APPLICATIONS

Simulated signals. Simulated signals with different characteristics were generated and the response of the lattice estimation algorithms noted. The simulated signals included white-noise driven autoregressive processes that were (1) stationary, (2) had linearly time-varying coefficients, (3) had step changes in coefficients, and (4) had impulse excitation at the step changes in coefficients.

For a stationary autoregressive process, the convergence of the least-squares lattice method is shown in Figure 5.10. An eighth-order fixed coefficient lattice filter driven by white noise generated the simulation data. The LSL algorithm with $\lambda = 0.99$ was used to compute the reflection coefficients. The first reflection coefficient converged in less than 50 samples and the first four reflection coefficients were near their correct values after 150 samples. Higher-order reflection coefficients approached their correct values after 250 samples.

When the simulated data were generated by a white-noise-driven second-order lattice with linearly time-varying coefficients, the adaptive nature was as seen in Figure 5.11. The two reflection coefficient estimates followed the actual parameter values. However, there was an increase in the variance of the estimate as the reflection coefficients approached zero or as the coefficient index increased. The previous experiment was repeated with piecewise-constant coefficients to generate the simulation data. The estimated reflection coefficient trajectory did not indicate that the model had step changes in the coefficients (see Figure 5.12).

The effect of the optimal weighting function γ was seen when the simulated data were generated by the same lattice with step changes in coefficients but also had a periodic impulse added to the white-noise-driving process at the instant of coefficient change. The presence of the impulses caused the estimates to readjust quickly to the new piecewise-constant values (see Figure 5.13). The impulse caused a sudden increase in the γ which allowed the estimates to focus on the new signal characteristics. Once the effect of the impulse has passed, the γ decreased so that convergence could take place.

Application for speech analysis. The most extensive use of the lattice filter has been for speech processing applications, including speech compression systems and stored vocabulary speech synthesis chips. Reflection coefficients and the lattice filter are well suited for speech processing for many reasons; the relationship to the acoustical tube model of the vocal tract, the advantageous quantization properties of the reflection coefficients, the finite-word-length arithmetic properties of the lattice filter, and the slowly time-varying nature of the reflection coefficients across speech sounds (making them amenable to interpolation).

Linear predictive coding (LPC) is a technique that has been used widely for low-bit-rate speech coding and fixed vocabulary speech synthesis. LPC uses a vocal tract model, the lattice filter parameterized by reflection coefficients, and an excitation model, periodic pulses for sounds produced by vocal chord oscillation (e.g., vowels) and white noise for hiss sounds. Short time segments of speech, typically

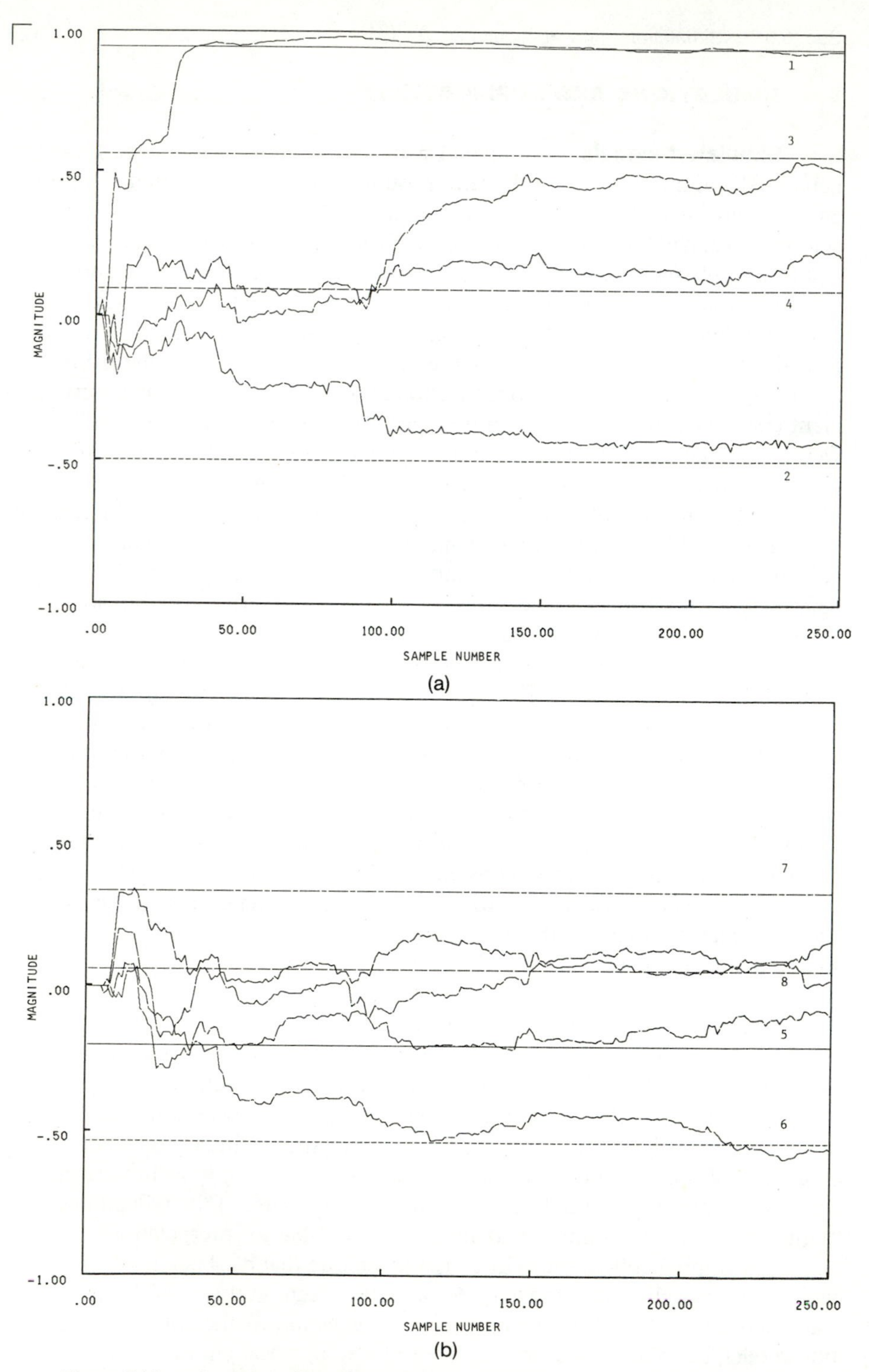

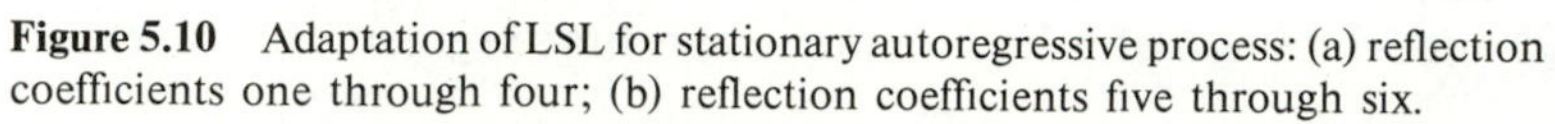

Figure 5.10 Adaptation of LSL for stationary autoregressive process: (a) reflection coefficients one through four; (b) reflection coefficients five through six.

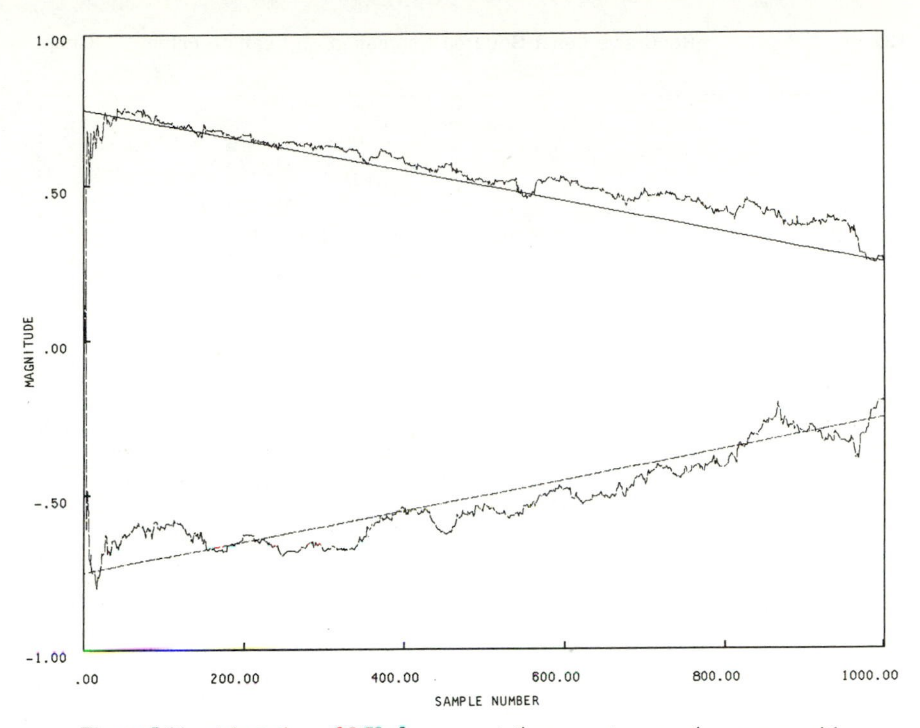

Figure 5.11 Adaptation of LSL for a nonstationary autoregressive process with linearly decreasing coefficients.

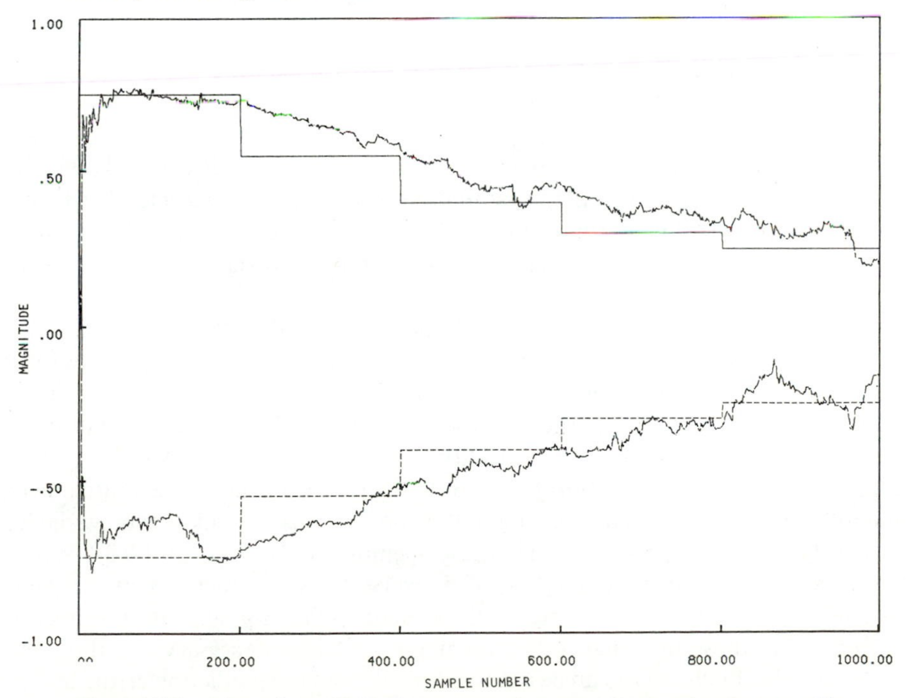

Figure 5.12 Adaptation of LSL for a nonstationary autoregressive process with step changes in coefficients.

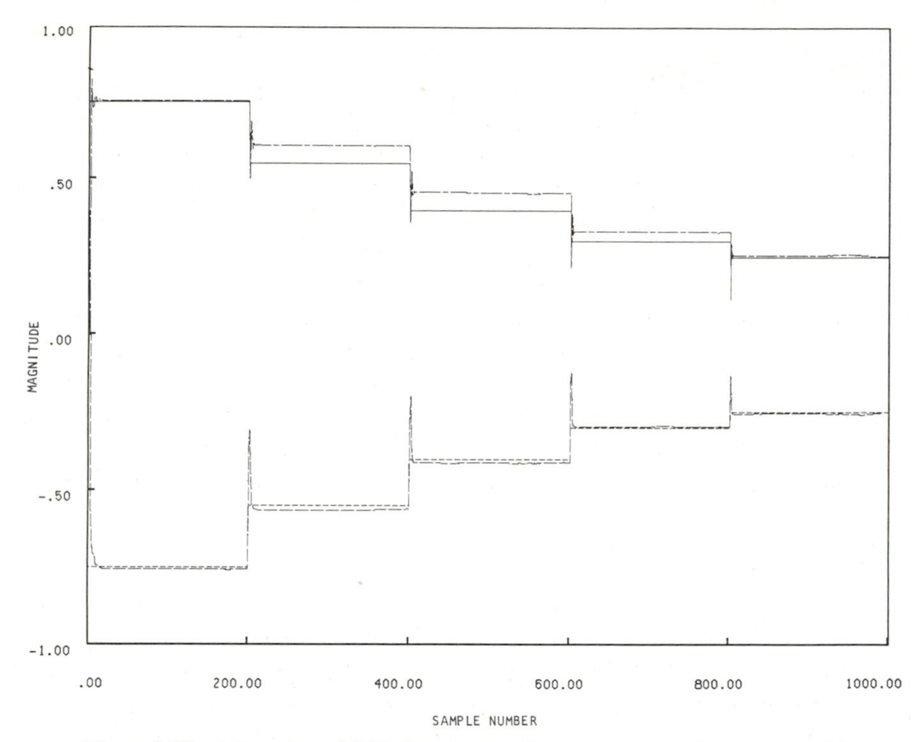

Figure 5.13 Adaptation of LSL for a nonstationary autoregressive process with step changes in coefficients. Input signal comprises white noise plus impulses which are synchronized with step changes.

20 ms, are characterized by 8 to 10 reflection coefficients, the pulse period (zero for noise), and an energy term. All the parameters can be quantized to a total of 48 bits per 20-ms interval. Using this compact description of sounds, speech synthesis integrated circuits have been developed that generate understandable speech using parameters stored in read-only memory.

Analyzing a spoken vowel sound by the LSL algorithm shows the properties of the likelihood variable. The time waveform [Figure 5.14(a)] clearly shows the periodic nature of this vowel. The LSL algorithm applied to this sound produces the forward prediction error shown in Figure 5.14(b). This relatively stationary sound produced fairly constant reflection coefficients (after convergence) [see Figure 5.14(c)]. The periodic jumps seen in all five reflection coefficients are due to the influence of the periodic opening of the vocal chords. The likelihood variable, γ, usually is small but increases when these openings occur [Figure 5.14(d)]. When the vocal chords open, a sudden pulse of air excites the vocal tract, which the likelihood variable interprets as a change in the structure of the signal. Determining the periodicity of these openings, called the pitch period, is necessary for the LPC speech model. Pitch pulses can be located directly from the prediction errors but do not always give accurate results. The periodicity is evident but not easily extracted from Figure 5.14(b). By combining the derivative of the likelihood variable γ with the prediction error sequence, a more easily discernible spike is generated at the onset of vocal chord oscillation [see Figure 5.14(e)]. This technique has been proposed as a pitch estimation method [Lee and Morf 1980(2)].

Since the recursive exact least-squares lattice algorithms can track quickly changing spectral characteristics, they can be used to differentiate the nature of transitional sounds [Turner 1982]. By exponential weighting of past data, the current estimate reflects the short-time signal characteristics. The beginnings of the words "bid" and "did" spoken by the same male speaker are shown in Figure 5.15(a) and (d). When analyzed by the SQNLSL with $\lambda = 0.98$, the reflection coefficients for the beginning of each word follow different trajectories corresponding to the different consonants. However, during the later vowel portion, the values are more similar [see Figure 5.15(b), (c), (e), and (f)]. The transitional part of the sounds is emphasized but the effects of the pitch pulses are also seen. The ability of the SQNLSL to differentiate these types of sounds may be useful in phoneme-based speech recognition systems.

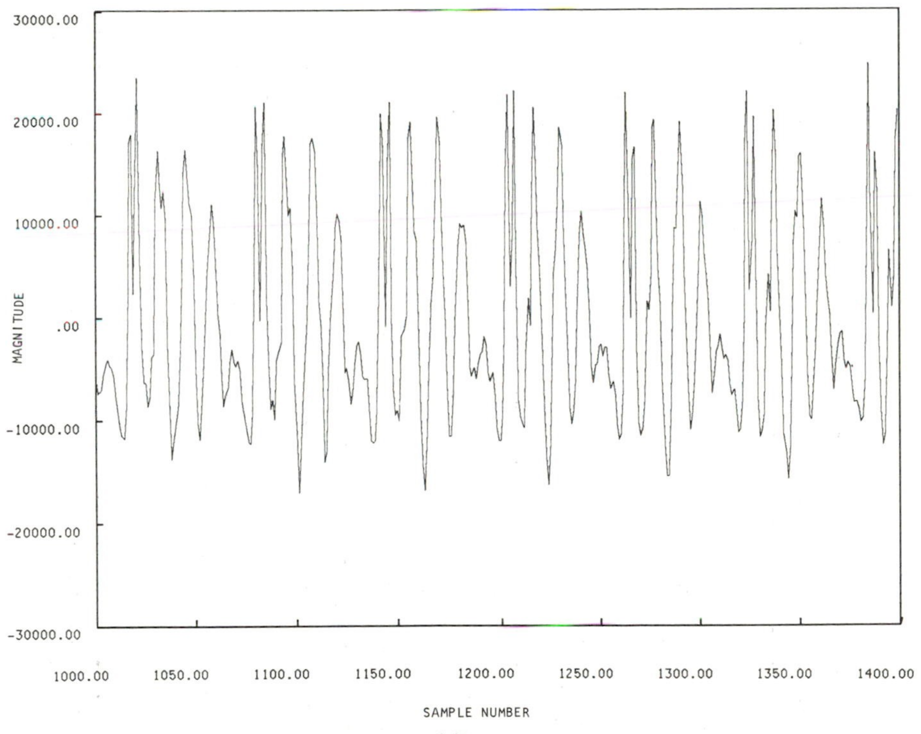

(a)

Figure 5.14 Adaptation of LSL algorithm on a spoken vowel: (a) time waveform for a typical vowel; (b) prediction error after fifth-order LSL filter; (c) adaptation of first five reflection coefficients; (d) likelihood variable γ; and (e) product of prediction error and derivative of likelihood variable (only positive part).

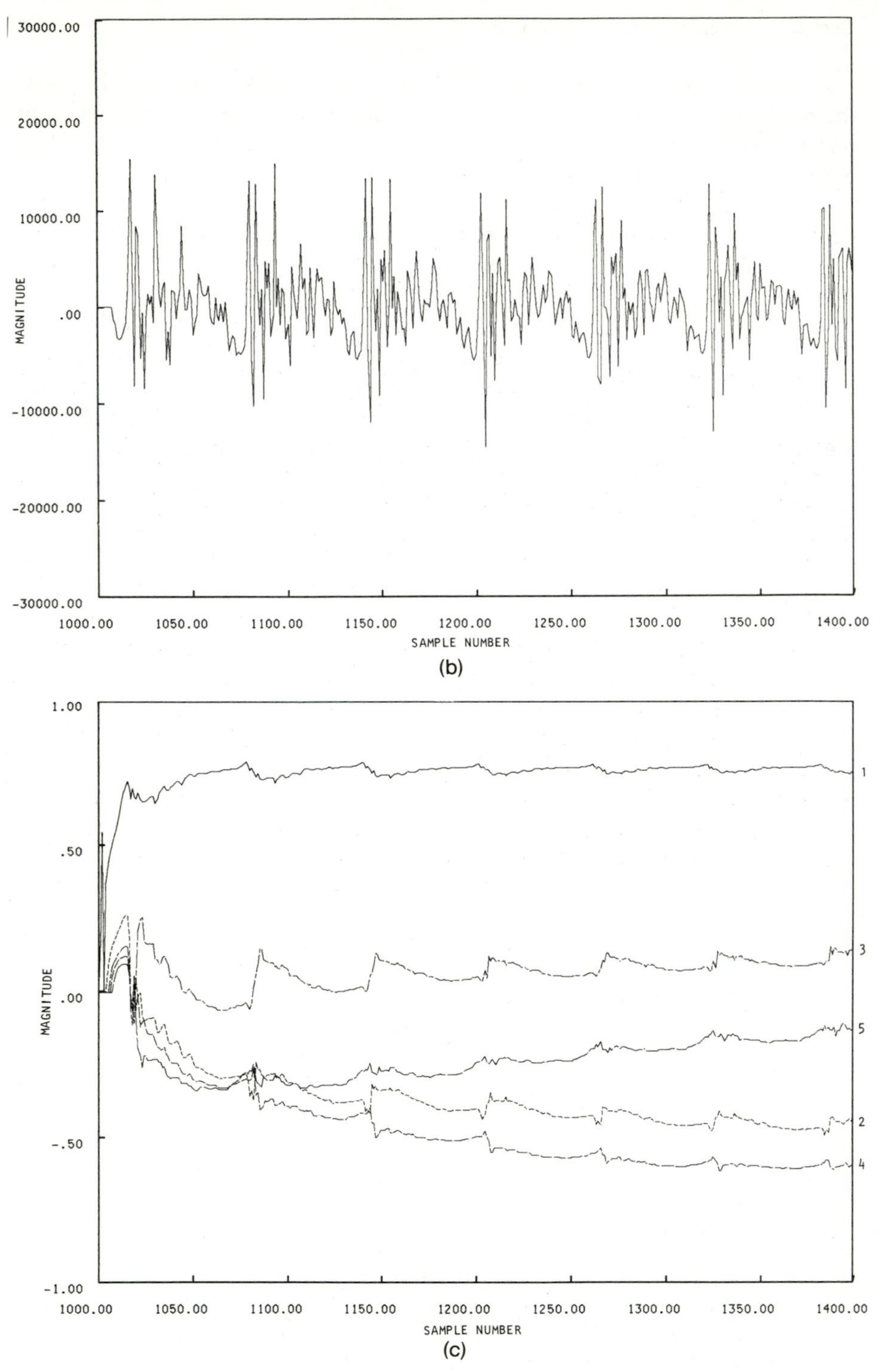

Figure 5.14 (cont.)

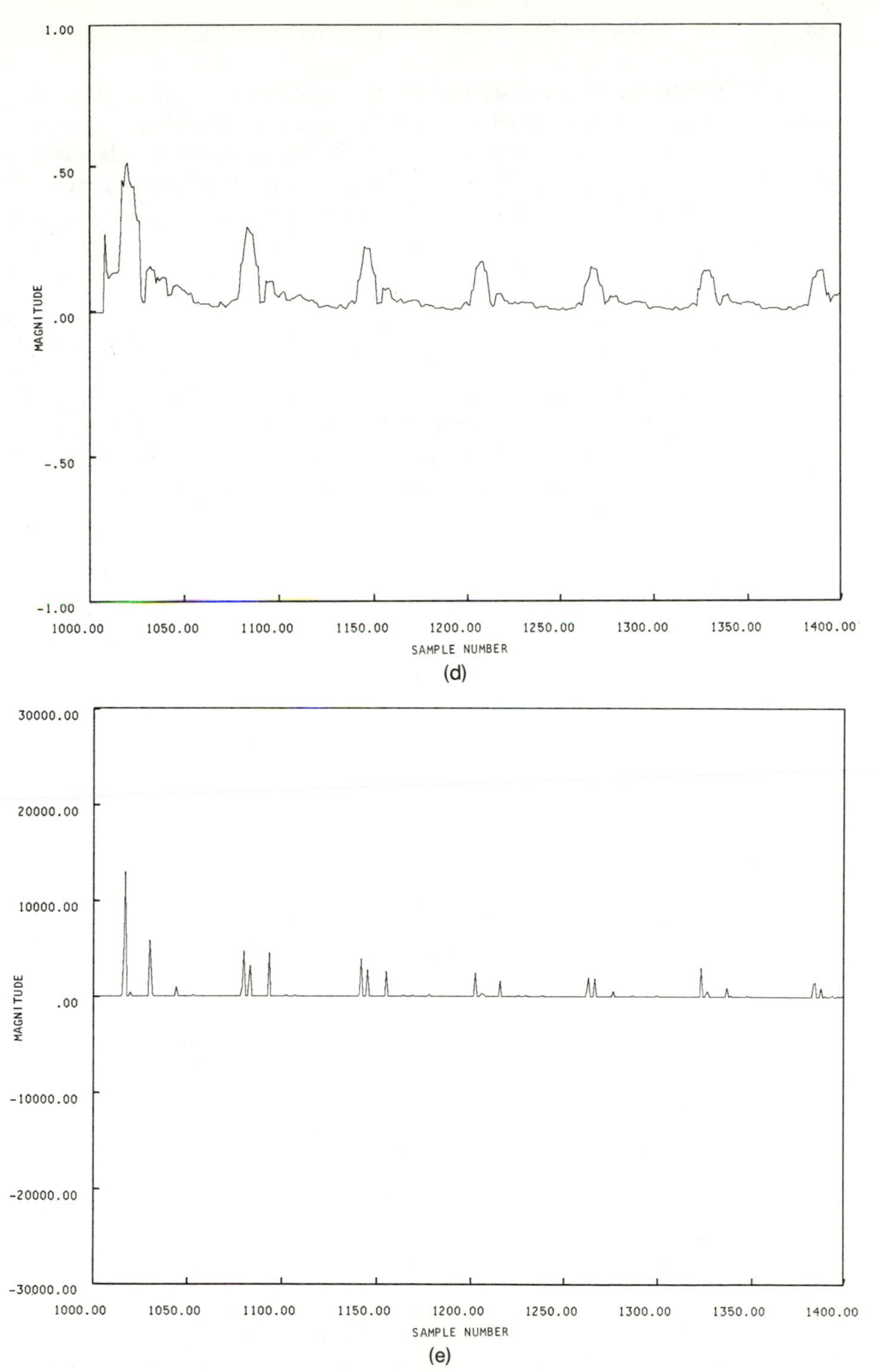

(d)

(e)

Figure 5.14 (cont.)

Application for channel equalization. The adaptive lattice filter offers substantial advantages for channel equalization where the orthogonalizing properties and fast-tracking characteristics are important. Tapped delay line adaptive gradient equalizers, although simple to implement, have a rate of convergence that depends on the ratio of the largest to smallest eigenvalues of the channel correlation matrix [Gitlin et al. 1973]. Self-orthogonalizing techniques have been proposed by [Gitlin and Magee 1977] and in lattice form by [Griffiths 1977, Griffiths and Medaugh 1979(1)]. The gradient lattice equalizer [Satorius and Alexander 1979(2)] and the LSL equalizer (see Section 5.6) [Satorius and Pack 1981] have been shown to provide very fast convergence. The lattice filter equalizers demonstrated fast convergence that was independent of the channel eigenvalue disparity ratios (see Figure 5.16 from [Satorius and Pack 1981]). Two simulated data channels with correlation matrices of eigenvalue disparity ratios (ratio of largest to smallest eigenvalues) of 11 and 21, respectively, were studied. An 11-tap equalizer

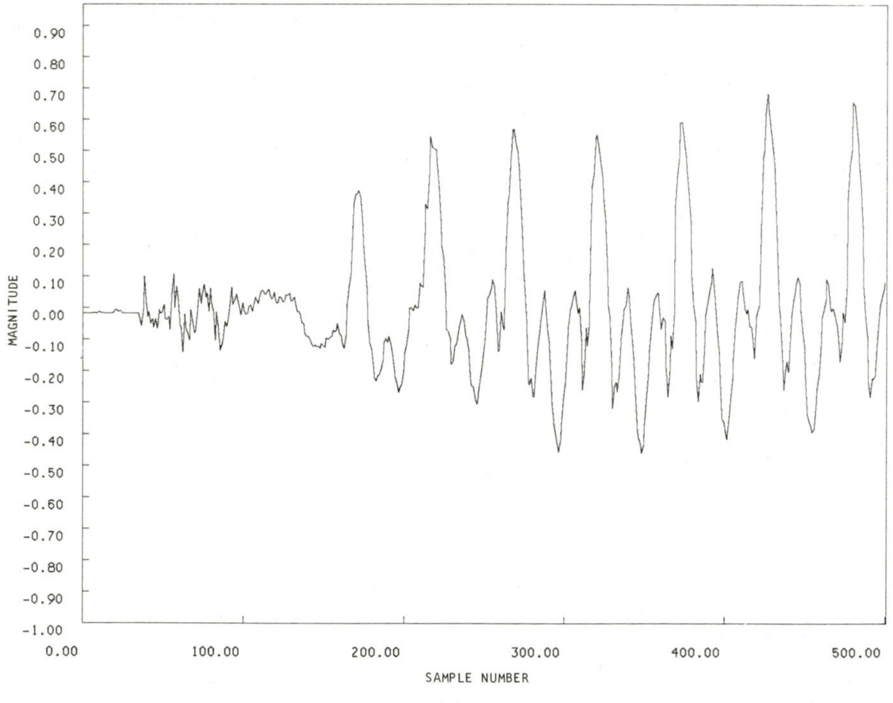

(a)

Figure 5.15 Comparison of two similar words beginning with stop consonants "b" and "d" (SQNLSL with $\lambda = 0.98$): (a) first 62.5 ms of the word "bid"; (b) first and second reflection coefficients for the word "bid"; (c) third and fourth reflection coefficients for the word "bid"; (d) first 62.5 ms of the word "did"; (e) first and second reflection coefficients for the word "did"; and (f) third and fourth reflection coefficients for the word "did."

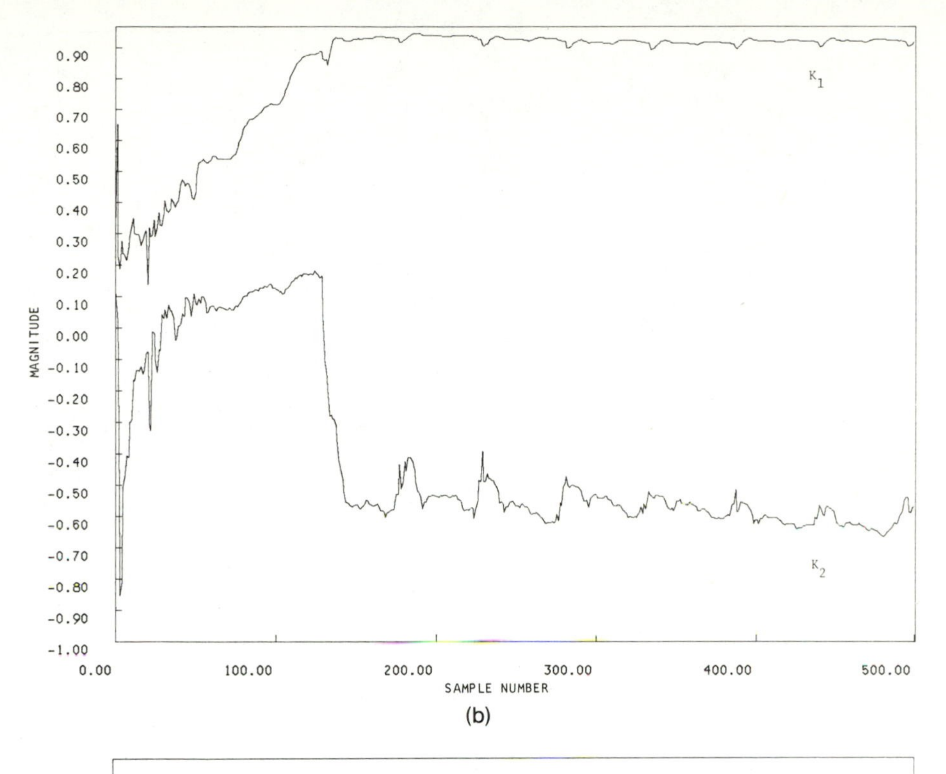

(b)

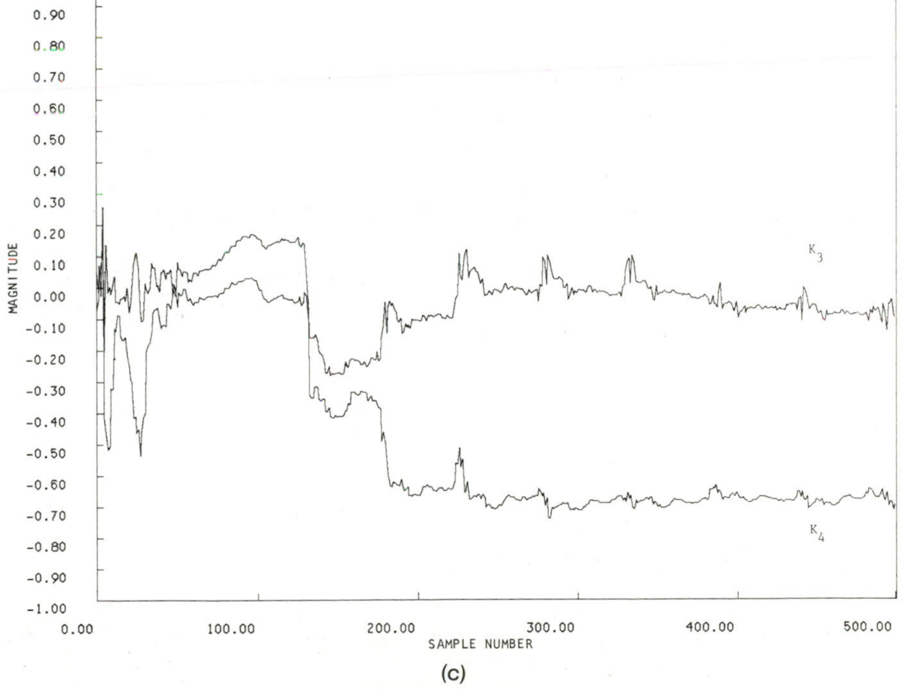

(c)

Figure 5.15 (cont.)

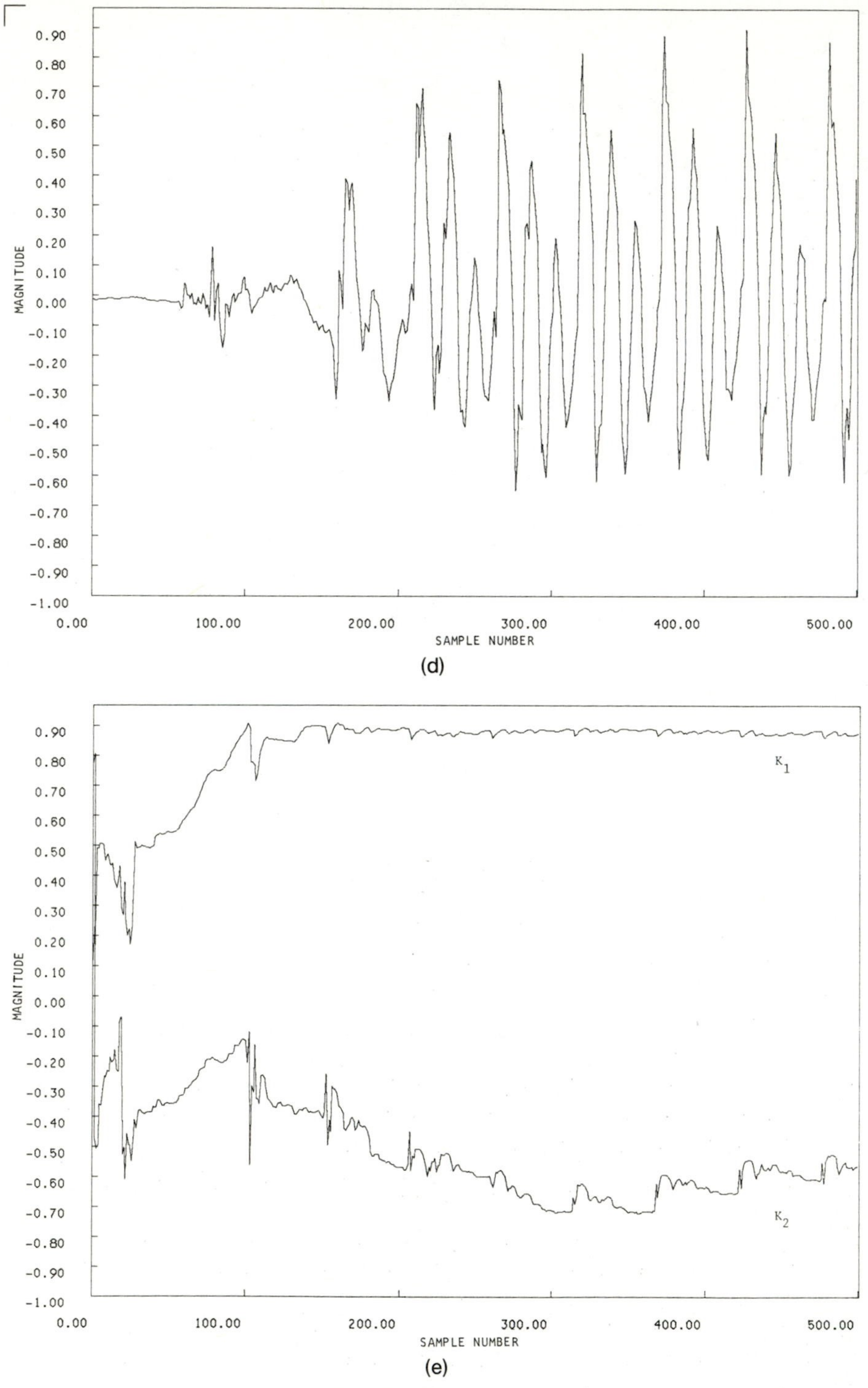

Figure 5.15 (cont.)

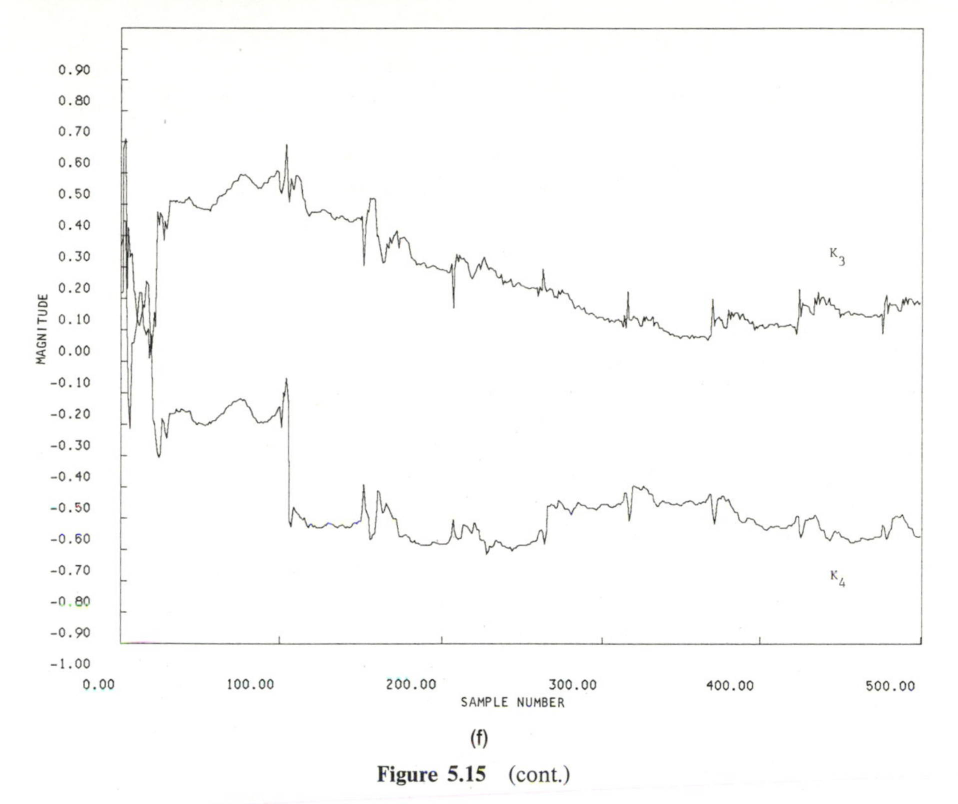

(f)

Figure 5.15 (cont.)

was implemented using the LMS gradient algorithm or adaptive lattice channel equalizer (ALCE) and the least-squares adaptive lattice equalizer algorithm (LSALE). The gradient tapped delay line equalizer has considerably slower convergence that depended on the eigenvalue ratio. The LSL equalizer converged in both cases in approximately 40 iterations, whereas the adaptive lattice equalizer required approximately 120 iterations to converge.

Application for electroencephalographic analysis. Electroencephalographic (EEG) data analyzed by autoregressive modeling can provide a better summary of EEG spectral information than can frequency-domain techniques such as FFTs. The reflection coefficients from the SQNLSL algorithm were studied to detect subtle changes in brain states as observed in EEG activity [Redington and Turner]. The data obtained from the left central EEG (C_1) response of an adult human subject monitored during sleep onset (sampled 60 times a second) is shown in Figure 5.17(a). A large change in activity appears near the beginning of the raw EEG data trace and is apparent in the reflection coefficients [Figure 5.17(b), (c), and (d)]. A second change in activity near the end of the trace is barely noticeable in the raw data, yet it is recognized in the activity of the higher-order reflection coefficient. The changes in reflection coefficients may reflect physiological transitions and provide a means of inferring presence or sequence of EEG brain states.

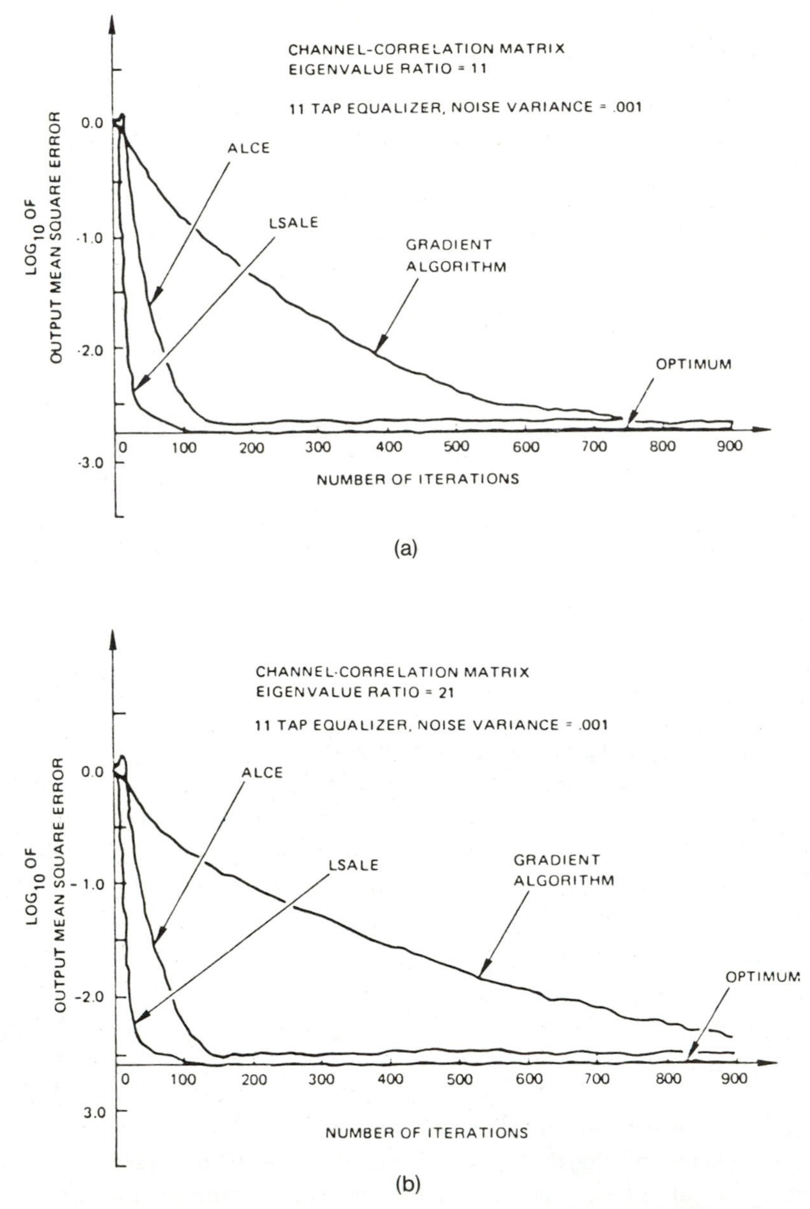

Figure 5.16 Equalizer convergence for two simulated channels: LMS gradient algorithm, or adaptive lattice channel equalizer (ALCE), and least-squares adaptive equalizer algorithm (LSALE): (a) equalizer convergence for eigenvalue ratio of 11; (b) equalizer convergence for eigenvalue ratio of 21. (After Satorius and Pack 1981, copyright © 1981 IEEE.)

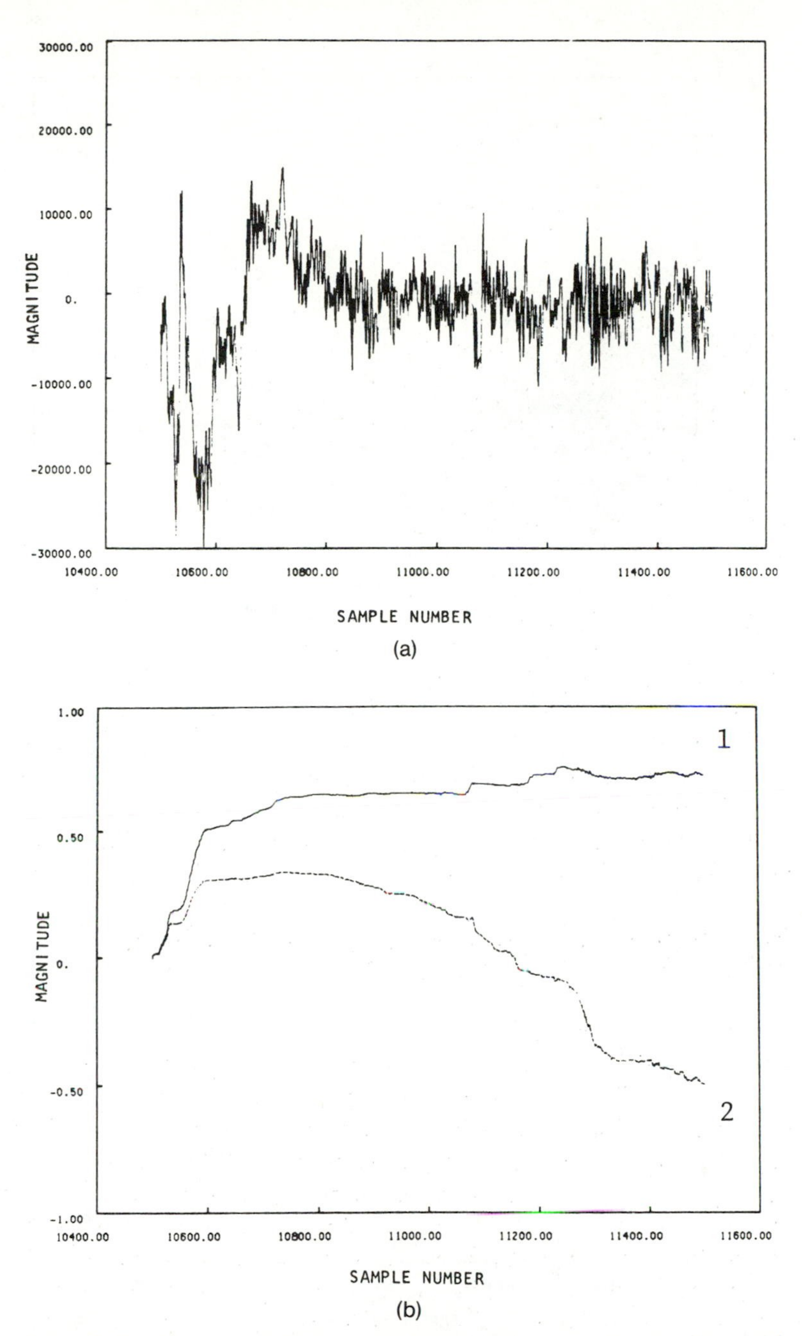

Figure 5.17 Analysis of electroencephalographic data by autoregressive modeling: (a) EEG data from central C_1 electrode; (b) first and second reflection coefficients; (c) third and fourth reflection coefficients; and (d) fifth and sixth reflection coefficients.

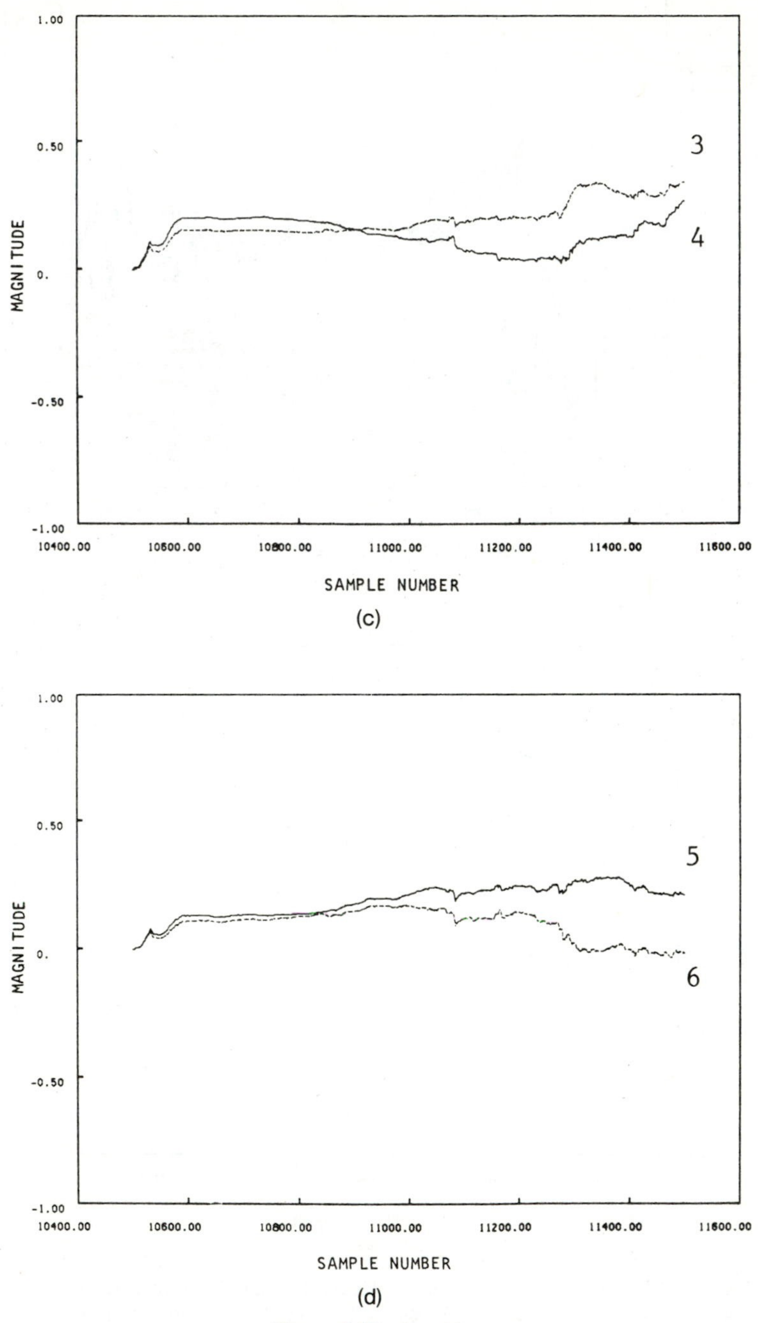

Figure 5.17 (cont.)

5.10 COMMENTS AND CONCLUSIONS

The lattice filter structure as a realization of a digital transfer function has several advantages; it is a cascade of identical sections, has a general insensitivity to round-off noise, and the reflection coefficients can be related to physical processes. The physical interpretation of reflection coefficients gives them intuitive appeal, particularly for speech signals. For adaptive estimation, the lattice structure is the natural form for an efficient solution to recursive least-squares problems. Lattice filters provide an orthogonalization or decoupling of the states of the input process. The stability of an all-pole model when expressed in lattice form can be determined by inspection.

The real advantage of the lattice structure lies in adaptive estimation and filtering. An N-stage lattice filter automatically generates all the outputs which would be generated by N different TDL filters with lengths from 1 to N. This allows dynamic assignment of any filter length that proves most effective at any instant of adaptive processing. When compared to the simpler adaptive transversal filter, the lattice filter has superior convergence properties and reduced sensitivity to finite-word-length effects.

Recursive lattice estimation algorithms allow the exact least-squares solution to be efficiently updated for each new time sample. The structure of this exact recursive approach is similar to the gradient lattice techniques; however, an optimal gain is calculated at every time sample. This optimal recursive solution has a complexity that is only slightly more than the gradient lattice solution. Consequently, the LSL and SQNLSL algorithms achieve extremely fast initial convergence and can track quickly time varying parameters. The SQNLSL has a very compact notation and normalizes all signals to unit variance at each stage. A single integrated circuit to execute this algorithm has been proposed.

However, as with all adaptive estimation procedures, there are various trade-offs to be made. The lattice structure involves more computation and is conceptually more complicated than the tapped delay line structure but has better convergence properties. The recursive least-squares lattice offers even better convergence than the gradient lattice, but again it is slightly more complex. For example, in the definition of the two reflection coefficients, there is a difference in the time subscripts of the normalizing covariances. In the stationary case, these terms are identical but in the LSL the difference is critical; in general, the algorithm will fail if this difference is overlooked [Satorius and Shensa 1980]. The SQNLSL allows a very short time constant to be applied to the sampled data so that the quickly time varying nature of the signal can be tracked. However, attempting to track transient speech sounds also tracks the pitch excitation signal. For processes that tend toward stationarity, the convergence properties of the gradient lattice and LSL lattice are similar [Honig 1983].

Many extensions to the basic recursive least-squares algorithm have been developed. Reviews of least-squares adaptive lattice filtering can be found in [Satorius and Shensa 1980, Friedlander 1982(3)]. Recursive ladder algorithms for ARMA modeling have been presented in [Lee et al. 1982]. The SQNLSL algorithm has been extended from the prewindowed data case presented here to the covariance data case in [Porat et al. 1982]. The problem of system identification has been addressed in [Porat and Kailath 1983]. A review of lattice filters for nonstationary processes was presented in [Kailath 1982].

There are other means to implement the lattice filter structure for estimation. The order-update recursions can also be obtained by using a Cholesky decomposition of the covariance matrix [Dickinson 1979(1), Dickinson and Turner 1979(2), Klein and Dickinson]. Alternatively, since a reflection coefficient is similar to a correlation coefficient, computationally simple techniques to estimate correlation coefficients can be applied to determining the reflection coefficients. Since the correlation of Gaussian random variables is related to the correlation of the hard-limited variables by an arcsin relationship, a very simple reflection coefficient approximation technique is possible [Turner et al. 1980]. This algorithm requires only a count of polarity changes in the prediction errors to estimate the reflection coefficients (assuming zero-mean-unit-variance Gaussian signals).

Overall, the adaptive lattice filter offers a compact algorithm for obtaining quickly converging estimates. The properties of the lattice filter and reflection coefficients motivate their use in many practical situations.

6

FREQUENCY-DOMAIN ADAPTIVE FILTERING

Earl R. Ferrara, Jr.

6.1 INTRODUCTION

In this chapter a class of adaptive filter algorithms is examined that transforms the input signal into the frequency domain before adaptive filtering. The transformations considered here are of a fixed nature [e.g., those based on the fast Fourier transform (FFT)], in contrast to the data-dependent orthogonalizing transforms described in Chapter 5. The adaptive algorithms considered here use the gradient descent algorithms reported in Chapter 3 or, as discussed in Section 6.6, modified gradient descent to adjust the filter coefficients (or "weights").

There are two principal advantages to frequency-domain implementations of adaptive filters. First, the amount of computation required to process a fixed amount of data can be greatly reduced compared with time-domain approaches. This reduction is accomplished by replacing convolution with a multiplication of transforms, as is done in "fast" convolution. Second, the convergence properties of the adaptive process can be improved over simple gradient descent.

In gradient descent algorithms, the weights converge to their optimal solution as a sum of exponentials, each exponential associated with a natural mode of the adaptive process. The time constants of these modes are inversely proportional to the eigenvalues of the input autocorrelation matrix. The mean-square error also decreases as a sum of exponentials whose time constants depend on the eigenvalues. For sufficiently long time-domain FIR filters, the eigenvalues of the input autocorrelation matrix are given approximately by uniformly spaced samples of the input power spectrum [Shensa 1979]. A heuristic interpretation of this result is that modes associated with areas of the spectrum having little power converge more slowly than those modes associated with frequencies having greater power. A large variation in the input power spectrum with frequency leads to highly disparate eigenvalues and therefore highly disparate time constants, some of which may be very long. Frequency-domain techniques, although based on steepest descent, can easily be modified to allow more uniform convergence of the modes of the adaptive process, thus improving convergence rate of the slower modes.

Use of the frequency domain results in block processing, in which a block of input data is processed simultaneously, producing a block of output. The nature of block processing requires that the filter coefficients be held fixed during the block. This process is in contrast to usual methods of time-domain adaptive processing, in which the filter coefficients may change at the input sampling rate. Although the filter coefficients are updated less often using frequency-domain approaches, they can be adjusted at each update with greater precision, since the gradient can be estimated using an entire block of data. The result is that adaptation can proceed just as rapidly and accurately in the frequency domain as it can in the time domain, with one exception. When the input autocorrelation matrix has highly disparate eigenvalues, stability considerations set an upper limit on the rate of adaptation that can be much slower than the corresponding limit for time-domain processing. Modified gradient techniques that effectively reduce the eigenvalue disparity can alleviate this problem. This technique is discussed in detail in Section 6.6.

In the following, uppercase symbols will denote frequency-domain variables, lowercase symbols stand for time-domain variables, and boldface will denote vectors or matrices. An asterisk will denote complex conjugate transpose. Define **F** to be a symmetric $N \times N$ matrix whose ith, kth element is $F_{ik} = \exp(-j2\pi ik/N)$, where j is the square root of -1. When **F** operates on a column vector of length N, the result is a column vector containing the DFT of the original vector. Similarly, $\mathbf{F}^{-1}$ is the inverse DFT operator. It can be shown that $\mathbf{F}^* = N\mathbf{F}^{-1}$ and $\mathbf{F}/\sqrt{N}$ is a unitary transformation. If **c** is a circulant matrix,† then $\mathbf{FcF}^{-1}$ is a diagonal matrix whose elements are the DFT of the first column of the circulant matrix [Gray 1972].

6.2 FREQUENCY-DOMAIN ADAPTIVE FILTER BASED ON CIRCULAR CONVOLUTION

One of the simplest frequency-domain adaptive filters is that shown in Figure 6.1 [Dentino et al. Bershad and Feintuch]. The input signal $x(n)$ and desired response $d(n)$ are accumulated in buffer memories to form N-point data blocks. They are then transformed by N-point FFTs. Each of the FFT outputs comprises a set of N complex numbers. The desired response transform values are subtracted from the input transform values at corresponding frequencies to form N complex error sig-

†A circulant matrix is a square matrix whose rows are obtained by successive right end-around shifts of the first row.

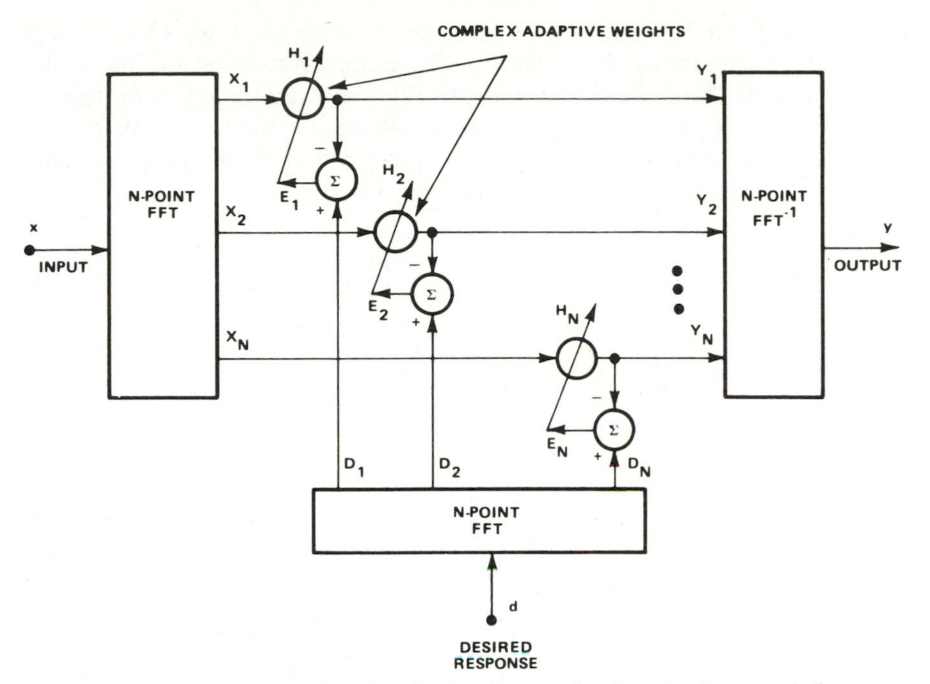

Figure 6.1 Frequency-domain adaptive filter performing circular convolution. (After Dentino et al., copyright © 1978 IEEE.)

nals. There are N complex weights, one corresponding to each spectral bin. Each weight is independently updated once for each data block. The weighted outputs are fed to an inverse FFT operator to produce the output signal $y(n)$.

The complex LMS algorithm [Widrow et al. 1975(1)] is used to update each weight. The ith complex weight for the ith frequency bin is updated according to

$$H_i(k+1) = H_i(k) + \mu E_i(k) X_i^*(k) \tag{6.1}$$

where μ is a constant that determines rate of convergence and stability of the adaptive process. For statistically stationary inputs, the weight-update equation (6.1) eventually minimizes the mean-square error at the ith frequency bin, provided that μ is chosen sufficiently small.

A substantial reduction in computation is obtained with this frequency-domain adaptive filter as compared with conventional time-domain adaptive filtering. This fact can be demonstrated by examining the number of multiply operations required to process a fixed amount of data. To produce N output data points with an N-tap time-domain LMS adaptive filter requires $2N^2$ real multiplies. To produce the same amount of output with this frequency-domain filter requires three N-point FFTs and $2N$ complex multiplies for the complex weighting and updating. For real input data, however, all transforms are symmetric, so that approximately half the weights can be discarded. Furthermore, for real data, an N-point FFT can be realized with an $N/2$-point FFT and $N/2$ complex multiplies

[Cooley et al.]. An $N/2$-point FFT takes approximately $(N/4)\log_2(N/2) - N/2$ complex multiplies for a radix-2 transform [Singleton]. Assuming four real multiplies per complex multiply yields $3N\log_2(N/2) + 4N$ real multiplies for the frequency-domain filter, compared with $2N^2$ multiplies for the time-domain filter. For large filters, the computational savings produced by the frequency-domain filter is substantial, as shown in the following listing.

N	Frequency-Domain Real Multiplies / LMS Real Multiplies
16	0.41
32	0.25
64	0.15
256	0.049
1024	0.015

Unfortunately, the frequency-domain filter of Figure 6.1 produces circular convolution [Oppenheim and Schafer], rather than linear convolution, of the input signal with the adaptive filter impulse response (the inverse FFT of the complex weights). The use of circular instead of linear convolution transforms a linear time-invariant filter into a periodic time-varying filter, whose output is periodically nonstationary for a stationary input [Pelkowitz] (see Appendix at end of Chapter 6). While the method of Figure 6.1 has found application in frequency-domain signal detection, its circular convolution property makes it less useful for general filtering applications. We shall see that the adaptive filter of Figure 6.1 minimizes the mean-square error between its output and the desired response, given that circular convolution is employed. The use of circular rather than linear convolution generally tends to shorten the effective impulse response length in order to reduce the effect of "wraparound" error due to circular convolution. The effect of circular convolution can be made small if the filter length (FFT size) can be chosen much larger than the effective nonzero length of the optimal impulse response for linear convolution. Although this substantially reduces the computational efficiency of the approach, its simplicity, coupled with the fact that the various weights are adapted independently, makes it attractive to consider.

To analyze this algorithm, define the frequency-domain weight vector for the kth block by

$$\mathbf{H}^{\mathrm{T}}(k) = [H_1(k) \quad H_2(k) \quad \cdots \quad H_N(k)] \tag{6.2}$$

and the diagonal matrix of input FFT coefficients by

$$\mathbf{X}(k) = \begin{bmatrix} X_1(k) & & & 0 \\ & X_2(k) & & \\ & & \ddots & \\ 0 & & & X_N(k) \end{bmatrix} \tag{6.3}$$

Similarly, let $\mathbf{Y}(k)$, $\mathbf{D}(k)$, and $\mathbf{E}(k)$ be vectors containing the frequency-domain output, desired response, and error for the kth block. Note that

$$\mathbf{Y}(k) = \mathbf{X}(k)\mathbf{H}(k) \tag{6.4}$$

and

$$\mathbf{E}(k) = \mathbf{D}(k) - \mathbf{Y}(k) \tag{6.5}$$

The frequency-domain weight update equation can be expressed as

$$\begin{aligned} \mathbf{H}(k+1) &= \mathbf{H}(k) + \mu\mathbf{X}^*(k)\mathbf{E}(k) \\ &= \mathbf{H}(k) + \mu\{\mathbf{X}^*(k)\mathbf{D}(k) - \mathbf{X}^*(k)\mathbf{X}(k)\mathbf{H}(k)\} \end{aligned} \tag{6.6}$$

It is useful to consider the equivalent time-domain operations implied by (6.6). Equation (6.6) can be transformed into the time domain to yield

$$\mathbf{h}(k+1) = \mathbf{h}(k) + \mu\{\boldsymbol{\chi}^{\mathrm{T}}(k)\mathbf{d}(k) - \boldsymbol{\chi}^{\mathrm{T}}(k)\boldsymbol{\chi}(k)\mathbf{h}(k)\} \tag{6.7}$$

where

$$\mathbf{h}(k) = \mathbf{F}^{-1}\mathbf{H}(k) \tag{6.8}$$

$$\mathbf{d}(k) = \mathbf{F}^{-1}\mathbf{D}(k) \tag{6.9}$$

and $\boldsymbol{\chi}(k)$ is a circulant matrix given by

$$\boldsymbol{\chi}(k) = \mathbf{F}^{-1}\mathbf{X}(k)\mathbf{F} \tag{6.10}$$

The first column of $\boldsymbol{\chi}(k)$ is the input vector $\mathbf{x}(k)$, since it is the inverse DFT of the diagonal elements of $\mathbf{X}(k)$. Therefore, the circulant matrix $\boldsymbol{\chi}(k)$ is given by

$$\boldsymbol{\chi}(k) = \begin{bmatrix} x(k) & x(k+N-1) & \cdots & x(k+1) \\ x(k+1) & x(k) & \cdots & x(k+2) \\ \vdots & \vdots & & \vdots \\ x(k+N-1) & x(k+N-2) & \cdots & x(k) \end{bmatrix} \tag{6.11}$$

Denoting the ith row of $\boldsymbol{\chi}(k)$ by $\mathbf{x}_i^{\mathrm{T}}(k)$, (6.7) becomes

$$\mathbf{h}(k+1) = \mathbf{h}(k) + \mu \sum_{i=1}^{N} \{d_i(k)\mathbf{x}_i(k) - y_i(k)\mathbf{x}_i(k)\} \tag{6.12}$$

where $y_i(k)$ is the ith element of the output vector

$$\mathbf{y}(k) = \boldsymbol{\chi}(k)\mathbf{h}(k) \tag{6.13}$$

The vector $\mathbf{y}(k)$ contains the elements of the kth output block of the filter. The elements of $\mathbf{y}(k)$ are obtained by circularly convolving the impulse response $\mathbf{h}(k)$ with rotated versions of the input vector $\mathbf{x}(k)$.

The equivalent weight-update equation in the time domain can be written

$$\mathbf{h}(k+1) = \mathbf{h}(k) + \mu \sum_{i=1}^{N} e_i(k)\mathbf{x}_i(k) \tag{6.14}$$

where $e_i(k) = d_i(k) - y_i(k)$. Note that (6.14) differs from the usual LMS algorithm in that, although adaptation is performed only once per block, gradient estimates are summed over an entire block of data before being used to update the weights.

Optimum weight vector. The optimum solution for $\mathbf{H}$, which minimizes the mean-square error between $\mathbf{y}(k)$ and $\mathbf{d}(k)$, can be determined (assuming that d and x are stationary) in the following manner. It is sufficient to minimize

$$NE[(\mathbf{d}(k) - \mathbf{y}(k))^*(\mathbf{d}(k) - \mathbf{y}(k))]$$
$$= E[(\mathbf{D}(k) - \mathbf{Y}(k))^*(\mathbf{D}(k) - \mathbf{Y}(k))] \tag{6.15}$$
$$= E[\mathbf{D}^*(k)\mathbf{D}(k)] - \mathbf{R}_{xd}^*\mathbf{H} - \mathbf{H}^*\mathbf{R}_{xd} + \mathbf{H}^*\mathbf{R}_{xx}\mathbf{H}$$

where

$$\mathbf{R}_{xd} = E[\mathbf{X}^*(k)\mathbf{D}(k)] \tag{6.16}$$

and

$$\mathbf{R}_{xx} = E[\mathbf{X}^*(k)\mathbf{X}(k)] \tag{6.17}$$

Note that $\mathbf{R}_{xx}$ is diagonal, with its ith diagonal element given by $E[X_i^*(k)X_i(k)]$. The ith element of $\mathbf{R}_{xd}$ is $E[X_i^*(k)D_i(k)]$. Taking the gradient of (6.15) with respect to $\mathbf{H}$ and setting it to zero yields the optimum frequency-domain weights:

$$\mathbf{H}_{\text{opt}} = \mathbf{R}_{xx}^{-1}\mathbf{R}_{xd} \tag{6.18}$$

Taking the inverse DFT of (6.18) yields the optimum time-domain weights for a circularly convolving filter:

$$\mathbf{h}_{\text{opt}} = \mathbf{r}_{xx}^{-1}\mathbf{r}_{xd} \tag{6.19}$$

where

$$\mathbf{r}_{xx} = \mathbf{F}^{-1}\mathbf{R}_{xx}\mathbf{F} \tag{6.20}$$

and

$$\mathbf{r}_{xd} = \mathbf{F}^{-1}\mathbf{R}_{xd}\mathbf{F} \tag{6.21}$$

The matrix $\mathbf{r}_{xx}$ is circulant, since $\mathbf{R}_{xx}$ is diagonal. The first row of $\mathbf{r}_{xx}$ is given by lags zero through $N - 1$ of the circular autocorrelation function of the input x. The circular autocorrelation function at lag i, $\phi_c(i)$, can be expressed in terms of the usual linear autocorrelation function $\phi(i)$ by

$$\phi_c(i) = \frac{N - i}{N}\phi(i) + \frac{i}{N}\phi(i - N) \tag{6.22}$$

A similar expression is obtained for the circular cross-correlation between d and x, which make up the elements of the vector $\mathbf{r}_{xd}$.

Convergence properties. Taking expected values of (6.6) and assuming stationary inputs, we have

$$E[\mathbf{H}(k + 1)] = E[\mathbf{H}(k)] + \mu\{\mathbf{R}_{xd} - \mathbf{R}_{xx}E[\mathbf{H}(k)]\} \tag{6.23}$$

where the usual assumption of uncorrelatedness between $\mathbf{X}(k)$ and $\mathbf{H}(k)$ has been made. Although this assumption is generally not strictly true, it is commonly made in adaptive filter analysis because of its simplicity, and has led to a number of useful results. We will make this simplifying assumption here in order to gain insight into the adaptive process.

Equation (6.23) implements the method of steepest descent and has been well studied [Widrow et al. 1976(2)]. The input autocorrelation matrix $\mathbf{R}_{xx}$, whose eigenvalues determine the stability and convergence rate of the adaptive process, is diagonal in this case. The eigenvalues of this matrix are therefore given by its diagonal elements, which are the powers of the DFT bins. If μ is chosen small enough, the expected value of the weight vector will converge to

$$\lim_{k\to\infty} E[\mathbf{H}(k)] = \mathbf{R}_{xx}^{-1}\mathbf{R}_{xd} = \mathbf{H}_{\text{opt}} \tag{6.24}$$

Thus the mean weight vector converges to the optimum weight vector. The condition for stability of the algorithm is

$$\mu < \frac{2}{\lambda_{\max}} \tag{6.25}$$

where $\lambda_{\max}$ is the maximum eigenvalue of $\mathbf{R}_{xx}$. The weights converge independently of each other since $\mathbf{R}_{xx}$ is diagonal. The time constant for the convergence of the pth weight is given by

$$\tau_p = \frac{1}{\mu\lambda_p} \quad \text{blocks} \tag{6.26a}$$

$$= \frac{N}{\mu\lambda_p} \quad \text{samples} \tag{6.26b}$$

where λ_p is the pth eigenvalue of $\mathbf{R}_{xx}$. Note that since $\mathbf{R}_{xx}$ is diagonal, its eigenvalues are simply its diagonal elements, and these are the powers in the FFT bins.

Misadjustment. Since the gradient is estimated using a finite amount of data, it has some error, resulting in random fluctuations of the filter coefficients about their optimal value. The result is a mean-square error greater than the minimum mean-square error. This excess mean-square error, normalized by the minimum mean-square error, is defined as the "misadjustment" [see (3.50)].

If it is assumed that the $\mathbf{x}_i(k)$ are uncorrelated and that $e_i(k)$ and $\mathbf{x}_i(k)$ are zero-mean Gaussian, an expression for the misadjustment of the adaptive process can be derived. Although these assumptions may not be strictly true, they are commonly made in order to simplify the analysis and will allow a misadjustment formula to be obtained that can be compared with that of conventional time-domain filters. Starting from (6.14) and following a misadjustment derivation similar to that of [Widrow et al. 1976(2)], the misadjustment M [see (3.91)] can be found to be

$$M = \frac{\text{excess mean-square error}}{\text{minimum mean-square error}} = \frac{\mu}{2N}\operatorname{tr}\{\mathbf{R}_{xx}\} \tag{6.27a}$$

$$= \frac{\mu NP}{2} \tag{6.27b}$$

where P is the power of the filter input. Note that the stability condition (6.25) requires that

$$M < \frac{\lambda_{\text{avg}}}{\lambda_{\max}} \tag{6.28}$$

which sets an upper limit on allowable misadjustment. Combining (6.26b) and (6.27) yields an expression for the adaptation time constant for the pth mode in terms of the misadjustment:

$$\tau_p = \frac{N}{2M}\frac{\lambda_{avg}}{\lambda_p} \qquad \text{samples} \tag{6.29}$$

The corresponding relationship for a time-domain LMS adaptive filter is identical in form to (6.29). Therefore, although adaptation in the frequency domain is performed only once per block, adaptation can proceed at the same rate as if the weights were adapted every sample, with no increase in the misadjustment (provided that $M < \lambda_{avg}/\lambda_{max}$). This is due to the fact that because of block processing, the individual gradient estimates at each sample are summed to provide a more accurate estimate of the gradient [Reed and Feintuch]. It should be noted that the converged solution and resultant minimum mean-square error of the frequency-domain filter is not identical to that of a conventional LMS filter since the former performs circular convolution.

6.3 ALGORITHMS FOR GENERAL ADAPTIVE FILTERING

Adaptive filters that allow linear convolution of the filter input and impulse response are more generally useful in filtering applications than are those that perform only circular convolution. Two frequency-domain adaptive filters are described in this section that allow linear convolution. One adaptive filter, denoted in the literature as block LMS or fast LMS, performs strictly linear convolution. It permits an efficient frequency-domain implementation while maintaining performance equivalent to that of the widely used LMS adaptive filter. The other adaptive filter considered in this section, the unconstrained frequency-domain LMS adaptive filter, allows either linear or circular convolution, whichever best minimizes the mean-square error.

6.3.1 Fast LMS Adaptive Filter

The block LMS adaptive filter [Clark et al. 1981] and the fast LMS adaptive filter [Ferrara 1980] are essentially identical frequency-domain implementations of the time-domain block LMS algorithm. In this algorithm, the data are grouped into N-point blocks, with the filter weights held constant over each block. During the kth block, the adaptive filter equations are

$$\begin{aligned}\mathbf{h}(k+1) &= \mathbf{h}(k) + \mu \sum_{i=0}^{N-1} e(kN+i)\mathbf{x}(kN+i) \\ &= \mathbf{h}(k) + \mu\mathbf{\nabla}(k)\end{aligned} \tag{6.30}$$

and

$$y(kN+i) = \mathbf{h}^{\mathrm{T}}(k)\mathbf{x}(kN+i) \qquad i = 0, 1, \ldots, N-1 \tag{6.31}$$

where $\mathbf{h}(k)$ is a vector containing the filter weights during the kth block:

$$\mathbf{h}^T(k) = [h_0(k) \quad h_1(k) \quad \cdots \quad h_{N-1}(k)] \tag{6.32}$$

and $\mathbf{x}(n)$ contains the N most recent filter input samples at time n:

$$\mathbf{x}^T(n) = [x(n) \quad x(n-1) \quad \cdots \quad x(n-N+1)] \tag{6.33}$$

The elements of $\mathbf{x}(n)$ can be viewed as the outputs of an N-tap tapped delay line. The error $e(n)$ is the difference between the desired response $d(n)$ and the filter output,

$$e(n) = d(n) - y(n) \tag{6.34}$$

The block LMS algorithm has properties that are identical, except for stability, to the conventional LMS adaptive filter, in which the filter weights are updated at the sampling rate. These properties will be discussed later.

The block LMS algorithm can be implemented in the frequency domain using the "overlap-save" method [Oppenheim and Schafer], resulting in a substantial reduction in computation over time-domain processing. A frequency-domain implementation employing the "overlap-add" method is also possible but results in more computations than are needed in the overlap-save method [Clark et al. 1983]. Although it is possible to implement the filter with any amount of overlap, the case of 50% overlap (block size equal to number of weights) is the most efficient [Clark et al. 1981] and will be presented here. The filter output equation (6.31) is a convolution between the filter input and impulse response, and can be computed efficiently using the overlap-save method. According to this method, the weights must be padded with N zeros, and $2N$-point FFTs must be used. Let $\mathbf{H}(k)$ be a vector of length $2N$ whose elements are the FFT coefficients of the zero-padded, time-domain weight vector

$$\mathbf{H}^T(k) = \text{FFT}[\mathbf{h}^T(k) \quad 0 \quad \cdots \quad 0] \tag{6.35}$$

$\mathbf{H}(k)$ is the frequency-domain weight vector. Let $\mathbf{X}(k)$ be a diagonal matrix whose elements are the $2N$-point transform of the $(k-1)$th and kth input blocks:

$$\mathbf{X}(k) = \text{diag}\{\text{FFT}[\underbrace{x(kN-N) \quad \cdots \quad x(kN-1)}_{(k-1)\text{th block}}\underbrace{x(kN) \quad \cdots \quad x(kN+N-1)}_{k\text{th block}}]\} \tag{6.36}$$

The convolution in (6.31) is realized by

$$\begin{aligned} \mathbf{y}(k) &= [y(kN) \quad \cdots \quad y(kN+N-1)]^T \\ &= \text{last } N \text{ terms of } \text{FFT}^{-1}\{\mathbf{X}(k)\mathbf{H}(k)\} \end{aligned} \tag{6.37}$$

Equation (6.37) gives the filter output values for the kth block. Note that an N-weight transversal filter in the time domain requires a $2N$-weight filter in the frequency domain.

To implement the weight-vector-update equation (6.30) in the frequency domain, notice that the jth element of $\nabla(k)$ can be rewritten as

$$\nabla_j(k) = \sum_{i=0}^{N-1} e(kN+i)x(kN+i-j), \qquad j = 0, 1, \ldots, N-1$$

so that the elements of $\nabla(k)$ are given by the cross-correlation of the error sequence with the filter input. $\nabla(k)$ can be computed using FFTs if we first compute the transform $\mathbf{E}(k)$ of the error sequence preceded by N zeros:

$$\mathbf{E}(k) = \text{FFT}[\underbrace{0 \cdots 0}_{N \text{ zeros}} \ \underbrace{d(kN) - y(kN) \cdots d(kN + N - 1) - y(kN + N - 1)}_{k\text{th error block}}]^{\text{T}} \tag{6.38}$$

and then compute

$$\nabla(k) = \text{first } N \text{ terms of } \text{FFT}^{-1}\{\mathbf{X}^*(k)\mathbf{E}(k)\} \tag{6.39}$$

Finally, the frequency-domain weight-vector-update equation is

$$\mathbf{H}(k + 1) = \mathbf{H}(k) + \mu\text{FFT}\begin{bmatrix} \nabla(k) \\ \left.\begin{matrix} 0 \\ \cdot \\ \cdot \\ \cdot \\ 0 \end{matrix}\right\} N \text{ zeros} \end{bmatrix} \tag{6.40}$$

If the last N values of the inverse transform of the initial weight vector $\mathbf{H}(0)$ are forced to zero, (6.40) is an exact implementation of (6.30) in the frequency domain.

Equations (6.35) to (6.40) define the fast LMS (FLMS) adaptive filter. A block diagram of the filter is shown in Figure 6.2. Double lines in Figure 6.2 denote parallel flow of frequency-domain data.

For each N-point block, the FLMS filter requires five $2N$-point FFTs and two $2N$-point complex multiplies. For real input data, all transforms are symmetric and require computation of only the first $N + 1$ terms. Furthermore, for real data, a $2N$-point FFT can be realized with an N-point FFT and N complex multiplies [Cooley et al.]. An N-point radix-2 FFT requires approximately $(N/2)\log_2(N) - N$ complex multiplies [Singleton] (a radix-4 FFT requires somewhat less computation). Therefore, the number of complex multiplies per block is $(5N/2)\log_2 N$ for the five FFTs and approximately $2N$ for the complex weighting and updating. To produce N output points with the conventional LMS adaptive filter requires $2N^2$ real multiplies. Assuming that one complex multiply is equivalent to four real multiplies yields the following ratio:

$$\frac{\text{FLMS real multiplies}}{\text{LMS real multiplies}} = \frac{5(\log_2 N) + 4}{N} \tag{6.41}$$

This ratio is computed for several values of N in the following tabulation:

N	$\dfrac{\text{FLMS Real Multiplies}}{\text{LMS Real Multiplies}}$
16	1.5
32	0.91
64	0.53
256	0.17
1024	0.053

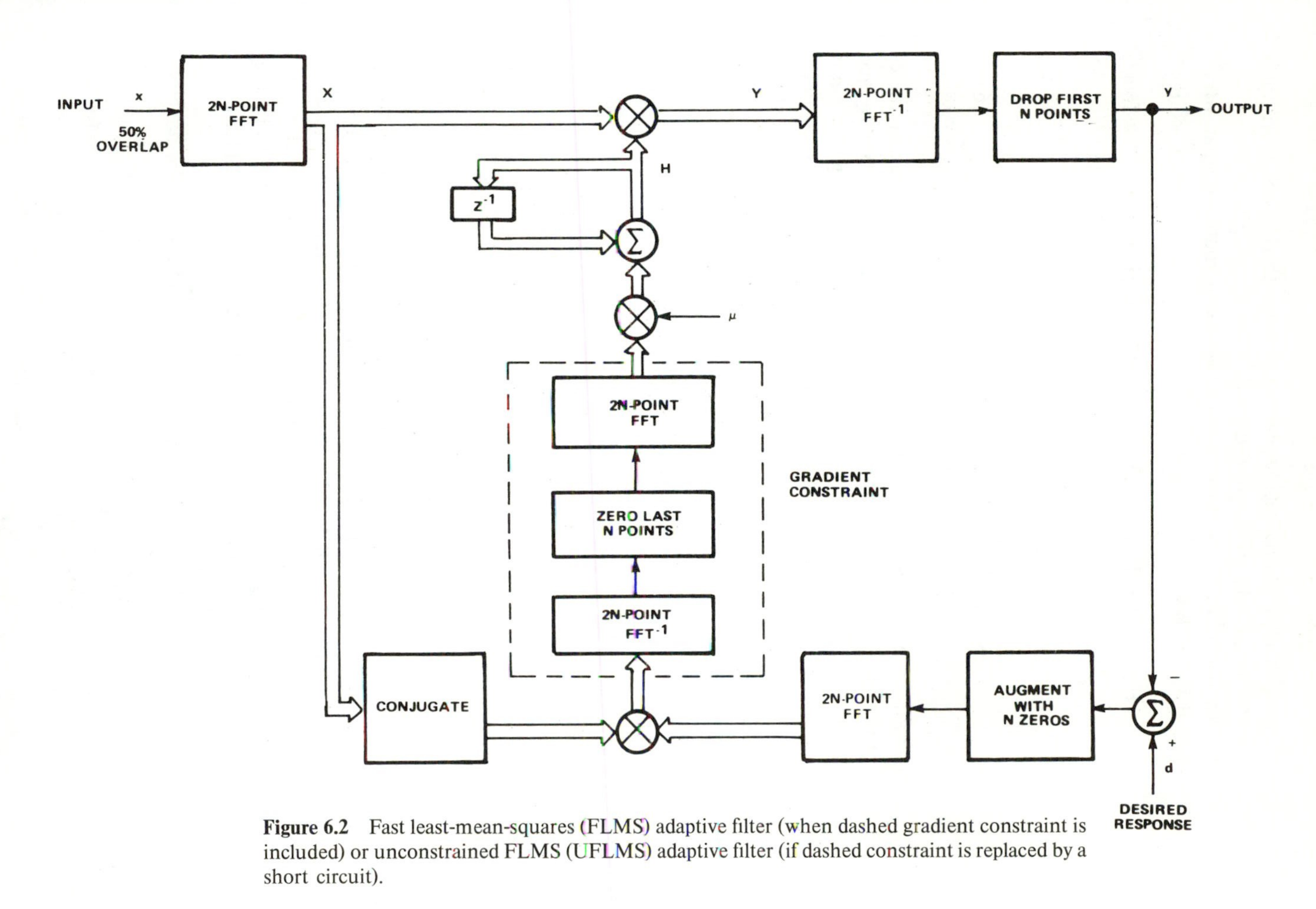

Figure 6.2 Fast least-mean-squares (FLMS) adaptive filter (when dashed gradient constraint is included) or unconstrained FLMS (UFLMS) adaptive filter (if dashed constraint is replaced by a short circuit).

For larger filters the computational savings gained by using the FLMS algorithm is substantial, even though five FFTs are required.

The FLMS algorithm can be written in the following matrix format, which will be useful in the sequel:

$$\mathbf{H}(k+1) = \mathbf{H}(k) + \mu\mathbf{F}\begin{bmatrix}\mathbf{I}_N & \mathbf{0}\\ \mathbf{0} & \mathbf{0}\end{bmatrix}\mathbf{F}^{-1}\mathbf{X}^*(k)\mathbf{E}(k) \tag{6.42}$$

$$\mathbf{E}(k) = \mathbf{F}\begin{bmatrix}\mathbf{0}\\ \mathbf{I}_N\end{bmatrix}(\mathbf{d}(k) - \mathbf{y}(k)) \tag{6.43}$$

$$\mathbf{y}(k) = [\mathbf{0} \quad \mathbf{I}_N]\mathbf{F}^{-1}\mathbf{X}(k)\mathbf{H}(k) \tag{6.44}$$

where $\mathbf{F}$ is a $2N \times 2N$ DFT matrix, and $\mathbf{I}_N$ is an $N \times N$ identity matrix.

Convergence properties. Since the FLMS algorithm is an exact implementation of the block LMS algorithm, it is sufficient to study the convergence properties of the latter (see [Clark et al. 1981] for detailed proofs). Using (6.30) and (6.31), a recursion for the expected value of the weight vector can be obtained under the assumption that $d(n)$ and $x(n)$ are stationary and that the $\mathbf{x}(n)$ are uncorrelated in time:

$$E[\mathbf{h}(k+1)] = E[\mathbf{h}(k)] + \mu N\{\mathbf{r}_{xd} - \mathbf{r}_{xx}E[\mathbf{h}(k)]\} \tag{6.45}$$

where

$$\mathbf{r}_{xx} = E[\mathbf{x}(n)\mathbf{x}^{\mathrm{T}}(n)] \tag{6.46}$$

and

$$\mathbf{r}_{xd} = E[d(n)\mathbf{x}(n)] \tag{6.47}$$

Applying the results for the method of steepest descent [Widrow et al. 1976(2)], it can be shown that

$$\lim_{k\to\infty} E[\mathbf{h}(k)] = \mathbf{r}_{xx}^{-1}\mathbf{r}_{xd} \tag{6.48}$$

provided that

$$\mu < \frac{2}{N\lambda_{\max}} \tag{6.49}$$

where $\lambda_{\max}$ is the maximum eigenvalue of $\mathbf{r}_{xx}$. Therefore, the converged weight vector is identical to that obtained with the conventional LMS algorithm. However, the condition for stability of the conventional LMS algorithm does not have the N in the denominator of (6.49), allowing faster adaptation. Therefore, the stability condition for the block LMS algorithm is more restrictive than that for the conventional LMS algorithm. This may be a problem when the eigenvalues of $\mathbf{r}_{xx}$ are highly disparate.

The time constants for the convergence of the N modes of the adaptive process can be shown to be

$$\tau_i = \frac{1}{\mu\lambda_i N} \qquad \text{blocks} \tag{6.50a}$$

$$= \frac{1}{\mu\lambda_i} \qquad \text{samples} \tag{6.50b}$$

These time constants are identical to that of the conventional LMS algorithm.

Assuming further that $e(n)$ and $\mathbf{x}(n)$ are zero-mean Gaussian, an expression for the misadjustment of the adaptive process can be found to be

$$M = \frac{\mu}{2} \operatorname{tr}\{\mathbf{r}_{xx}\} \tag{6.51a}$$

$$= \frac{\mu NP}{2} \tag{6.51b}$$

which is identical to that for the conventional LMS algorithm. Therefore, to achieve the same steady-state error in either algorithm, μ should be chosen the same, resulting in identical convergence rates for either algorithm. This may not always be possible because of the more restrictive stability conditions for the block LMS algorithm. The stability requirement limits the misadjustment of the block LMS algorithm to satisfy

$$M < \frac{\lambda_{\text{avg}}}{\lambda_{\text{max}}} \tag{6.52}$$

Since the desired misadjustment would normally be under 0.1, the restriction (6.52) is only a problem for the case of highly disparate eigenvalues. The restriction becomes less of a problem as the data overlap between successive FFTs is increased, although this results in less efficient processing since fewer output samples are obtained from each iteration.

6.3.2 Unconstrained Frequency-Domain LMS Adaptive Filter

The FLMS algorithm of the preceding section required five FFTs per processed block; two of them were needed to impose a time-domain constraint in which the last half of the time-domain weights were forced to zero. This was necessary in order to implement strictly linear convolution between the filter input and impulse response. In the unconstrained frequency-domain LMS (UFLMS) adaptive filter [Mansour and Gray], this constraint is removed, which produces a simpler adaptive filter that can implement either linear or circular convolution, whichever best minimizes the mean-square error. Allowing the filter the freedom to implement circular convolution would be acceptable in a number of applications—those for which we do not care how the filter input is used to minimize the mean-square error (e.g., noise canceling [Widrow et al. 1975(2)] or multichannel signal enhancing [Ferrara and Widrow 1981]). In fact, it is even possible that the mean-square error may be reduced by allowing some circular convolution. However, a problem arises if we attempt to use UFLMS for adaptive prediction or line enhancement [Treichler 1979]. In these applications, past values of an input signal are used by an adaptive filter to predict the signal's current value. Obviously, the adaptive filter should not be able to use the signal's current value to predict it. However, when block processing is used, as in the UFLMS algorithm, the adaptive filter has access, not only to past input values, but also to many current and future values. Unless a constraint is placed on the weights, the adaptive filter can use the current and future values to minimize the mean-square error. This is disastrous for line enhancing since it produces a filter output that is nearly identical to the filter input, with little enhancement of spectral lines. The foregoing comments apply only when predicting ahead some number of samples that is less than the FFT size; for longer prediction intervals there is no problem, since the data in corresponding desired response and filter input blocks are disjoint.

The block diagram of the UFLMS adaptive filter is identical to that in Figure 6.2 except that the constraint shown in the dashed line is removed, which eliminates two of the FFTs used in the FLMS adaptive filter. The ratio of UFLMS real multiplies to conventional LMS real multiplies is therefore

$$\frac{\text{UFLMS real multiplies}}{\text{LMS real multiplies}} = \frac{3(\log_2 N) + 4}{N} \tag{6.53}$$

This ratio is tabulated below for several values of N.

N	UFLMS Real Multiplies / LMS Real Multiplies
16	1.0
32	0.59
64	0.34
256	0.11
1024	0.033

By removing the zero-forcing constraint in the FLMS algorithm [(6.42) to (6.44)] the UFLMS algorithm can be expressed as

$$\mathbf{H}(k+1) = \mathbf{H}(k) + \mu \mathbf{X}^*(k)\mathbf{E}(k) \tag{6.54}$$

$$\mathbf{E}(k) = \mathbf{F}\begin{bmatrix} \mathbf{0} \\ \mathbf{I}_N \end{bmatrix}(\mathbf{d}(k) - \mathbf{y}(k)) \tag{6.55}$$

$$\mathbf{y}(k) = [\mathbf{0} \quad \mathbf{I}_N]\mathbf{F}^{-1}\mathbf{X}(k)\mathbf{H}(k) \tag{6.56}$$

Optimum weight vector. The optimum weights that minimize the mean-square error between the filter output and desired response are now computed. The mean-square error ζ expressed as a function of the weight vector $\mathbf{H}$ is

$$\begin{aligned}\zeta &= E[(\mathbf{d}(k) - \mathbf{y}(k))^*(\mathbf{d}(k) - \mathbf{y}(k))] \\ &= \frac{1}{2N}E\left\{(\mathbf{d}(k) - \mathbf{y}(k))^*[\mathbf{0} \quad \mathbf{I}_N]\mathbf{F}^*\mathbf{F}\begin{bmatrix} \mathbf{0} \\ \mathbf{I}_N \end{bmatrix}(\mathbf{d}(k) - \mathbf{y}(k))\right\}\end{aligned} \tag{6.57}$$

where we have used the fact that $\mathbf{F}^*\mathbf{F} = 2N\mathbf{I}_{2N}$. Therefore,

$$2N\zeta = E\{[\mathbf{D}(k) - \mathbf{FwF}^{-1}\mathbf{X}(k)\mathbf{H}]^*[\mathbf{D}(k) - \mathbf{FwF}^{-1}\mathbf{X}(k)\mathbf{H}]\} \tag{6.58}$$

where $\mathbf{D}(k)$ is the $2N$-point DFT of $\mathbf{d}(k)$ preceded by N zeros,

$$\mathbf{D}(k) = \mathbf{F}\begin{bmatrix} \mathbf{0} \\ \mathbf{d}(k) \end{bmatrix} \tag{6.59}$$

and $\mathbf{w}$ is a $2N \times 2N$ windowing matrix

$$\mathbf{w} = \begin{bmatrix} \mathbf{0} & \mathbf{0} \\ \mathbf{0} & \mathbf{I}_N \end{bmatrix}$$

Expanding (6.58) results in

$$\begin{aligned} 2N\zeta = {} & E[\mathbf{D}^*(k)\mathbf{D}(k)] - E[\mathbf{D}^*(k)\mathbf{WX}(k)]\mathbf{H} - \mathbf{H}^*E[\mathbf{X}^*(k)\mathbf{W}^*\mathbf{D}(k)] \\ & + \mathbf{H}^*E[\mathbf{X}^*(k)\mathbf{WX}(k)]\mathbf{H} \end{aligned} \tag{6.60}$$

where

$$\mathbf{W} = \mathbf{FwF}^{-1} \tag{6.61}$$

Setting the gradient of (6.60) with respect to $\mathbf{H}$ equal to zero yields the optimum weight vector,

$$\mathbf{H}_{\text{opt}} = \mathbf{R}_{xx}^{-1}\mathbf{R}_{xd} \tag{6.62}$$

where

$$\mathbf{R}_{xx} = E[\mathbf{X}^*(k)\mathbf{WX}(k)] \tag{6.63}$$

and

$$\mathbf{R}_{xd} = E[\mathbf{D}(k)\mathbf{W}^*\mathbf{X}^*(k)] \tag{6.64}$$

The equivalent time-domain weight vector can be found to be

$$\mathbf{h}_{\text{opt}} = \mathbf{r}_{xx}^{-1}\mathbf{r}_{xd} \tag{6.65}$$

where

$$\mathbf{r}_{xx} = E[\boldsymbol{\chi}^*(k)\mathbf{w}\boldsymbol{\chi}(k)] \tag{6.66}$$

$$\mathbf{r}_{xd} = E\left\{\boldsymbol{\chi}^*(k)\begin{bmatrix} \mathbf{0} \\ \mathbf{d}(k) \end{bmatrix}\right\} \tag{6.67}$$

and $\boldsymbol{\chi}(k)$ is a circulant matrix defined by

$$\boldsymbol{\chi}(k) = \mathbf{F}^{-1}\mathbf{X}(k)\mathbf{F} \tag{6.68}$$

Convergence properties. Using an argument similar to that in Section 6.2, the UFLMS algorithm satisfies

$$\lim_{k\to\infty} E[\mathbf{H}(k)] = \mathbf{H}_{\text{opt}} \tag{6.69}$$

provided that

$$\mu < \frac{2}{\lambda_{\max}} \tag{6.70}$$

where $\lambda_{\max}$ is the maximum eigenvalue of $\mathbf{R}_{xx}$ or $\mathbf{r}_{xx}$. The time constants of the $2N$ modes of the adaptive process are given by

$$\tau_p = \frac{1}{\mu\lambda_p} \qquad \text{blocks} \tag{6.71a}$$

$$= \frac{N}{\mu\lambda_p} \qquad \text{samples} \tag{6.71b}$$

where λ_p is the pth eigenvalue of $\mathbf{R}_{xx}$ or $\mathbf{r}_{xx}$. Under a Gaussian assumption stated in Section 6.1, the misadjustment can be found to be

$$M = \frac{\mu}{2N} \operatorname{tr}\{\mathbf{R}_{xx}\} \tag{6.72a}$$

$$= \mu NP \tag{6.72b}$$

where P is the power of the filter input.

To ensure stability, the misadjustment is bounded by

$$M < \frac{\lambda_{\text{avg}}}{\lambda_{\max}} \tag{6.73}$$

where λ_{avg} is the average eigenvalue of $\mathbf{R}_{xx}$.

6.4 CHANNEL EQUALIZATION

The rate of data transmission over a dispersive communication channel is limited by intersymbol interference—distortion of data pulses by the channel so as to overlap other transmitted pulses. To alleviate the effects of intersymbol interference, it is common to equalize the channel with an adaptive equalizer as shown in Figure 6.3. The parameters of the adaptive equalizer are chosen to minimize the mean-square error between its output and some desired output. Quite a number of techniques have been proposed for adaptive equalization [Lucky 1973]. Two frequency-domain approaches are considered here. One technique performs equalization using isolated test pulses; the other uses random data sequences. Although other frequency-domain approaches could be used for channel equalization, the methods presented here are specifically for channel equalization.

Figure 6.3 Communication channel equalization.

6.4.1 Isolated Pulse Equalization

In this method [Walzman and Schwartz], a training mode is used in which an isolated test pulse (or a series of nonoverlapping test pulses) is transmitted through the channel and used as input to an adaptive equalizer. The equalizer is adjusted so that its output is as close as possible to a desired output (e.g., the transmitted test pulse shape). At this point, the training mode is completed and transmission of information at a higher data rate may begin. The training mode may be repeated periodically to account for changing channel characteristics.

Let $\mathbf{x}(k)$ be an M-vector containing samples of the channel output produced by the kth test pulse; it will also be the input to the equalizer. Let $\mathbf{h}(k)$ contain the M-point impulse response of the equalizer and $\mathbf{y}(k)$ contain the N-point result of the convolution of the equalizer input and impulse response $(N \geq 2M - 1)$. Finally, let $\mathbf{d}(k)$ contain the N-point desired output for the kth pulse (it might not depend on k), and let $\mathbf{D}(k)$ contain its N-point FFT coefficients.

Since $\mathbf{x}(k)$ and $\mathbf{h}(k)$ are finite sequences, their linear convolution can be computed using FFTs in the following manner. Let $\mathbf{X}(k)$ be a diagonal matrix whose entries are given by the N-point FFT coefficients of $\mathbf{x}(k)$ padded with $N - M$ zeros, that is,

$$\mathbf{X}(k) = \text{Diag}\{\text{FFT}[\mathbf{x}^{\text{T}}(k) \quad \underbrace{0 \cdots 0}_{N-M \text{ zeros}}]\} \tag{6.74}$$

Define $\mathbf{H}(k)$ to be the N-point FFT of the zero-padded impulse response,

$$\mathbf{H}^{\text{T}}(k) = \text{FFT}[\mathbf{h}^{\text{T}}(k) \quad \underbrace{0 \cdots 0}_{N-M \text{ zeros}}] \tag{6.75}$$

Then $\mathbf{y}(k)$ is given by

$$\begin{aligned} \mathbf{y}(k) &= \text{FFT}^{-1}\{\mathbf{Y}(k)\} \\ &= \text{FFT}^{-1}\{\mathbf{X}(k)\mathbf{H}(k)\} \end{aligned} \tag{6.76}$$

We wish to choose $\mathbf{h}$ to minimize

$$E[(\mathbf{d}(k) - \mathbf{y}(k))^*(\mathbf{d}(k) - \mathbf{y}(k))] \tag{6.77}$$

which is equivalent to choosing $\mathbf{H}$ to minimize

$$E[(\mathbf{D}(k) - \mathbf{Y}(k))^*(\mathbf{D}(k) - \mathbf{Y}(k))] \tag{6.78}$$

subject to the constraint that the last $N - M$ elements of $\mathbf{h}$ are zero. The optimum value of $\mathbf{H}$ can be calculated iteratively, provided that the initial value of $\mathbf{H}$ satisfies the constraint, by

$$\mathbf{H}(k + 1) = \mathbf{H}(k) + \mu\mathbf{P}(k)\mathbf{X}^*(k)[\mathbf{D}(k) - \mathbf{Y}(k)] \tag{6.79}$$

where $\mathbf{P}$ is a constraint matrix given by

$$\mathbf{P} = \mathbf{F}\begin{bmatrix} \mathbf{I}_M & \mathbf{0} \\ \mathbf{0} & \mathbf{0} \end{bmatrix}\mathbf{F}^{-1} \tag{6.80}$$

In (6.80) $\mathbf{F}$ is the DFT matrix and $\mathbf{I}_M$ is an $M \times M$ identity matrix. Equation (6.79) is a linearly constrained gradient descent algorithm. The constraint can be implemented by transforming $\mathbf{X}^*(k)[\mathbf{D}(k) - \mathbf{Y}(k)]$ into the time domain, zeroing the last $N - M$ values, and transforming back into the frequency domain. The equalization algorithm is illustrated in Figure 6.4.

The block diagram of Figure 6.4 is similar to that of the fast LMS adaptive filter (Figure 6.2), the difference being that the output is not windowed in the time domain before being fed back to adapt the weights. This windowing can be eliminated while preserving linear convolution only because the input data sequence was finite and thus could be padded with zeros before convolving it with the impulse response. Elimination of this window is computationally advantageous since the weights can be updated without actually computing the output FFT^{-1}. When the training mode is completed and transmission of information is begun, the weight adaptation is disabled and the output must be computed using the overlap-save technique.

Convergence properties. The convergence behavior of linearly constrained gradient descent algorithms such as (6.79) has been analyzed [Frost]. If the diagonal matrix $\mathbf{R}_{xx}$ is defined by

$$\mathbf{R}_{xx} = E[\mathbf{X}(k)\mathbf{X}^*(k)] \tag{6.81}$$

the time constants of the adaptive process are determined by the eigenvalues of $\mathbf{P}^{\mathrm{T}}\mathbf{R}_{xx}\mathbf{P}$; that is, the time constant for the pth mode is

$$\tau_p = \frac{1}{\mu\sigma_p} \qquad \text{blocks} \tag{6.82}$$

where σ_p is the pth eigenvalue of $\mathbf{P}^{\mathrm{T}}\mathbf{R}_{xx}\mathbf{P}$. If μ is chosen so that

$$0 < \mu < \frac{2}{\sigma_{\max}} \tag{6.83}$$

the mean weight vector converges to its optimum value given by

$$\mathbf{H}_{\text{opt}} = \{\mathbf{I}_N - \mathbf{R}_{xx}\mathbf{W}(\mathbf{W}^*\mathbf{R}_{xx}\mathbf{W})^{-1}\mathbf{W}^*\}\mathbf{R}_{xx}^{-1}\mathbf{R}_{xd} \tag{6.84}$$

where

$$\mathbf{R}_{xd} = E[\mathbf{X}^*(k)\mathbf{D}(k)] \tag{6.85}$$

and

$$\mathbf{W} = \mathbf{F}\begin{bmatrix} \mathbf{0} \\ \mathbf{I}_{M-N} \end{bmatrix} \tag{6.86}$$

The nonzero eigenvalues of $\mathbf{P}^{\mathrm{T}}\mathbf{R}_{xx}\mathbf{P}$ are bounded between the smallest and largest eigenvalues of $\mathbf{R}_{xx}$, that is,

$$\lambda_{\min} \le \sigma_{\min} \le \sigma_i \le \sigma_{\max} \le \lambda_{\max} \tag{6.87}$$

where $\lambda_{\min}$ and $\lambda_{\max}$ are the smallest and largest eigenvalues of $\mathbf{R}_{xx}$. Since $\mathbf{R}_{xx}$ is diagonal, its eigenvalues are easily obtained and are given by the mean-square values of the FFT coefficients. A sufficient condition for stability of the adaptive process is

$$0 < \mu < \frac{2}{\lambda_{\max}} \tag{6.88}$$

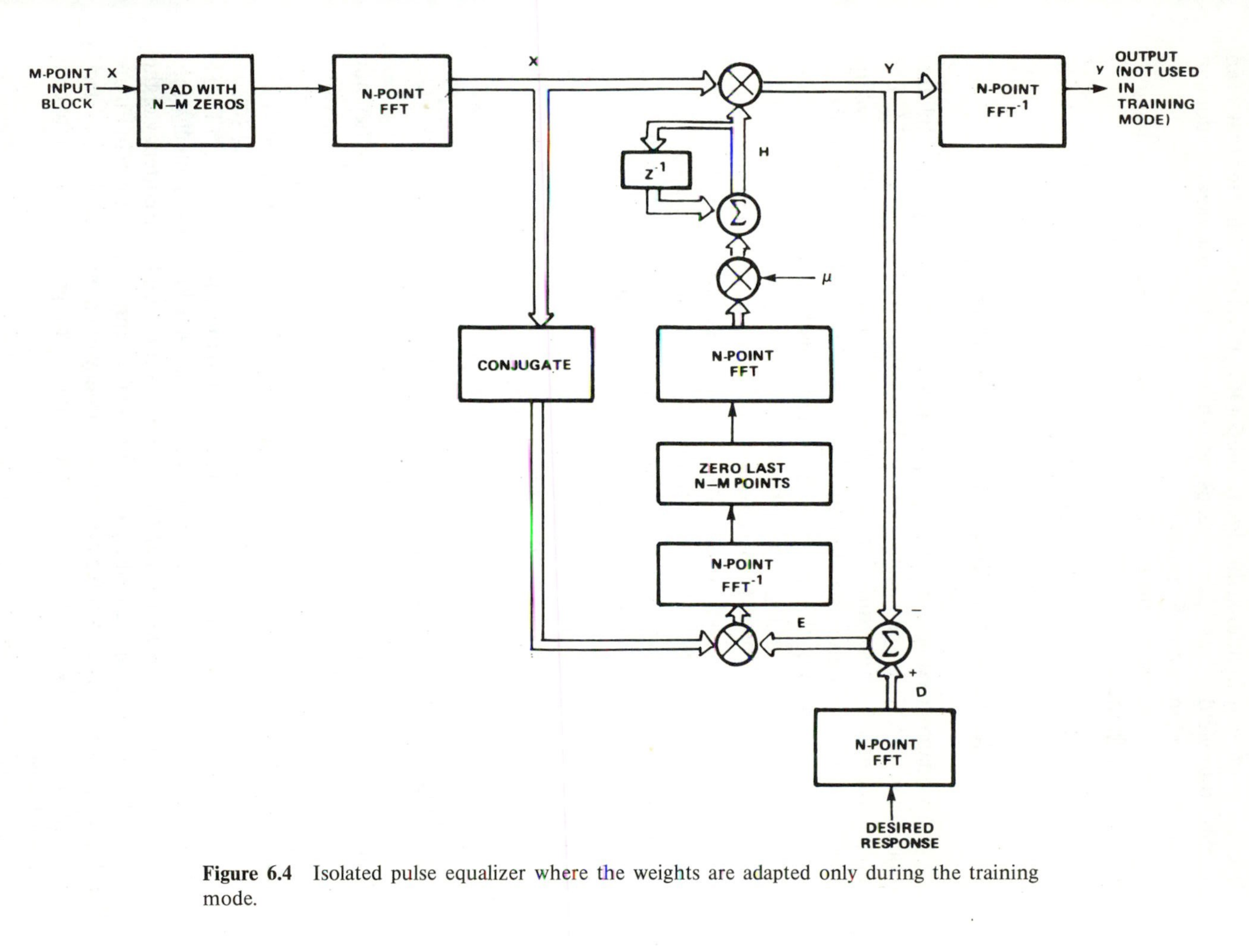

Figure 6.4 Isolated pulse equalizer where the weights are adapted only during the training mode.

Because of the ease in finding the eigenvalues of $\mathbf{R}_{xx}$, it has been claimed that μ can be chosen to yield a faster convergence rate than with comparable time-domain equalizers [Walzman and Schwartz].

Using an approach similar to that in the preceding sections, the misadjustment can be found to be

$$M = \frac{\mu}{2N} \operatorname{tr}\{\mathbf{P}^{\mathrm{T}}\mathbf{R}_{xx}\mathbf{P}\} \leq \frac{\mu}{2N} \operatorname{tr}\{\mathbf{R}_{xx}\} \tag{6.89}$$

6.4.2 Random Data Sequence Equalization

An adaptive equalizer has been proposed for use with pulse-amplitude modulation (PAM) or phase-shift keying (PSK) systems [Maiwald et al.]. The input to the equalizer is assumed to be the (possibly complex) output of a receiving filter sampled at the pulse repetition rate. The desired response for the equalizer is obtained by estimating the transmitted sequence with a decision circuit after equalization as shown in Figure 6.5. The equalizer structure differs from that of Figure 6.4 in that overlap-add is performed at the equalizer output in order to produce linear convolution with an infinitely long input sequence, allowing information transmission to occur concurrently with equalizer adaptation. However, the equalizer output that is fed back for the adaptation process is not overlap-added. In fact, adaptation proceeds exactly as it did for the isolated pulse equalizer, that is, as if the input were composed of disjoint blocks separated by zeros. Although the mean-square error between the properly overlapped equalizer output and the desired response is not being minimized by this approach, encouraging results have been reported.

6.5 TRANSMULTIPLEXER ADAPTIVE FILTER

Transmultiplexers are used in telecommunication networks to translate efficiently between time-division multiplexing (TDM) and frequency-division multiplexing (FDM) [Bellanger and Daguet]. Some of the techniques used in transmultiplexing can be used to implement an adaptive filter in an efficient manner [Copeland]. For our purposes, the effect of a transmultiplexer is shown in Figure 6.6(a) (actual implementation will be discussed later). The input to the transmultiplexer (TM) is filtered by a bank of N complex bandpass filters, each having essentially zero response outside a bandwidth of $2f_s/N$, where f_s is the sampling frequency of the input data. Adjacent bandpass filters may overlap in frequency, as shown in Figure 6.7. The filter outputs are shifted down in frequency to baseband and can be decimated by a factor of $N/2$ without loss of information. This process is similar to converting FDM to TDM.

The effect of the inverse transmultiplexer (similar to TDM-to-FDM conversion) is shown in Figure 6.6(b). Each input channel is interpolated to increase the sampling rate by a factor of $N/2$ and then processed by another bank of bandpass

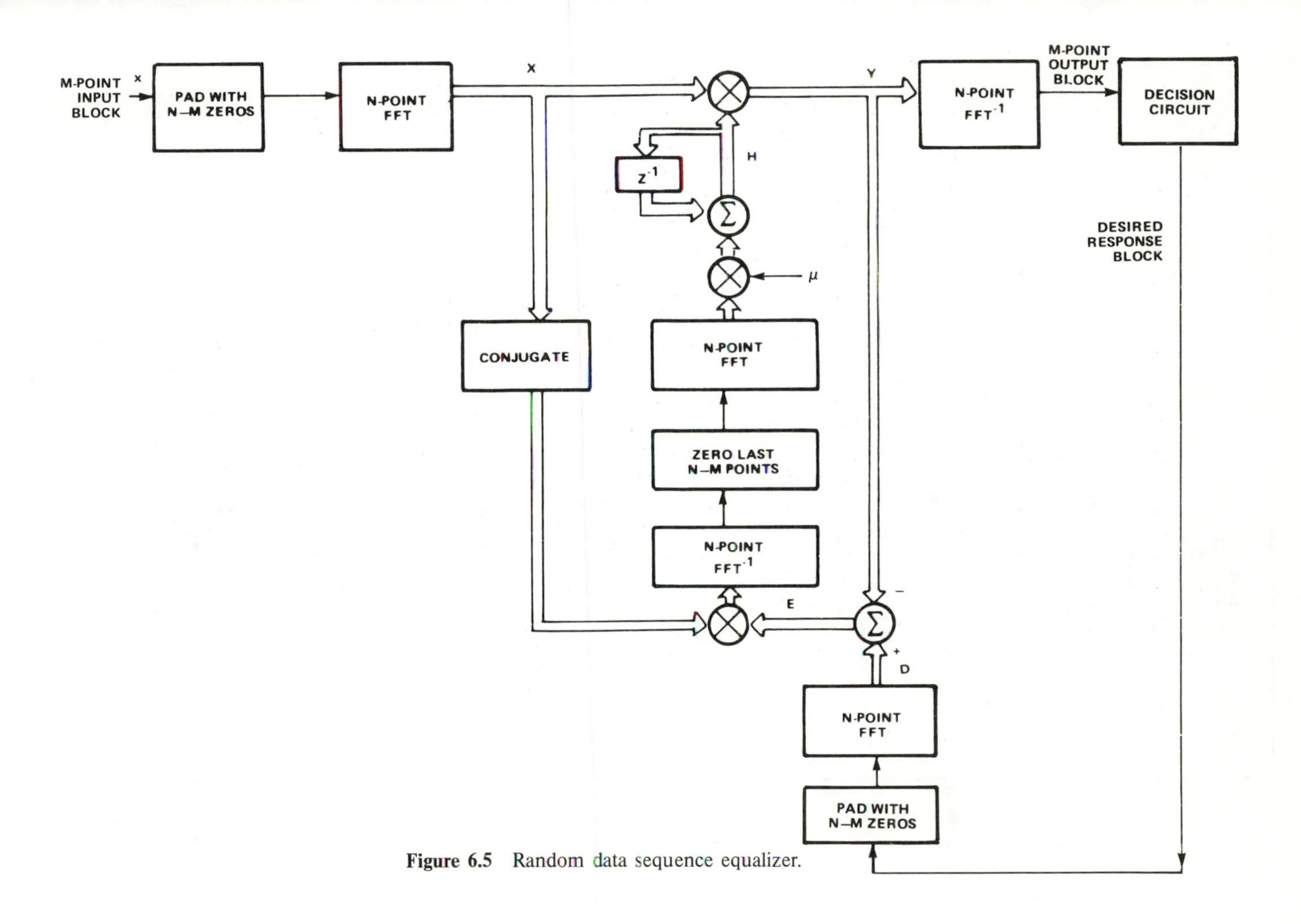

Figure 6.5 Random data sequence equalizer.

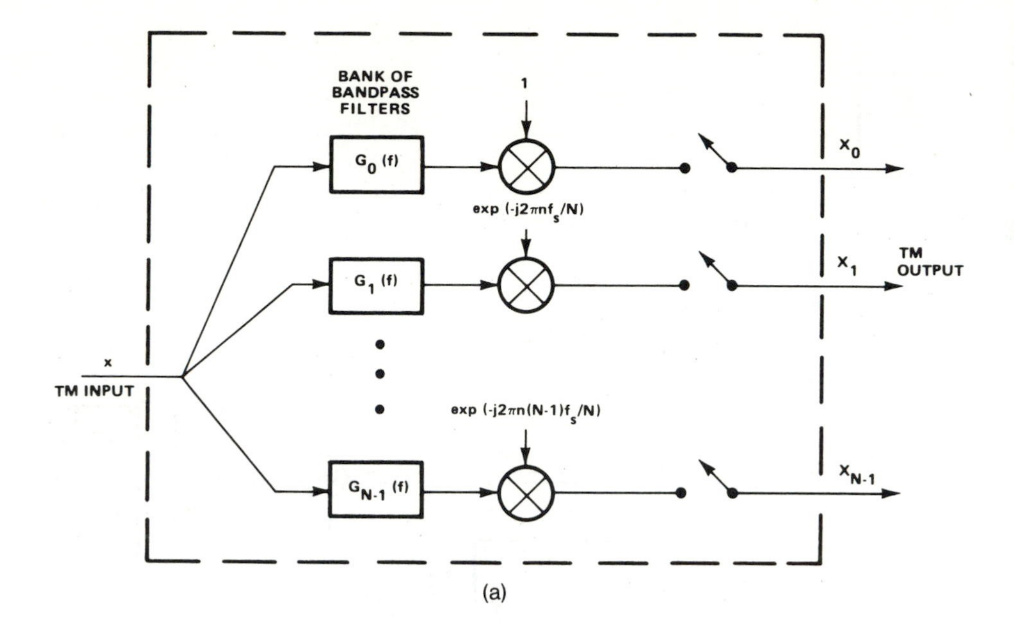

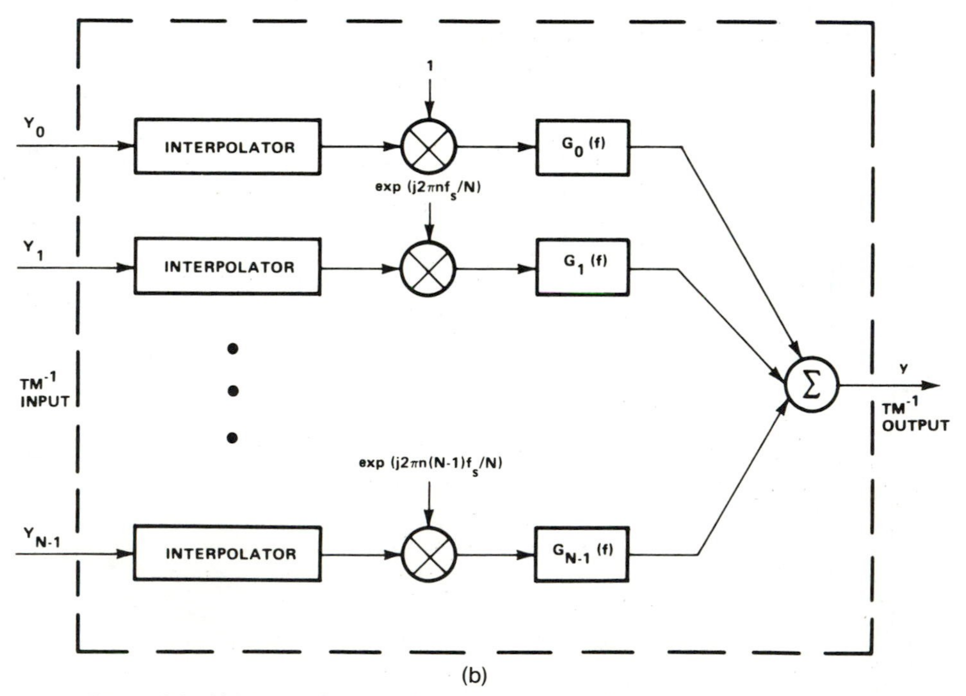

Figure 6.6 (a) Model of transmultiplexer operation; (b) model of inverse transmultiplexer operation.

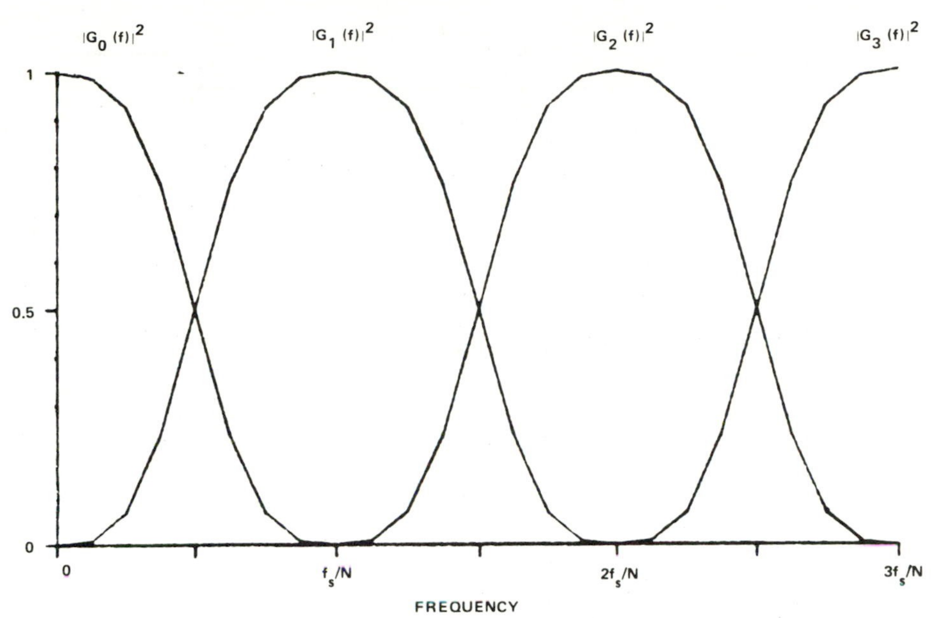

Figure 6.7 Possible magnitude-squared transfer functions for the bandpass filter bank.

filters. The bandpass filter outputs are then summed. For the processes described in Figure 6.6 to truly be inverses of each other, the bandpass filters must satisfy

$$\sum_{i=0}^{N-1} |G_i(f)|^2 = 1 \tag{6.90}$$

General filtering can be performed using transmultiplexers by suitably weighting the outputs of the TM, which are then processed by an inverse TM. If the ith TM output is weighted by H_i, the resulting overall transfer function obtained is (except for a delay)

$$H(f) = \sum_{i=0}^{N-1} H_i |G_i(f)|^2 \tag{6.91}$$

Since only adjacent bandpass filters may overlap, design of many types of fixed filters could be achieved rather easily by setting the desired gain of each frequency bin. However, the primary advantage of the approach is in adaptive filtering.

Figure 6.8 shows a block diagram of an adaptive filter using transmultiplexers [Copeland]. The filter weight for the ith frequency bin is adapted to minimize the mean-square error between the output for that bin, $Y_i(k)$, and the desired response for that bin, $D_i(k)$, using the LMS algorithm:

$$H_i(k+1) = H_i(k) + \mu E_i(k) X_i^*(k) \tag{6.92}$$

where

$$E_i(k) = D_i(k) - Y_i(k) \tag{6.93}$$

and

$$Y_i(k) = H_i(k) X_i(k) \tag{6.94}$$

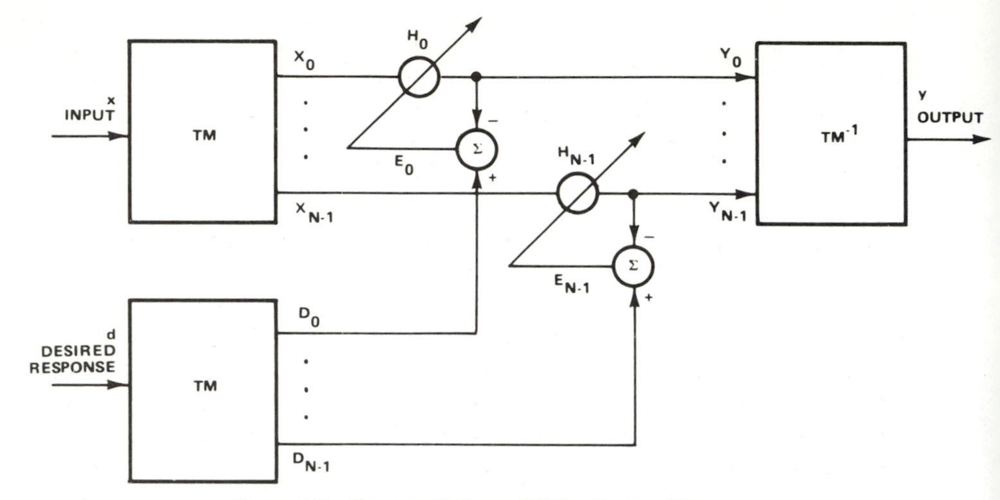

Figure 6.8 Transmultiplexer (TM) adaptive filter.

The convergence properties of each channel are the same as that for a one-tap LMS filter. For statistically stationary inputs and μ sufficiently small, the mean value of the ith weight will converge to

$$\lim_{k\to\infty} E[H_i(k)] = \frac{E[D_i(k)X_i^*(k)]}{E[X_i(k)X_i^*(k)]} \tag{6.95}$$

This weight is not, in general, the optimum weight setting to minimize the mean-square error between $d(k)$ and $y(k)$ unless the bandpass filters have no frequency overlap and have unity gain in their passband. Even so, quite reasonable results can be obtained for many scenarios, since overlap is confined to adjacent bins. Several experiments have shown that the transfer function obtained is practically indistinguishable from optimum when the optimum varies little over a bin width. Even when the optimum transfer function contained narrow notches, near-optimum performance was obtained with the adaptive filter. Optimum performance can be approached as closely as desired by reducing the amount of overlap between adjacent bandpass filters, although the complexity of the transmultiplexers is increased.

The optimum weights can be determined in the following manner. Let $\mathbf{H}$ be a vector containing the weights and let $\mathbf{\Gamma}(f)$ be a vector whose ith element is $|G_i(f)|^2$. Then, according to (6.91), the filter transfer function is

$$H(f) = \mathbf{\Gamma}^{\mathrm{T}}(f)\mathbf{H} \tag{6.96}$$

We wish to choose $\mathbf{H}$ to minimize

$$E\int_0^{f_s} |Y(f) - D(f)|^2\, df = E\int_0^{f_s} |X(f)\mathbf{\Gamma}^{\mathrm{T}}(f)\mathbf{H} - D(f)|^2\, df \tag{6.97}$$

Expanding out (6.97) and setting its gradient with respect to $\mathbf{H}$ equal to zero results in

$$\mathbf{H}_{\text{opt}} = \left[\int_0^{f_s} E[\mathbf{\Gamma}(f)X(f)X^*(f)\mathbf{\Gamma}^T(f)]\, df \right]^{-1} \int_0^{f_s} E[\mathbf{\Gamma}(f)X^*(f)D(f)]\, df \qquad (6.98a)$$

$$\equiv \mathbf{R}_{xx}^{-1}\mathbf{R}_{xd} \qquad (6.98b)$$

The ith element of the second integral above can be shown to be

$$\int_0^{f_s} E[G_i^*(f)X^*(f)D(f)G_i(f)]\, df = E[X_i^*(k)D_i(k)] \qquad (6.99)$$

If the bandpass filters are disjoint in frequency and have unity gain in their passband (zero elsewhere), $\mathbf{R}_{xx}$ is diagonal and its ith element is given by $E[X_i(k)X_i^*(k)]$. Therefore, under these assumptions, the proposed adaptive filter converges to the optimum solution. However, these assumptions will not be strictly satisfied in a practical system and some loss in performance from optimum will generally be obtained.

A transmultiplexer producing the effect of Figure 6.6(a) can be implemented in the following manner [Coker]. Let $g(n)$ be the M-point impulse response of a lowpass filter whose transfer function $G(f)$ satisfies

$$G(f) \approx 0 \qquad \text{for } |f| > \frac{f_s}{N} \qquad (6.100)$$

and

$$\sum_{i=0}^{N-1} \left| G\left(\frac{f - if_s}{N}\right)\right|^2 \approx 1 \qquad (6.101)$$

The bandpass filter $G_i(f)$ of Figure 6.6(a) will be $G(f - if_s/N)$, the frequency-shifted low-pass filter. It will be convenient to let N be a power of 2 and $M = LN$ for some small positive integer $L \geq 2$. The bigger the value of L, the less the overlap can be between adjacent bandpass filters. The bandpass filter shapes of Figure 6.7 were obtained with $L = 3$. Figure 6.9 shows the transmultiplexer implementation for $L = 3$. The input is commutated into two banks of three-tap filters whose filter weights are obtained from the original low-pass filter impulse response. When the upper commutator reaches the top of the filter bank, an N-point FFT^{-1} of the filter bank outputs is computed. When the lower commutator reaches the middle of its filter bank, another FFT^{-1} is computed. The FFT^{-1} outputs are interleaved to produce a sampling rate for each output channel that is $2/N$ times the input sampling rate.

The inverse transmultiplexer can be implemented as shown in Figure 6.10 for $L = 3$. The input channels are processed by an FFT^{-1} whose outputs go into a bank of N $2L$-tap filters whose filter weights are again determined by the original low-pass impulse response. The output commutator sweeps down $N/2$ filters from the top. At this time, a new FFT^{-1} is computed and updates all N filters. The commutator continues sweeping down until it is ready to return to the top filter, at which time another FFT^{-1} must be computed. Thus two FFTs are computed for every N points of output obtained. Because of this, each filter need only compute every other output.

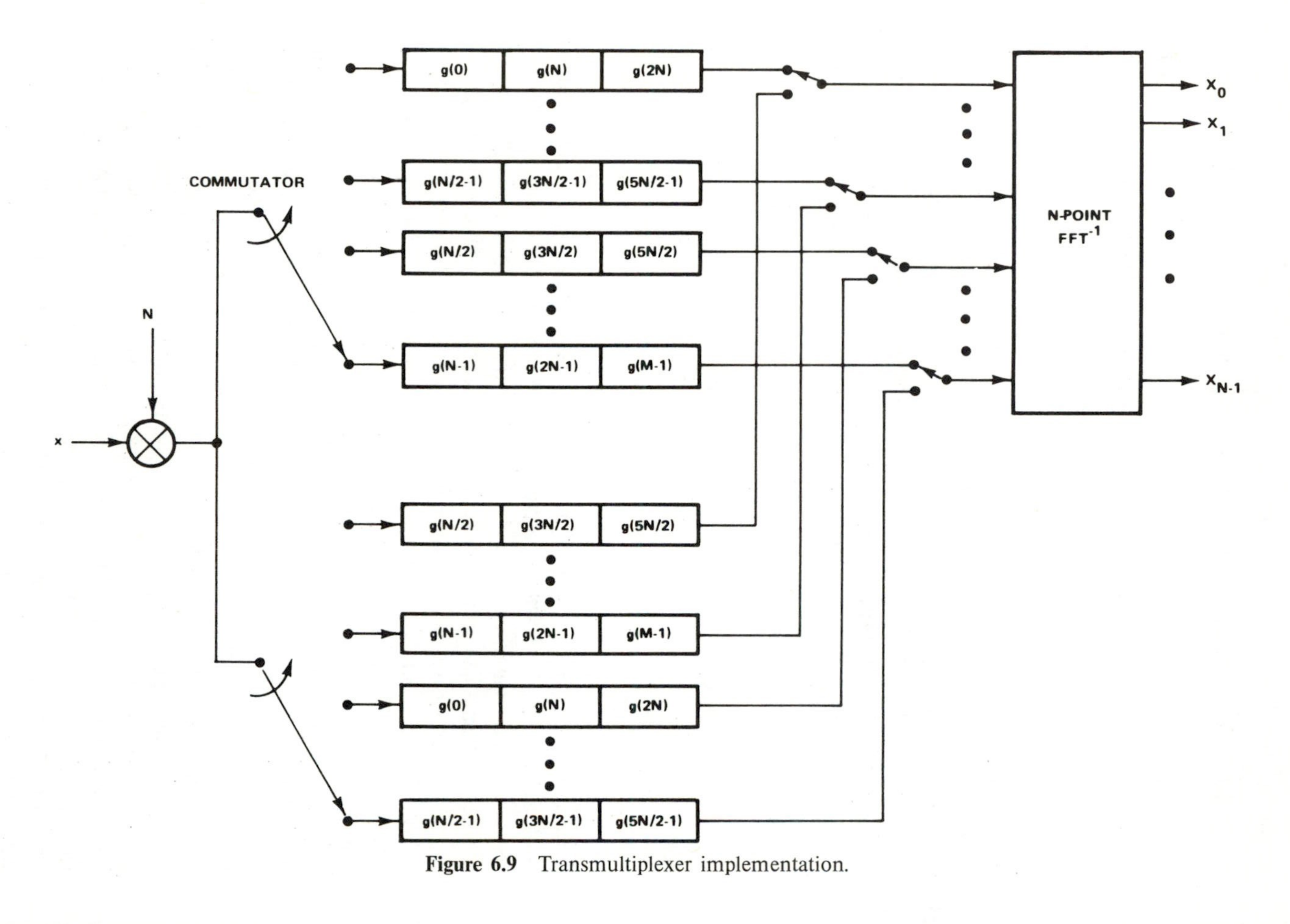

Figure 6.9 Transmultiplexer implementation.

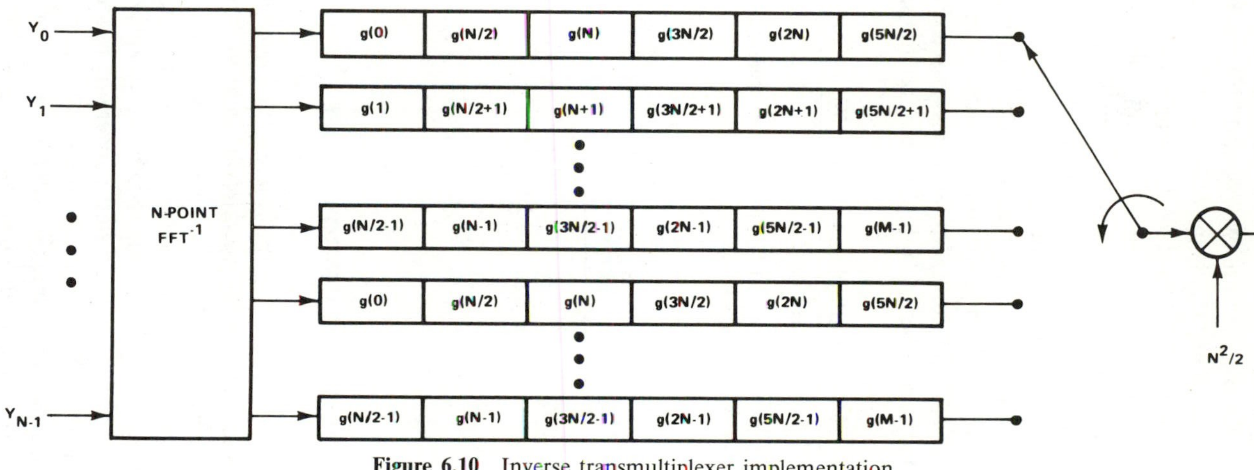

Figure 6.10 Inverse transmultiplexer implementation.

The number of multiplies needed to compute N real output points for the transmultiplexer adaptive filter can now be computed. The input, desired response, and output transmultiplexers require two N-point FFTs each. As stated earlier, an N-point FFT can be computed using an $N/2$-point FFT and $N/2$ complex multiplies. Therefore, a total of $(3N/2)\log_2(N/2)$ complex multiplies are needed for the FFTs. Since the weights should be symmetric, only half of them need to be used. Weighting and weight updating then take $2N$ complex multiplies per N processed points. Computation for the transmultiplexer input and output filters takes $2M = 2LN$ real multiplies each for the input, desired response, and output. The input and output scaling can be absorbed into the transmultiplexer filter coefficients. Assuming that one complex multiply equals four real multiplies yields $6N\log_2N + 2N + 6LN$ real multiplies per N processed data points. This figure can be compared to the number of multiplies required for an N-tap LMS adaptive filter:

$$\frac{\text{TM real multiplies}}{\text{LMS real multiplies}} = \frac{3(\log_2 N) + 1 + 3L}{N} \tag{6.102}$$

This ratio is computed for $L = 3$ in the following table. The amount of computation is slightly more than that of UFLMS.

N	$\dfrac{\text{TM Real Multiplies}}{\text{LMS Real Multiplies}}$ $(L = 3)$
16	1.4
32	0.78
64	0.44
256	0.13
1024	0.04

6.6 CONVERGENCE RATE IMPROVEMENT

In this section methods for altering the convergence rate of the various modes of an adaptive process are discussed. Generally, the modes of an adaptive process converge at different rates, with the rate of each mode determined by the associated eigenvalue of the input autocorrelation matrix. The eigenvalues, and thus the convergence rates, can be vastly different when the input power spectrum varies greatly with frequency. To keep the fastest converging modes from becoming unstable, the convergence rates of some of the other modes may be unacceptably slow, resulting in an inability to track nonstationarities in the input data. It would generally be desirable to make the convergence rates of the modes more equal.

The convergence rate problem is particularly bothersome for the case of block processing employed by the frequency-domain adaptive algorithms discussed in this chapter. In this case the filter weights are updated only once per block and, even though a very good estimate of the gradient may be obtained by averaging data over the entire block, the updates must be in small enough increments to ensure stability of the adaptive process. For example, when using the FLMS algorithm with N-point blocks, 2μ must be chosen less than $1/N\lambda_{max}$ to attain stability; with sample-by-sample updates, it need only be less than $1/\lambda_{max}$. Choosing the adaptation rate to satisfy the block processing stability limit is not a problem when the eigenvalues are nearly equal, as discussed in Section 6.3, since the maximum μ would be determined by reasonable misadjustment rather than stability. However, when the eigenvalues are highly disparate, stability requires that adaptation proceed much more slowly than is necessary to achieve reasonable performance.

In either the adaptive transmultiplexer or the circular adaptive filter of Section 6.2, the weights are adapted independently from each other so that each weight is associated with one mode of the adaptive process. Since the modes are easily accessible, it is easy to alter their convergence rates. Since each weight corresponds to a one-tap LMS adaptive filter, the convergence time for the ith weight, assuming stationary inputs, is inversely proportional to $\mu\lambda_i$, where λ_i is the input power seen by that weight (λ_i is also the ith eigenvalue of the input autocorrelation matrix). In order to make all the modes converge at the same rate, we could make μ be different for each weight according to

$$\mu_i = \frac{\alpha}{p_i} \tag{6.103}$$

where p_i is an estimate of the input power seen by the ith weight. If the p_i are good estimates for the powers, the weights all converge at the same rate with a time constant of

$$\tau = \frac{N}{\alpha} \quad \text{samples} \tag{6.104}$$

For the circular adaptive filter the misadjustment for N weights would then be

$$M = \frac{\alpha}{2} \tag{6.105}$$

These convergence properties would be obtained provided that the environment is stationary and good estimates of the powers are available at the start of adaptation. If the environment is nonstationary, or if the power seen by each weight is unknown, a recursive update of the power estimates can be used. A simple recursion is

$$p_i(k) = \gamma p_i(k-1) + (1-\gamma)|X_i(k)|^2 \tag{6.106}$$

where $X_i(k)$ is the input to the ith weight at time k and γ is chosen between zero and 1. This corresponds to using an exponentially weighted average of the magnitude squared of the input values:

$$p_i(k) = (1-\gamma) \sum_{m=0}^{\infty} \gamma^m |X_i(k-m)|^2 \tag{6.107}$$

The ith weight is then adapted according to

$$H_i(k+1) = H_i(k) + \frac{\alpha}{p_i(k)} E_i(k) X_i^*(k) \tag{6.108}$$

A particularly good choice for α is $1 - \gamma$, for then the weights are chosen at each point in time to minimize an exponentially weighted average of the square error [Falconer and Ljung 1978] given by

$$\sum_{m=0}^{\infty} \gamma^m |E_i(k-m)|^2 \tag{6.109}$$

It should be noted that when p_i changes in time, either because of nonstationarity or transients in the power estimates following startup of the adaptive process, convergence is not exponential as in gradient descent algorithms because the gradient estimate in the weight-update equation has been modified by the multiplication of the various time-varying μ_i values. For stationary inputs, the weights converge in the mean to the same final solution obtained with identical μ_i values for each weight (except when the weights are constrained, as discussed below), provided that the modes are all stable. Stability is ensured by choosing $\alpha = 1 - \gamma$, for then the weights always minimize a weighted least-squares criterion.

To see the effect of recursively updating the power estimates on adaptive filter convergence, it is useful to consider a step change in input power to one of the weights. For a step increase in input power of the ith mode, the effective $\mu_i\lambda_i$ product is initially big, since the power estimate is close to the input power before the increase. Then $\mu_i\lambda_i$ decreases to its steady-state value as the power estimate converges to the input power after the step increase. Thus the response of the adaptive filter is fastest right after the step. For a step decrease in power, just the opposite is true—$\mu_i\lambda_i$ starts out small following the step, then gradually increases.

In the UFLMS adaptive filter, adaptation of the weights is not uncoupled, since the error used to adapt each weight depends on all the weights. However, we shall see that the weight adaptation becomes approximately uncoupled when there are a sufficiently large number of weights, provided that the input is stationary with no periodic components. This allows the time constants of the adaptive process to be set approximately to desired values by choosing a different μ_i for each weight as discussed previously.

The DFT coefficients of a stationary random process are essentially uncorrelated provided that the power spectrum of the process changes slowly over the bandwidth of a single DFT bin [Hodgkiss and Nolte]. If there are no periodic components present (including DC), this condition can be satisfied by choosing a large enough DFT size, which, for the UFLMS algorithm, is equal to the number of weights. (For a periodic signal, the DFT coefficients become uncorrelated only if the DFT size is a multiple of the period.) If the DFT coefficients are uncorrelated, the frequency-domain autocorrelation matrix $\mathbf{R}_{xx}$, given by (6.63), is diagonal. This is so since the ith, jth element of $\mathbf{R}_{xx}$ is

$$(\mathbf{R}_{xx})_{ij} = W_{ij} E[X_i^*(k) X_j(k)] \tag{6.110}$$

where $X_i(k)$ is the ith DFT coefficient of the kth input block. A diagonal $\mathbf{R}_{xx}$ matrix implies that the expected values of the frequency-domain weights are uncoupled during adaptation, the eigenvalues of the $\mathbf{R}_{xx}$ matrix are given by its diagonal elements, and the ith time constant is $\tau_i = 1/(\mu_i \lambda_i)$. Computer simulation has shown that the UFLMS algorithm, with a different μ_i for each weight determined by (6.103) and (6.106), converges faster than LMS for the case of highly disparate eigenvalues [Mansour and Gray]. (See also Figure 7.27.)

In the FLMS adaptive filter, the last half of the gradient estimate is set to zero in the time domain to ensure that half of the time-domain weights are zero. The gradient modification discussed above must be applied after the gradient constraint or the converged weights will be biased away from their optimal values. The gradient, and therefore the convergence rate, can be modified without biasing the final solution, in the following way. Recall that FLMS is an exact implementation of the block LMS algorithm. To equalize the convergence rates exactly, it is necessary to multiply ∇ in (6.30) or (6.40) by the inverse of the $N \times N$ time-domain autocorrelation matrix $\mathbf{r}_{xx}$. If the N-point DFT coefficients of the filter input are essentially uncorrelated, $\mathbf{r}_{xx}$ is essentially circulant [Pearl]. The inverse of a circulant matrix is also circulant. Thus we wish to multiply ∇ by a circulant matrix which is approximately $\mathbf{r}_{xx}^{-1}$, which is most easily accomplished in the frequency domain. The N-point DFT of ∇ is computed and its ith element is divided by an estimate of the power of the ith coefficient of an N-point DFT of the filter input. Then an N-point DFT^{-1} is computed. The result is padded with N zeros and a $2N$-point DFT is taken and used to update the frequency-domain weights. Performing these operations requires a considerable amount of additional computation. The simpler approach of modifying the gradient in the frequency domain before applying the constraint has been proposed [Picchi and Prati]. Although this causes the converged solution for the weights to be biased, the bias was found to be insignificant, at least for certain scenarios.

The channel equalization algorithms of Section 6.4 also require a constraint to be imposed on the gradient. Comments on convergence rate improvement are therefore similar to those for FLMS.

Figure 6.11 compares the convergence rates of some of the adaptive filters described in this chapter for $N = 256$. The desired response in these experiments was a white noise sequence of unit power. The adaptive filter input was obtained by passing this white noise sequence through a three-tap digital filter whose impulse response is 0.33, 0.63, and 0.33. A different μ was chosen for each frequency bin according to (6.103) and (6.106) with $\alpha = 0.9$ and $\gamma = 1 - \alpha$. The power estimates in (6.106) and the filter weights were initially set to zero. Figure 6.11 shows the mean-square error between the filter output and desired response computed over successive 256-point blocks. According to this figure, the TM and circular adaptive filters converge quickly, although the steady-state performance of the circular adaptive filter is poorer since it must perform circular convolution. The UFLMS filter achieves good steady-state performance but converges more slowly since the UFLMS weights are not strictly uncoupled. Although not shown in Figure 6.11,

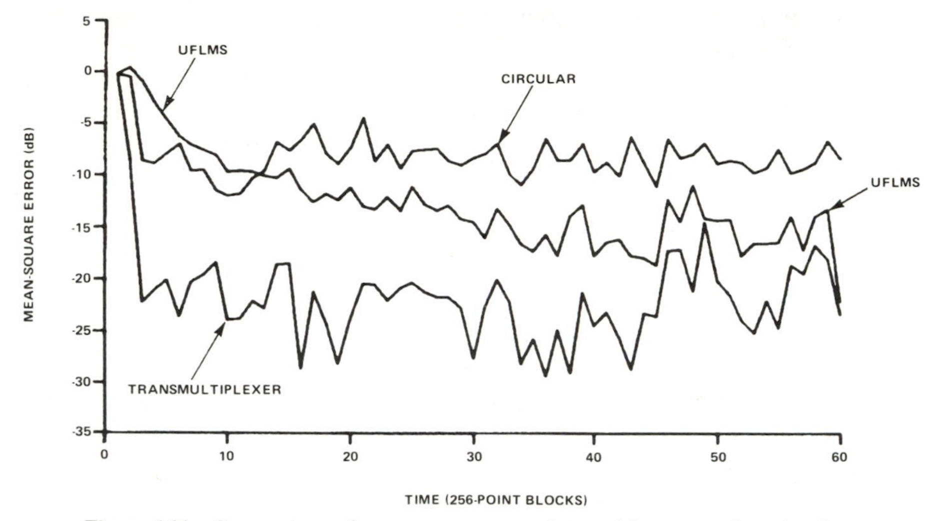

Figure 6.11 Comparison of convergence rates of several frequency-domain adaptive algorithms.

the FLMS adaptive filter, with gradient modification applied before the constraint as in [Picchi and Prati], exhibited convergence properties nearly identical to that of UFLMS.

We mention briefly another frequency-domain method that has been proposed to improve adaptive convergence rate but that does not result in reduced computation compared with time-domain processing. Narayan et al. and Bitmead and Anderson decompose the input signal into its frequency components using a bank of N bandpass filters implemented with a sliding DFT; the DFT is updated recursively for each new input sample. A different μ can be used to adjust the convergence rate separately for each frequency bin. The bandpass filter outputs are not decimated, so that the computation required to process a fixed amount of data is not reduced compared with time-domain processing, although the convergence time may be reduced. If the bandpass filter outputs are decimated by a factor of N, the circular adaptive filter of Section 6.2 is obtained.

6.7 SUMMARY

Several frequency-domain approaches to adaptive filter implementation have been presented. Adaptation in the frequency domain can result in substantial reduction in computation and convergence time compared with time-domain processing. Of the frequency-domain approaches, the FLMS adaptive filter is closest in performance to the conventional time-domain LMS adaptive filter and has the advantage of easily imposed time-domain constraints on its impulse response. The UFLMS filter is simpler to implement than FLMS and, except in certain situations described in Section 6.3.2, provides steady-state performance at least as good as FLMS. Frequency-domain adaptive filters employing circular convolution are the simplest to implement but can yield poorer steady-state performance than adaptive filters allowed to perform linear convolution. A transmultiplexer adaptive filter has moderate complexity, near-optimum steady-state performance, and the advantage that the frequency-domain weights can be adapted independently.

The block processing inherent in most frequency-domain adaptive filters sets an upper limit on speed of adaptation that can result in much slower convergence than with time-domain processing when the input correlation matrix has highly disparate eigenvalues. To keep the fastest converging modes from becoming unstable, the convergence rates of some of the other modes may be unacceptably slow. This problem can be alleviated with the gradient modification techniques described in Section 6.6. The convergence rates of the various modes can be made identical for the transmultiplexer and circular adaptive filters; for FLMS and UFLMS this can be achieved only approximately.

6.8 APPENDIX: LINEAR VERSUS CIRCULAR CONVOLUTION

When the FFTs of two sequences are multiplied together and the inverse FFT taken of the product, the result is the circular, rather than the customary linear, convolution of the two sequences. This appendix contains a brief discussion of the difference between these two types of convolution.

Figure 6.12 shows two sequences that we wish to convolve, $x(n)$ and $h(n)$. To produce the linear convolution, $h(n)$ is reversed in time about the origin [Figure 6.13(b)], then shifted sequentially to the right to form the convolution output [Figure 6.13(c)]. For each shift, x and h are multiplied point for point, and these four products summed to produce an output sample.

Figure 6.13(d) and (e) illustrate the circular convolution obtained using N-point FFTs (in this example the block length N is set equal to 8). First, the N samples of $h(n)$ are reversed in time modulo N to produce $h(-n \text{ modulo } N)$, as in Figure 6.13(d). These N samples are then rotated sequentially to produce the N output samples shown in Figure 6.13(e). Due to the wraparound of h, the circular convolution differs from the linear convolution. However, some of the output samples are the same. This occurred only because the last $N/2$ samples of $h(n)$ were zero; if $h(n)$ had been chosen arbitrarily, the two convolutions would have been different in all but one of the output samples. For arbitrary $h(n)$, each output of the circular convolution is a weighted sum of all N samples of $x(n)$.

The overlap-save method is one way of using FFTs to produce the linear convolution of a long data sequence $x(n)$ with a relatively short sequence $h(n)$. If the short sequence is of length M, choose the FFT size to be $N > M$, and pad $h(n)$ with enough zeros to make it of length N. The sequence $x(n)$ is partitioned into sections of length N, each overlapping the adjacent section by $M - 1$ samples. The

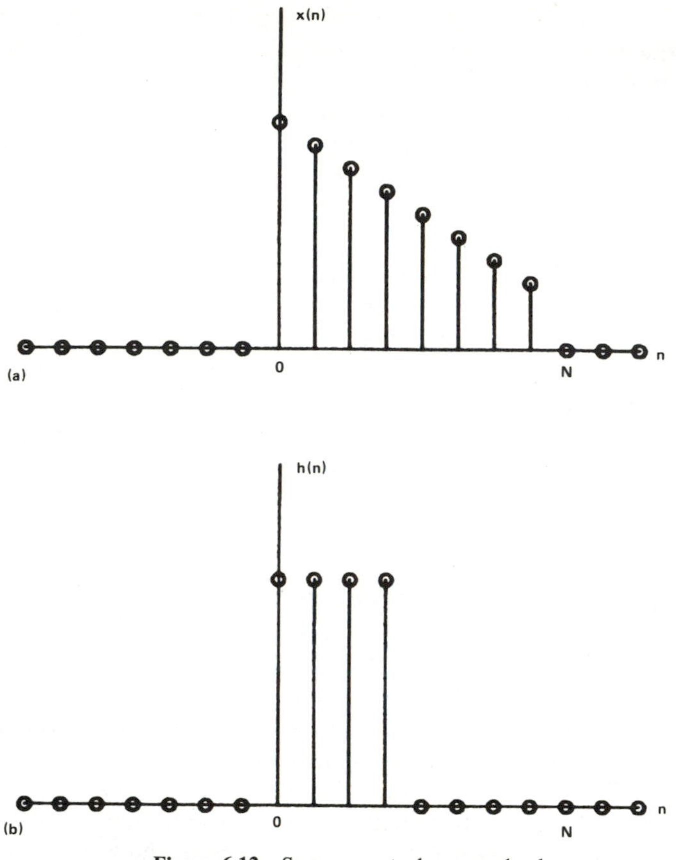

Figure 6.12 Sequences to be convolved.

reason for the overlap will become apparent later. The FFT of one of these sections is computed and multiplied by the FFT of $h(n)$. The inverse FFT of the product is an N- point circular convolution. The last $N - M + 1$ samples of the circular convolution are identical to those that would be obtained had we implemented linear convolution. We keep these samples and throw away the rest. Since we have produced $N - M + 1$ output samples, the sequence $x(n)$ must be shifted by this amount before computing the next block of output. This produces an overlap of $M - 1$ samples between successive sections. The normal optimum is the case where $N = 2M$, which achieves 50% overlap in the time-domain samples. An alternative technique is overlap-add, where the incorrect products are added between successive frames to obtain correct output samples [Clark et al. 1983].

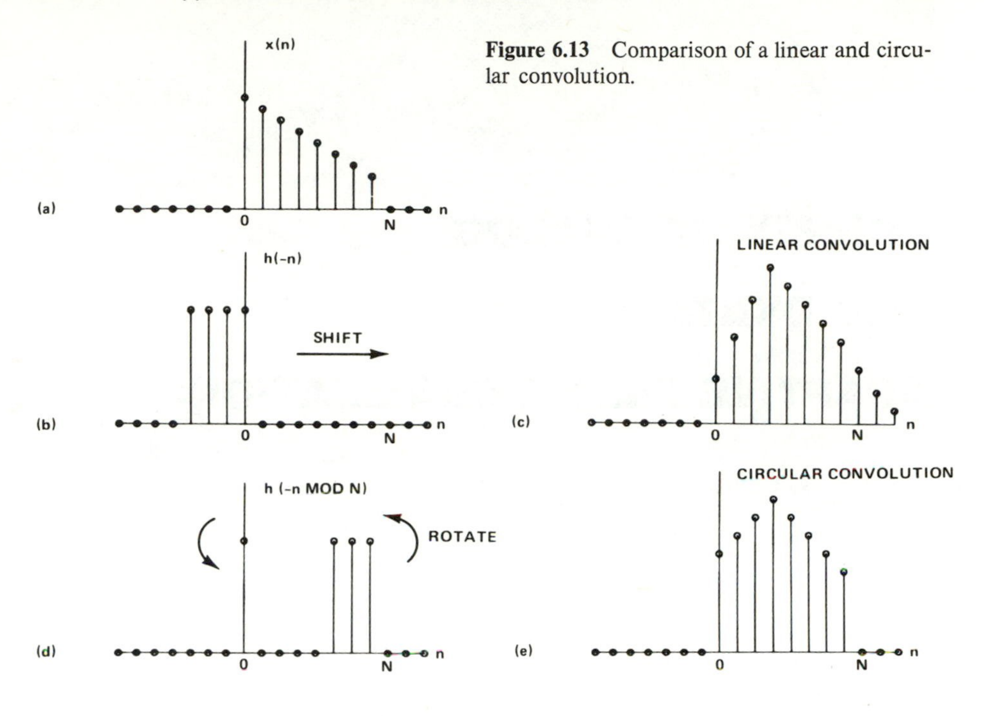

Figure 6.13 Comparison of a linear and circular convolution.

7

SURVEY OF ANALOG AND DIGITAL ADAPTIVE FILTER REALIZATIONS

Colin F. N. Cowan and Peter M. Grant

7.1 INTRODUCTION

Although the theory of adaptive systems has been well known for many years, through their use in control systems and antennas, it is only in recent years that technology advances have made feasible the construction of sampled data adaptive filters. In this chapter the various techniques for implementing this class of filter are surveyed [Cowan et al. 1979]. Some specific system designs are described and experimental results are shown to indicate some of the operational characteristics of adaptive filters.

The discussion is split broadly into two areas, concerned with digital and sampled-data analog implementations. In Section 7.2 a review of digital design techniques is given, principally showing the design restrictions imposed by current integrated-circuit limitations. From this it is possible to extrapolate to feasible structures which may be used once submicron very large scale integrated (VLSI) circuit feature sizes become readily available in silicon. This aspect is covered in Section 7.5.

Section 7.3 is concerned with the use of sampled-data analog structures for adaptive filter design. These filters are predominantly dependent on the use of charge-coupled-device (CCD) [Boyle and Smith, Amelio et al.] and bucket-brigade-device (BBD) [Weckler and Walby] technologies. Examples on the application of analog adaptive filters in telephony are given in Chapter 8. The shrinking of feature sizes in silicon processing is not presently likely to provide any advantage in these analog technologies and it may therefore be presumed that the results presented in this chapter represent the realizable limits of the performance which can be expected from this particular approach. Section 7.4 reports results on other analog adaptive filters based on surface acoustic wave devices which offer increased bandwidth over CCD and BBD approaches. It will be seen that although the analog systems offer great advantages in terms of power, weight, and size, they have fundamental limitations in dynamic range and accuracy which may not be easily overcome. It seems likely that the next generation of adaptive filter design lies in the use of digital techniques. Hence Section 7.5, which describes future device trends, is devoted entirely to this approach.

7.2 DIGITAL IMPLEMENTATIONS

In this section a review of currently feasible digital adaptive filter designs is given. The first part concentrates on classical design techniques which rely heavily on the use of linear digital multipliers. At present these designs must be implemented using standard TTL-based systems or large custom-designed integrated circuits, if reasonable bandwidth is to be achieved. The latter part of this section is devoted to a consideration of memory-based digital systems which do not make use of high-accuracy digital multipliers. These systems are based on the use of residue number systems (RNS) [Soderstrand et al. 1980(1)] and distributed arithmetic architectures [Cowan and Mavor 1981(2)].

7.2.1 Classical Digital Design

Given the limitations of currently available technology, only adaption algorithms which are relatively simple in a computational sense may be executed in digital terms at real-time bandwidths. A generalized block diagram of an adaptive filter is represented in Figure 7.1. Here, the major computational load is imposed by the need to perform the linear convolution of the input signal samples with the stored filter weights (the finite impulse response filter) with a marginal overhead being provided by the adaption algorithm. To provide updates for the weight value

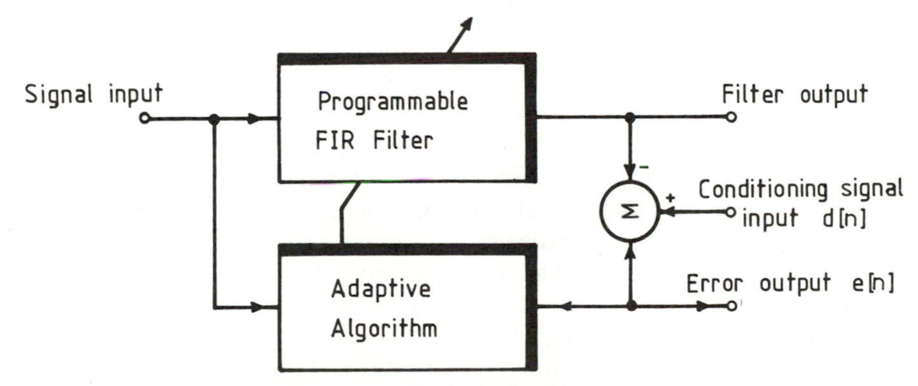

Figure 7.1 Generalized adaptive filter structure.

estimates for the filter the adaption algorithm must be supplied with samples of the signal input and the output error from the estimation process. This information permits implementation of any of the algorithms derived in Chapter 3.

Before considering the implementation of the adaption process itself we shall examine one common approach for implementing the FIR filter section. This technique is illustrated in Figure 7.2 [Mavor et al. 1977]. It uses a single digital multiplier with vectors representing the incoming signals and the filter weights stored in digital memory. This implementation with digital memories is a development of the initial versions of this realization, which was based on analog (bulk delay) memories. Each pair of inputs (signal and weight) is presented to the multiplier sequentially and the multiplication product is accumulated. Therefore, for a filter length of N points the multiplier must operate at N times the sampling rate of the filter. This, obviously, provides a fundamental limitation to the available time-bandwidth product, given a target sampling frequency. The time-bandwidth product may be extended by cascading two or more such filters without affecting the sampling frequency. This, however, results in the use of further multipliers, therefore increasing the cost and power consumption of the system. A number of alternative schemes for filter implementation may be used which have various trade-offs in terms of overall circuit efficiency. Some of these techniques are reviewed at the end of this section.

When considering the adaption algorithm that should be used with this filter structure, it is important that no serious degradation in filter throughput or bandwidth should result from the added computational load to calculate the update values for the filter weights. Using current technologies this has usually led to implementations that make use of the stochastic gradient search least-mean-squares (LMS) [Widrow et al. 1975(2)] adaption algorithms, which have been dealt with theoretically in Chapter 3. The basic LMS adaption algorithm is given by

$$\mathbf{H}(n+1) = \mathbf{H}(n) + 2\mu e(n)\mathbf{S}(n) \tag{7.1}$$

where $\mathbf{H}(n)$ = vector of filter weights at time n
$e(n)$ = estimation error at time n
$\mathbf{S}(n)$ = vector of input signal samples stored in the filter at time n
μ = convergence factor which is less than unity and determines convergence time and the final accuracy of the converged solution

There are a number of possible simplifications of this algorithm which may produce considerable advantages in terms of hardware efficiency. However, we shall initially consider the straightforward implementation of the algorithm given in (7.1) (the linear LMS algorithm). A block diagram of such a system is shown in Figure 7.3, which uses one extra digital multiplier in order to implement the multiplication of signal and error in the LMS algorithm. This particular choice of implementation depends on the desire to maintain the sampling rate of the original FIR filter. However, if a faster multiplier is available or the time-bandwidth product is reasonably low, the original multiplier used in the FIR filter may be further multiplexed to perform the multiplication needed by the adaption algorithm.

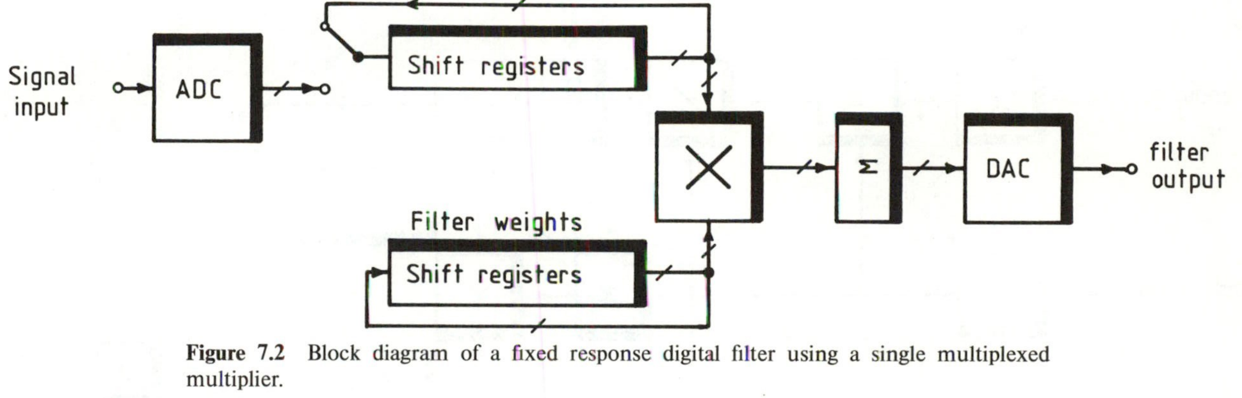

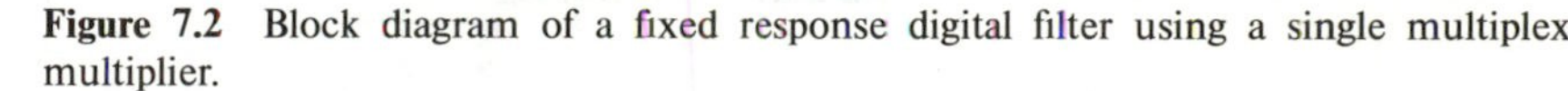

Figure 7.2 Block diagram of a fixed response digital filter using a single multiplexed multiplier.

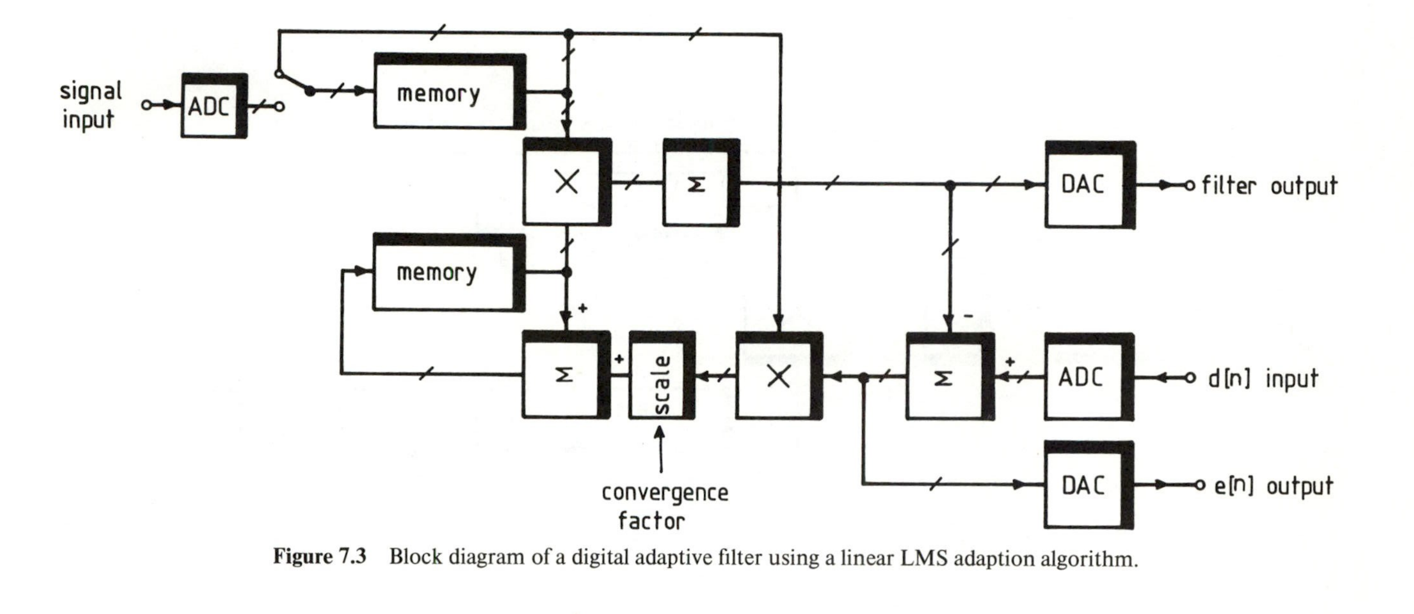

Figure 7.3 Block diagram of a digital adaptive filter using a linear LMS adaption algorithm.

This type of consideration, involving the number and usage of digital multipliers, is the more obvious of the trade-offs involved in the actual design of the hardware adaptive filter. However, a consideration which is not so obvious is that of the number of bits necessary to store the digital signals and weights in the filter. The effect of quantization on normal fixed filter design is fairly well understood and the actual word length for the signal channel may be easily derived from dynamic range requirements. It is even possible to reduce the amount of storage required (and indeed simplify the multiplication process) by using either A-law or μ-law compressive coding, used in pulse-code modulation.

The main problem in determining word lengths occurs in consideration of the accuracy required to store the individual weights, taking into account the adaptive update process. Looking at the form of the filter shown in Figure 7.3, if we assume the input signals to be quantized as 8-bit words and that the weights used in the filtering operation are also 8 bits wide, the filter multiplication produces a 16-bit word. After accumulation this will be further extended, according to the filter order. Normally, the output word would be truncated down to about 8 or 10 bits, with the position of the output word being determined by the application. The significant part of the output word will differ depending on whether the filter is a matched filter, an inverse filter, or other type. Using this particular architecture it is not, therefore, possible to define a completely general-purpose processor for these reasons. Rather, it is necessary to know the application in order to define the significant regions within the data words.

Assuming that the output signal word has been appropriately truncated to yield an 8-bit output word, it is then subtracted from the conditioning signal input and the result multiplied by the signal word. This multiplication produces yet another 16-bit word, which may, once more, be rounded to an 8-bit residual. The result must then be multiplied by the convergence factor, which is done either using another digital multiplication or by scaling by a power of 2 (scaling by an integer power of 2 is usually preferred to linear multiplication, as it is a much simpler hardware operation). The actual width of the update word after this scaling will be dependent on the chosen range of convergence factors, which will be determined by the desired convergence times and converged noise performance. The size of this word is, essentially, the same as the depth to which the filter weights must be stored. Typically, to take full advantage of the performance of the 8-bit arithmetic in the filter, this will require that the filter weights be stored to a depth of between 16 and 24 bits. This is sufficient to ensure that when the filter is used in a cancellation mode the updates caused by the excess error signal do not corrupt the filter weights. Results showing these effects are presented in Section 7.2.2. Filter architectures typical of this type of design procedure have been discussed in some detail in [Neissen and Willim, Morgan and Craig, South et al.]. In addition to these digital FIR adaptive filter structures, gradient lattice hardware has also been reported [Friedlander 1982(3)]. Figure 7.4 shows the block diagram of a lattice equalizer which is based on a single multiplexed 12-bit parallel digital multiplier, with 24-bit accuracy for coefficient storage to minimize the algorithm noise [Rutter

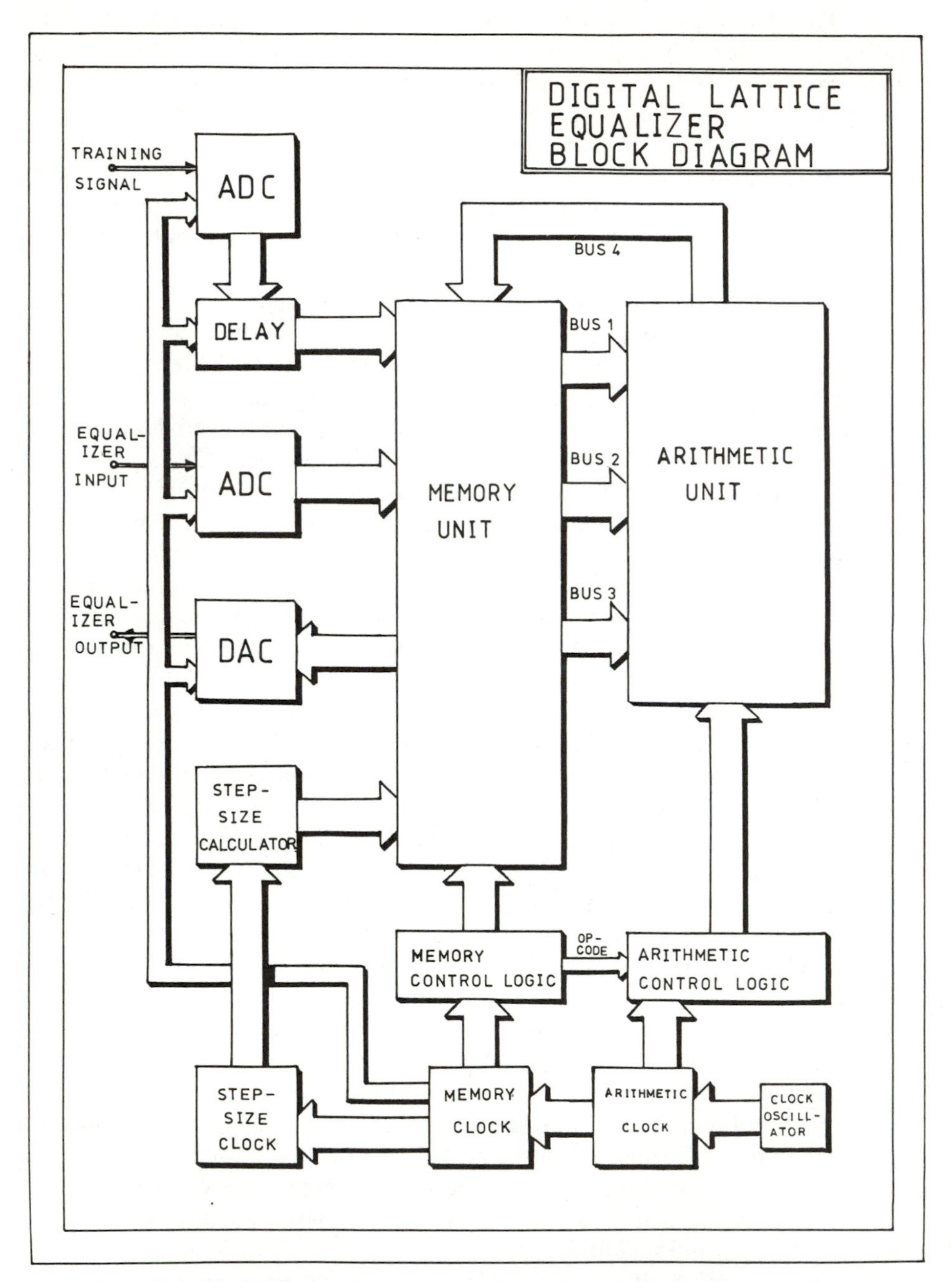

Figure 7.4 Block diagram showing the digital hardware realization of an adaptive lattice filter. (Copyright © 1982 IEE.)

et al. 1983(1) and (2)]. This processor, which comprises 160 standard TTL integrated circuits, implements directly the fixed μ gradient lattice filter discussed in Chapter 5. The eight multiplications per filter stage are implemented in the 2.2-MHz-cycle-rate arithmetic unit and Schottky RAM-based memory units. For a 16-stage lattice equalizer this gives an input sample rate of 17kHz, providing a usable bandwidth of 5 to 8 kHz.

7.2.2 Digital Adaptive Filters Using Simplified Algorithms

In this section the use of simplified LMS algorithms is described with particular reference to their impact on hardware design. In particular, the standard linear LMS algorithm may be degraded in three ways, which are outlined in Table 7.1. All these degradations involve the clipping (reduction to single bit numbers representing sign information only) of one or both of the linear signals involved in the calculation of the updated filter weights. This automatically produces an enhancement in hardware efficiency, due to the fact that the previous linear multiplication may now be replaced by a simpler operation using exclusive-or gates.

TABLE 7.1 LMS-Type Algorithms Showing Simple Degradation Alternatives

Algorithm	Update Product	Convergence Dependence
Linear	$s(t)e(t)$	Input signal power
Clipped	$\text{sgn}[s(t)]e(t)$	Signal amplitude
Pilot	$s(t)\ \text{sgn}[e(t)]$	Signal amplitude
Zero forcing	$\text{sgn}[s(t)]\ \text{sgn}[e(t)]$	Zero crossings

The most commonly used of these algorithms is the clipped LMS adaption algorithm [Moschner], given by

$$\mathbf{H}(n+1) = \mathbf{H}(n) + 2\mu e(n)\ \text{sgn}[\mathbf{S}(n)] \tag{7.2}$$

Here only the signal, $s(n)$, is clipped, meaning that the multiplier determining the update product in Figure 7.3 may be replaced by a number of exclusive-or gates (the number of gates required is equal to the word length of the error). The convergence characteristics of this algorithm have been thoroughly investigated [Moschner] and it has been shown that the performance of this algorithm is approximately comparable to the linear LMS algorithm [Weiss and Mitra]. In the case of input signals which are random with Gaussian statistics, the convergence of the clipped LMS algorithm is approximately 25% slower. However, when the input signals are highly deterministic the clipped LMS algorithm has been shown to converge more quickly than the linear version. In addition, the convergence rate of the clipped algorithm is dependent on input signal amplitude rather than signal power (as in the case of the linear algorithm).

The photograph in Figure 7.5 shows a hardware module constructed to implement a 64-point adaptive transversal filter using the clipped LMS algorithm [Cowan et al. 1979]. This system used 8-bit signal quantization and incorporated a single 8-bit by 8-bit parallel digital multiplier. The filter had a sampling frequency of 8 kHz (i.e., compatible with voice-bandwidth processing). In this particular processor the convergence factor was not programmable; instead, it had a hard-wired convergence factor of 2^{-8} with weight vector storage of 16 bits, where the 8 most significant bits were connected to the multiplier. The output signal range was switchable to accommodate use of the filter in the situation of either matched or inverse filtering (Section 1.1.2). A few experimental results from this system are presented here to demonstrate the performance available with this simple type of system.

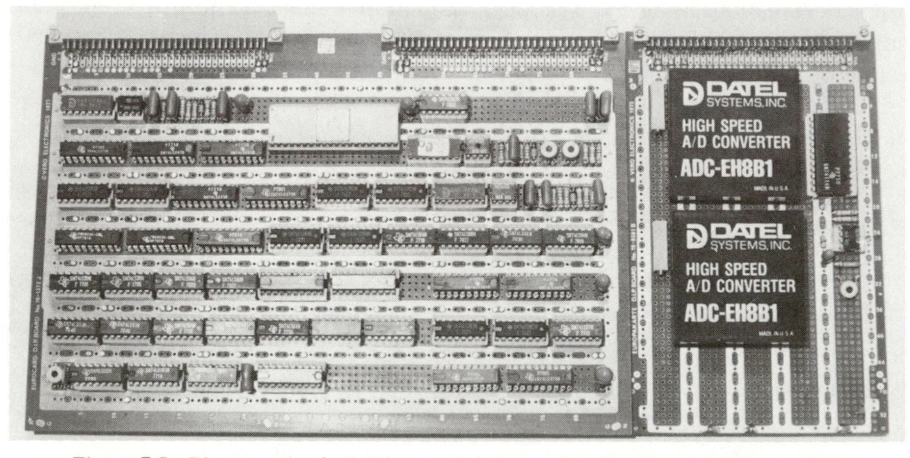

Figure 7.5 Photograph of a hardware processor using the clipped LMS adaption algorithm to implement a 64-point adaptive transversal filter.

The result in Figure 7.6 shows the convergence of the filter to purely sinusoidal inputs. Here the input signal, $s(n)$, is a continuous sinusoid and initially the training signal, $d(n)$, is zero. Therefore, the system is trained to notch out the input sinusoid at the filter output. The training signal is then switched to become an exact replica of the input signal, $s(n)$. The oscilloscope trace in Figure 7.6(d) shows the initial increase in the error signal and its decrease toward zero as the filter output converges [Figure 7.6(c)].

The result shown in Figure 7.7 demonstrates an interesting application in speech processing. Here, the input signal is a segment of speech corrupted by a highly deterministic signal. In the instance shown this is actually a continuous tone, but it could be music or a highly periodic background noise. This combined signal is applied to both inputs of the adaptive filter. The speech signal varies its characteristics too rapidly for the filter to track it adaptively. Therefore, the speech signal appears, virtually uncorrupted, at the filter error output. However, the

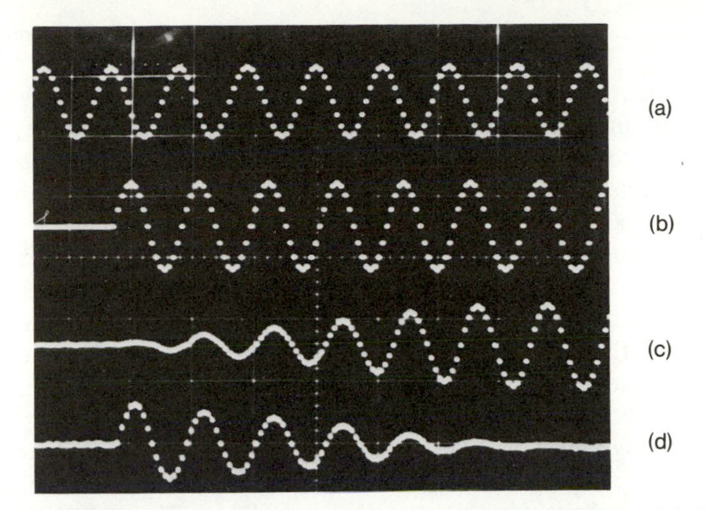

Figure 7.6 Experimental result showing the convergence of the 64-point digital adaptive filter to sinusoidal inputs. (a) Signal input; (b) training signal; (c) filter output; (d) error output. Horizontal scales 2 ms per division, vertical scales linear. (Copyright © 1979 IEE.)

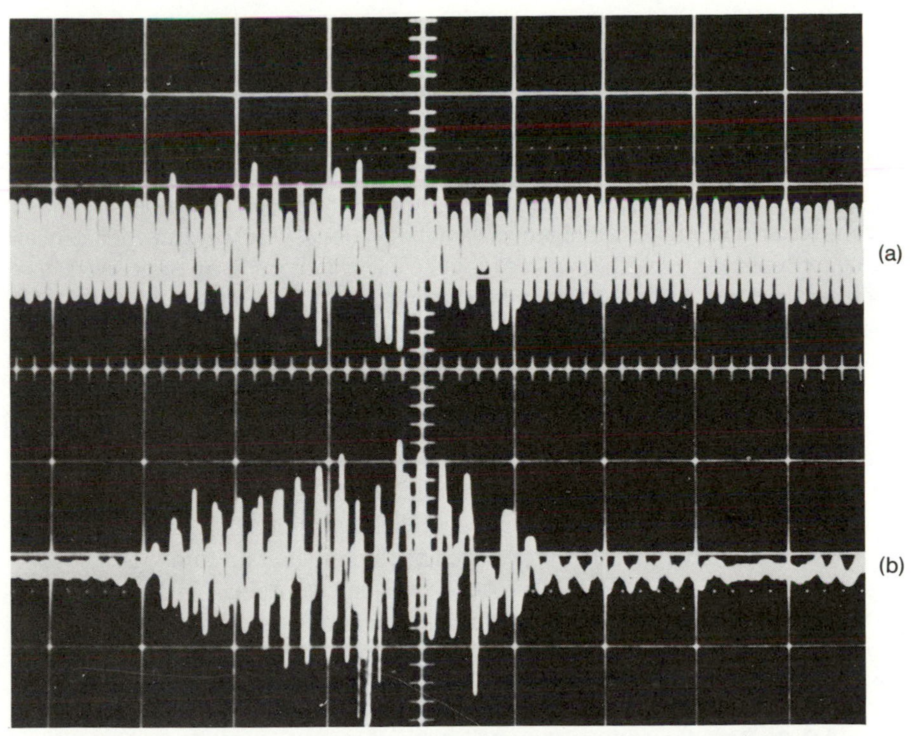

Figure 7.7 Result showing the use of the digital adaptive filter to cancel a tone on a speech signal. (a) Corrupted input signal; (b) cancellation output. Horizontal scale 20 ms per division, vertical scale linear. (Copyright © 1981 E. W. Communications Inc.)

periodic interference is relatively stationary and the filter easily tracks it and cancels the periodic interference on the error output. This result is shown in Figure 7.7(b), where the error output contains the original speech input uncorrupted by the background interference. The periodic interference is separately obtainable at the filter output.

The other two algorithm degradations shown in Table 7.1 have characteristics which are somewhat more difficult to predict than in the clipped LMS case [Mitra and Sondhi 1975]. The zero-forcing algorithm was proposed by [Lucky 1966] before the LMS algorithm had been fully documented. It was one of the first reported implementations of an equalizer, as it was the simplest to realize due to its hardware efficiency.

Although most digital systems of this type have been implemented as hardwired digital systems, there have been some attempts to implement fully monolithic adaptive digital filters [Verhoeckx et al., Duttweiler and Chen 1980]. These processors have been constructed specifically to overcome the problems of echo on the telephone network and further details of this type of design may be found in Chapter 8, which describes adaptive filter applications in telecommunications systems.

7.2.3 Digital Adaptive Filters Using Memory Access Techniques

The techniques for the implementation of digital adaptive filters in the preceding two sections rely on the use of parallel digital multipliers as the main processing block. This operation, of linear digital multiplication, is a common bottleneck encountered in all digital signal processing applications. One way in which explicit digital multiplication may be avoided is by the use of table look-up techniques; that is, rather than physically performing a multiplication the results are accessed from memory using the operands as addresses. For two n-bit inputs and a $2n$-bit product the memory size is $2n \cdot 2^{2n}$, which for $n = 8$ requires a 1-Mbit memory.

It is therefore obvious that a more subtle approach to the problem must be adopted. This can be done in two ways. The first technique which we shall consider is that of the distributed arithmetic filter [Peled and Liu 1974] and the second is a number-theoretic technique using residue number systems (RNS) [McClellan and Rader]. The distributed arithmetic approach (considered in detail in Section 7.2.4) uses the additional structure provided by the convolution sum to simplify the multiply operations. The RNS approach (considered in detail in Section 7.2.5) makes use of number theory to split multiplications of large numbers into a number of smaller operations, hence making the operation more tractable in terms of memory access.

7.2.4 Distributed Arithmetic Adaptive Filters

It is necessary before considering the structure of adaptive filters making use of distributed arithmetic to examine the distributed arithmetic approach to fixed filtering. It was indeed for the implementation of fixed-response biquadratic digital filters that the distributed arithmetic filter was first proposed by [Peled and Liu 1974].

Consider, first, the convolution product defined by

$$y = \mathbf{S}^{\mathrm{T}}\mathbf{H} \tag{7.3}$$

where y = filter output
$\mathbf{S}$ = column vector of the input signal values
$\mathbf{H}$ = column vector of the filter weights

and the superscript T refers to matrix transposition. The time index has been deleted for convenience in this treatment.

If the input signals are coded as offset binary numbers, we may define a matrix $\mathbf{B}$ such that the rows of $\mathbf{B}$ are the offset binary digits representing the corresponding signal elements in $\mathbf{S}$. The convolution in (7.3) may then be redefined as

$$y = (\mathbf{BX})^{\mathrm{T}}\mathbf{H} \tag{7.4}$$

where $\mathbf{X}$ is a column vector of the first K negative integer powers of 2. Rearranging (7.4) gives us

$$y = \mathbf{X}^{\mathrm{T}}(\mathbf{B}^{\mathrm{T}}\mathbf{H}) \tag{7.5}$$

Note that the transposition of $\mathbf{B}$ now implies that the weights are multiplied by bits of the same significance from different signal words and appropriate weighting is applied later by the premultiplication by $\mathbf{X}^{\mathrm{T}}$.

Since the multiplications involved in the bracketed term in (7.4) are all simple binary operations, the computation involved in the convolution has been considerably simplified. However, a further simplification is possible if the rows of the matrix $\mathbf{B}^{\mathrm{T}}$ are simply used as the address vector for a ROM that contains a set of partial product results corresponding to all possible multiplications of the rows of $\mathbf{B}^{\mathrm{T}}$ with the known vector $\mathbf{H}$. This then yields the fixed distributed arithmetic form shown in Figure 7.8.

It may be seen how this filter may be made adaptive if the LMS algorithm of (7.1) is substituted into (7.5), giving

$$y = \mathbf{X}^{\mathrm{T}}\mathbf{B}^{\mathrm{T}}(\mathbf{H} + 2\mu e\mathbf{BX}) \tag{7.6}$$

$$y = \mathbf{X}^{\mathrm{T}}\mathbf{B}^{\mathrm{T}}\mathbf{H} + 2\mu e\mathbf{X}^{\mathrm{T}}\mathbf{B}^{\mathrm{T}}\mathbf{BX} \tag{7.7}$$

Assuming the input signals to be random with a Gaussian statistical distribution, the product $\mathbf{B}^{\mathrm{T}}\mathbf{B}$ is a purely diagonal matrix in which the diagonal terms are N (the filter order). This means that an adaptive distributed arithmetic algorithm may be defined as

$$P'(x) = P(x) + 2\mu Ne(t)2^{-x} \tag{7.8}$$

where

$$y = \sum P'(x) \tag{7.9}$$

that is, $P(x)$ is a partial product evaluated from an input bit plane with significance 2^{-x}. This algorithm, first proposed by [Cowan and Mavor 1981(2)], may be inter-

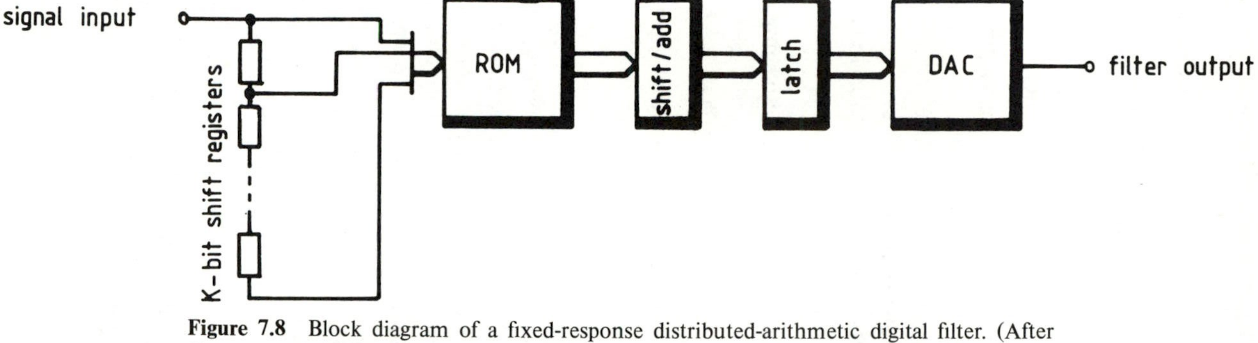

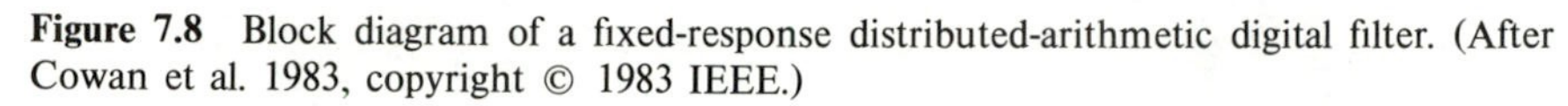

Figure 7.8 Block diagram of a fixed-response distributed-arithmetic digital filter. (After Cowan et al. 1983, copyright © 1983 IEEE.)

preted in hardware terms by replacing the ROM in Figure 7.8 by a RAM and incorporating a minimal amount of updating hardware. An example of such a system is illustrated in Figure 7.9.

Computer simulations have been carried out to empirically evaluate the performance of this structure and hardware modules have also been constructed [Cowan et al. 1983]. A few practical results are presented here to indicate the sort of performance available with this algorithm.

The result shown in Figure 7.10 shows traces of the mean-square error (MSE) plotted against the number of algorithm iterations for purely sinusoidal inputs. The simulated filter in this case was an eight-point transversal filter with 8-bit input signal quantization. A number of convergence curves are plotted in Figure 7.10 showing the performance as the convergence factor varies from 2^{-3} to 2^{-5}. It can be seen that the convergence time increases in the expected manner.

The second result, shown in Figure 7.11, is from a prototype hardware module constructed using standard transistor-transistor logic (TTL) integrated circuits [Cowan et al. 1983]. The filter, again, is an eight-point transversal filter with 8-bit input signal quantization. The sampling rate of this particular module was modestly set at 16 kHz. The result shown in Figure 7.11 demonstrates the cancellation that is achievable using the distributed arithmetic filter approach. The $d(n)$ input signal [Figure 7.11(b)] is a composite signal made up of two equal-amplitude sinusoidal inputs at 400 Hz and 2 kHz [Figure 7.11(e)]. The $s(n)$ input [Figure 7.11(a)] is an unsynchronized tone at 2 kHz. After adaption the filter reproduces the 2-kHz tone on the filter output [Figure 7.11(c)] at the correct phase and amplitude to subtract coherently from the $d(n)$ input, leaving only the 400-Hz tone on the $e(n)$ output [Figure 7.11(d)]. The spectrum of the error output is shown in Figure 7.11(f), demonstrating that the higher frequency has been canceled by about 36 dB. The convergence factor in this case was 2^{-20}.

This type of approach to adaptive filter design is potentially useful in a number of ways. First, it may be easily implemented on microprocessor-based equipment since the only operations used are memory access, shift, and addition. Such systems can be easily operated at voice bandwidths and offer a good alternative to more costly multiplier-based systems, reported earlier in Figure 7.3, or custom IC designs. Second, using hardwired systems the potential bandwidth available using this design is very much higher than the 40 kHz of the specific system reported here. Using the design approach shown in Figure 7.9, a sampling rate of up to 2.5 MHz should be feasible and a more advanced design using Schottky devices would easily perform at video bandwidths, for short-length (e.g., $N = 16$) filters.

A final potentially useful aspect of this structure is its suitability for recursive filter design. As the original distributed arithmetic architecture was configured for use in designing recursive filters, the structure is ideally suited to reconfiguration in this way with relatively minor modifications. Thus it is also potentially applicable to the designs reported in Chapter 4.

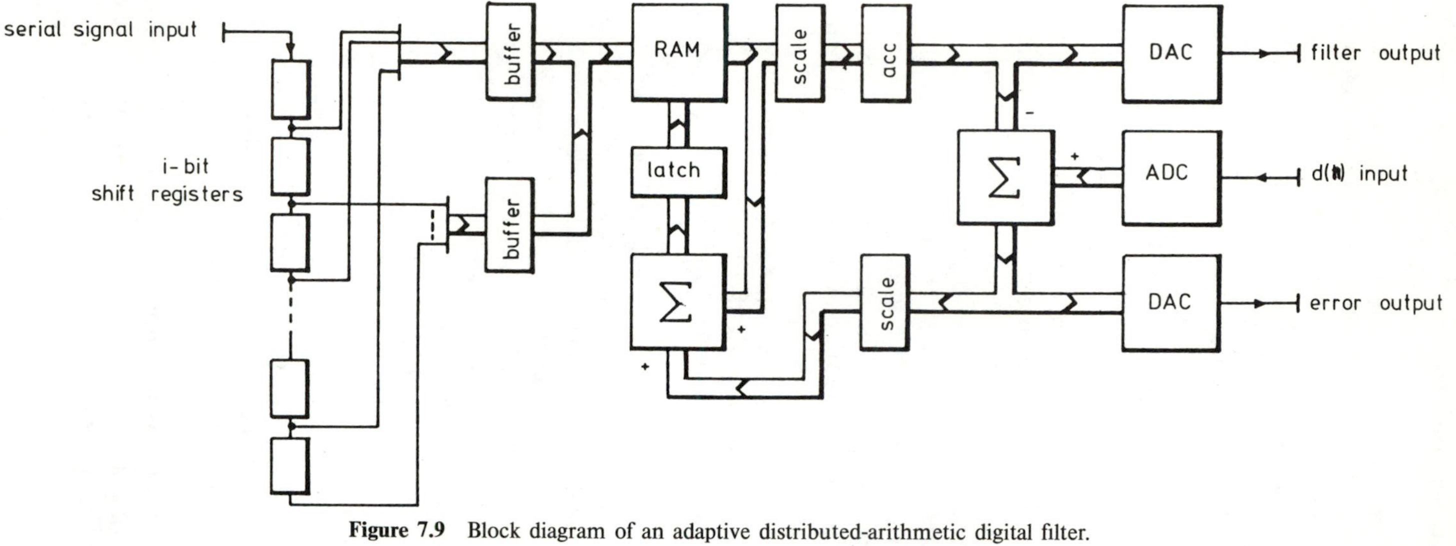

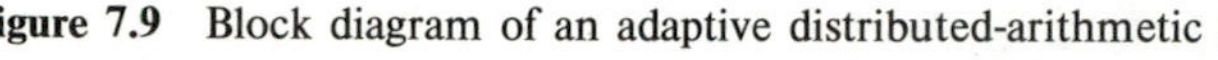

Figure 7.9 Block diagram of an adaptive distributed-arithmetic digital filter.

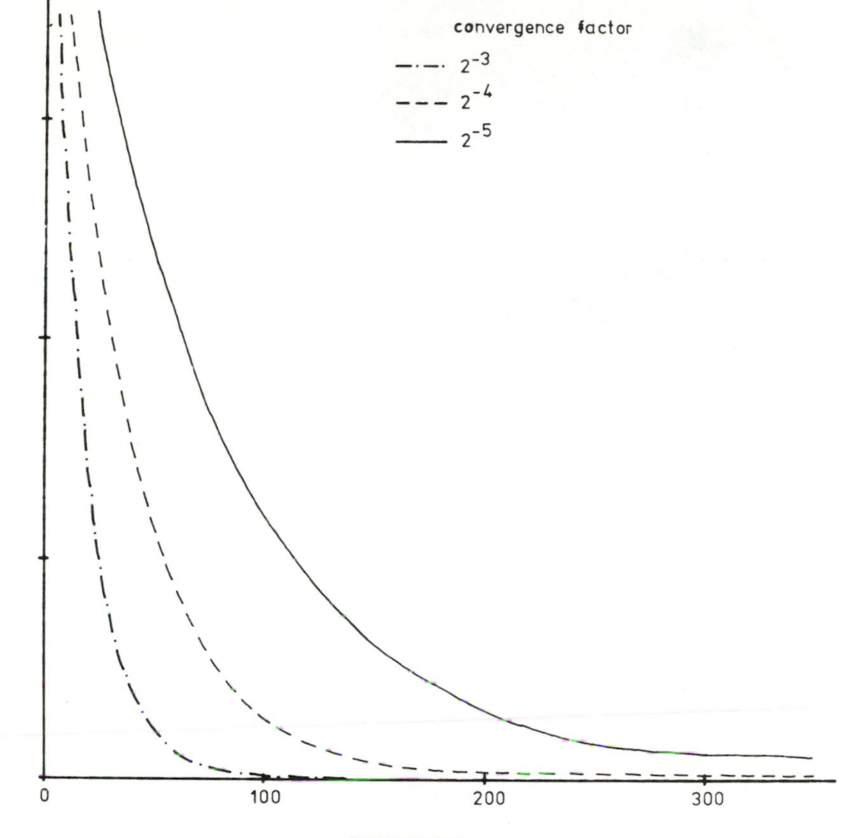

Figure 7.10 Plot of the convergence characteristic of the mean-square error output of an adaptive distributed-arithmetic filter as the convergence factor varies from 2^{-3} to 2^{-5}. (Copyright © 1981 IEE.)

7.2.5 Residue Number Systems

In contrast to the distributed arithmetic systems discussed in the preceding section, RNS-based systems do not actually require a modification of the filter architecture itself. Rather, they operate by reducing the complexity of large, high-accuracy, digital multiplications through the use of number-theoretic transforms.

The technique [Jenkins and Leon 1977] relies on the fact that a number may be divided into a set of smaller numbers which may be operated on independently and the result may then be transformed to yield the result that would have been obtained if a straightforward arithmetic approach had been adopted.

The RNS approach to arithmetic operations uses number subsets which are relatively prime moduli residues. That is, if we have an integer I, the individual residues, R_i, are defined by a set of moduli, p_i, according to the following coding

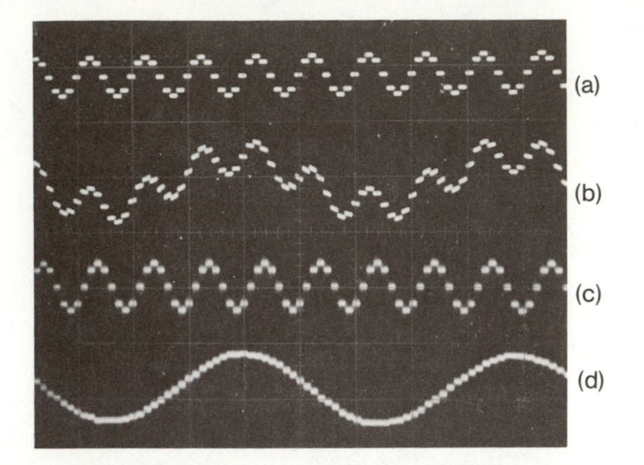

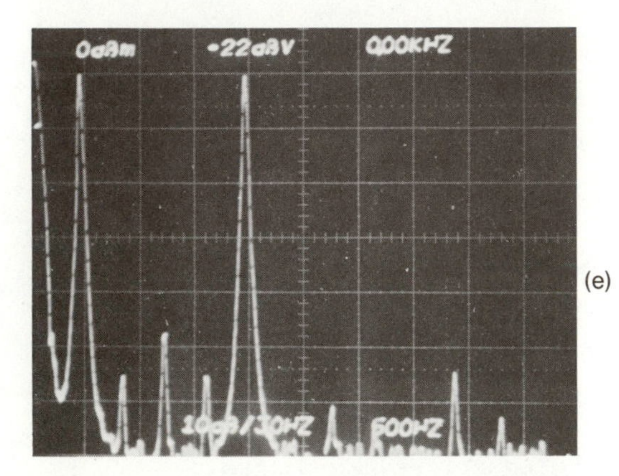

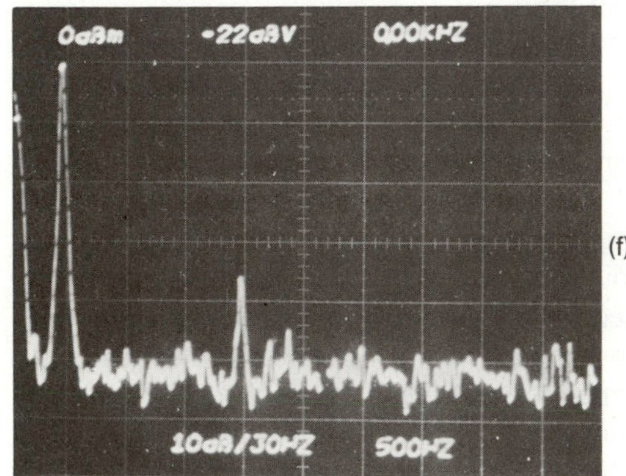

Figure 7.11 Experimental result showing the cancellation performance of the distributed-arithmetic adaptive filter. (a) Signal input; (b) $d(n)$ input; (c) filter output; (d) error output; (e) $s(n)$ spectrum; (f) $e(n)$ spectrum. (a)–(d) Horizontal scale 500 μs per division, vertical scale linear. (e) and (f) Horizontal scale 500 Hz per division, vertical scale 10 dB per division.

practice:

$$w = 0.5(\prod p_i - 1) \tag{7.10}$$

$$R_i = |I|_{\text{mod}\, p_i}, \qquad I \in [0, w] \tag{7.11a}$$

$$R_i = p_i - |I|_{\text{mod}\, p_i}, \qquad I \in [-w, 0] \tag{7.11b}$$

The residues, R_i, may then be operated on independently subject to the modulus arithmetic rules relevant to the particular modulus used, p_i. The result of these operations will be a set of numbers, y_i, which may be converted back to normal integers using the Chinese remainder theorem [McClellan and Rader]:

$$y = |\sum \bar{p}_i |\bar{p}_i^{-1} y_i|_{p_i}|_M \tag{7.12}$$

where $M = \prod p_i$ and $|\bar{p}_i \bar{p}_i^{-1}|_{p_i} = 1$. To take a trivial example of the savings involved in terms of memory space using this approach, consider the situation where the numbers being dealt with are all integers in the range 0 to 35. Using straightforward arithmetic this would require 6-bit words and making use of memory access to multiply two such words would require 4K bits of memory for each output bit. Using the RNS approach to the problem the input numbers would be split into two moduli subsets using moduli of $p_1 = 5$ and $p_2 = 7$. Each of these may be represented by 3-bit words and hence the multiplications, if performed using memory access techniques, require memory sizes of 64 bits for each output bit for each modulus subset (i.e., a total of 128 bits per output bit). There is obviously an overhead incurred in coding and decoding the numbers involved which has not been taken into account here. However, a considerable saving in required memory space is involved in using this technique.

The use of RNS techniques in the design of fixed-response digital filters has been well documented [Jenkins and Leon 1977, Jullien, Soderstrand 1977(2), Soderstrand and Fields 1977(1)] with particular reference to the use of RNS in finite impulse response filters. A major disadvantage of the RNS filter is its inability to handle word overflow in the residue moduli [Jullien]. This must be counteracted by using overall word lengths which are sufficient to ensure that arithmetic overflow is not possible. This is a particular problem when considering recursive filters and often requires the use of base extension algorithms in the structure [Jullien, Jenkins 1979]. These algorithms involve recoding the residue sets in such a way that word growth may be accommodated. This operation is also required in performing multiplications, although this may also be handled using the square-law multiplier [Soderstrand and Fields 1977(1)]. A block diagram of such a multiplier implemented using an RNS approach is shown in Figure 7.12, where k and p are the word sizes corresponding to the moduli p_1 and p_2.

The basic principles under consideration so far in this section have been concerned with the general construction of digital filters using RNS structures. However, specific designs for adaptive digital filters using RNS have been proposed by [Soderstrand et al. 1980(1), Soderstrand and Vigil 1980(2)] which make specific use of microprocessors in the filter construction.

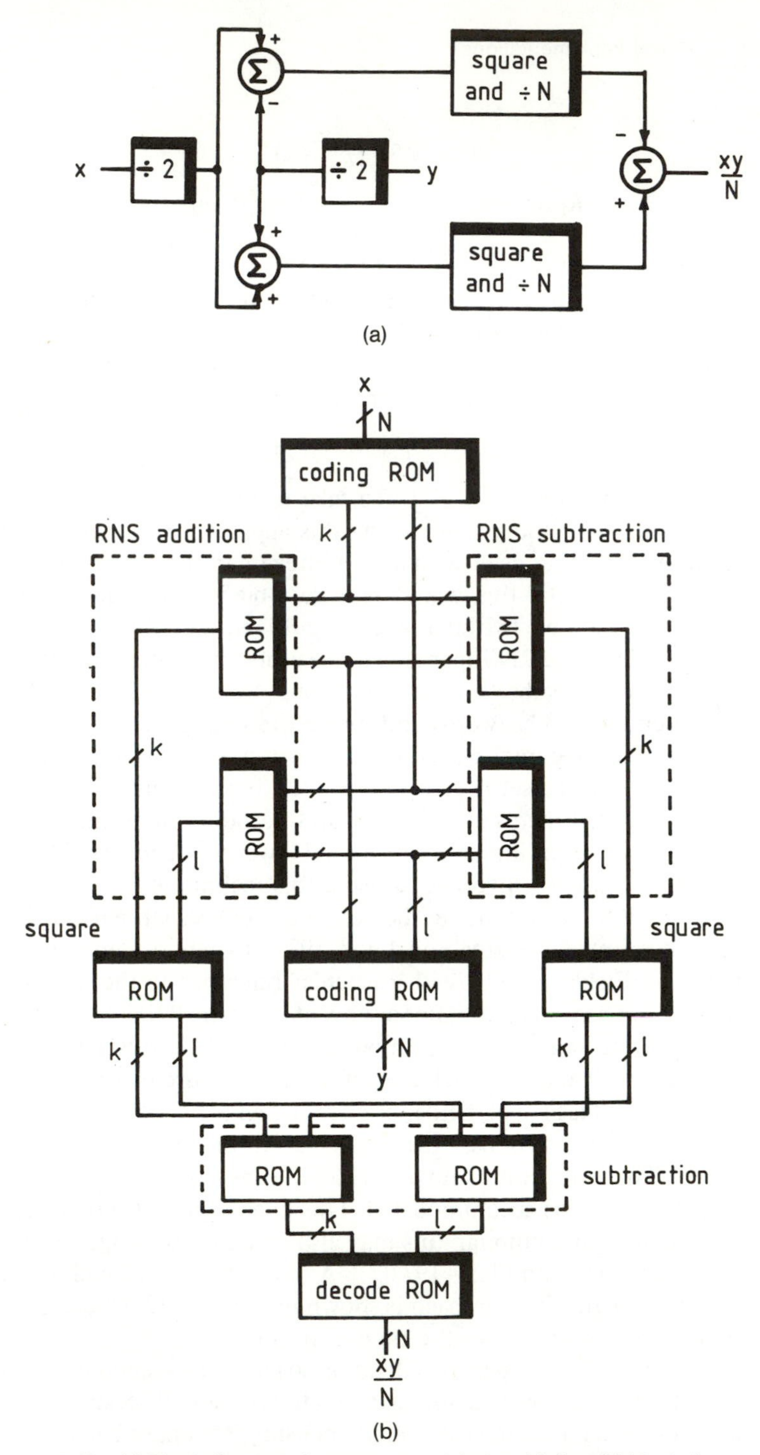

Figure 7.12 (a) Block diagram of a square-law digital multiplier; (b) block diagram of a square-law digital multiplier using residue number systems (RNS).

A major restriction regarding the use of microprocessors in digital signal processing has been the low sampling rates available. In the RNS-based systems this is avoided, to some extent, since 8-bit arithmetic may be used throughout (provided that coding and decoding occurs in hardware), hence maintaining the maximum speed advantage using 8-bit bus structures in currently available microprocessors. Full advantage of the flexibility of the microprocessor-based approach may be taken by using the "totally adaptive" algorithms [Soderstrand and Vigil 1980(2)]. In this case the system adaptively adjusts not only the filter coefficients, but also the filter format. That is, the adaptive algorithm may reconfigure the filter to be all-zero, all-pole, or any combination of poles and zeros. The filter does this by stepping through each available form of filter and allowing it to converge. Once all the filter formats have been tried, the one yielding the lowest converged error is chosen as the optimum filter. The adaptive algorithms chosen by Soderstrand for this filter were the LMS algorithm for the purely transversal filter and the Stearns–White algorithm [White 1975, Stearns and Elliott 1976] for the recursive forms.

This form of filter has an advantage over the distributed arithmetic format considered in Section 7.2.4 in that the filter is inherently linear in nature. This is because the RNS filter does not alter the underlying structure of the normal digital filter and therefore the adaptive algorithms still operate directly on the canonical filter coefficients. This means that the extra degrees of freedom available using the distributed arithmetic approach are not available using RNS. However, the convergence times of the RNS-based filter will be the same as those for a straightforward digital filter using hardware multipliers. Hence it possesses the distinct advantage that large word sizes may be efficiently handled using standard microprocessor hardware.

7.3 ANALOG SAMPLED-DATA ADAPTIVE FILTERS

In this section the construction of adaptive filters using sampled-data analog structures is investigated. These systems depend mainly on the use of charge-coupled devices (CCDs), bucket-brigade devices (BBDs), and sample-and-hold-based technologies. The background material necessary to understand the operation of CCD and BBD delay lines is not covered in this text, as several books are available which cover this topic comprehensively [Beynon and Lamb]. Instead, CCDs and BBDs will be referred to only in the context of their operation as analog delay lines, and the effects of cumulative errors in these devices will be considered only in terms of their final effects on signals rather than considering the mechanisms themselves. This approach is adopted as it is not intended that this survey should include a detailed examination of silicon technologies.

7.3.1 Charge-Coupled-Device Implementations

The CCD is a device which has been under investigation for over a decade for use in systems involving delay or convolution processes. These structures have been used successfully in the construction of video delay lines [Wen], programmable transversal filters [Denyer and Mavor 1979(2), Pelgrom et al., Weckler and Walby], Fourier transform processors [Kapur et al., Bailey et al.], and most notably for the construction of imaging devices [Mavor 1979].

The aspect of CCD technology of most interest to us here is their use in the design of programmable transversal filters (PTFs). In this section some of the techniques for implementing PTFs using CCDs are reviewed and the ways in which these filters may be made adaptive are considered. Many of the techniques discussed here are also applicable to the case of BBD and other sample-and-hold-based technologies. However, the specific use of CCD technology to implement completely monolithic adaptive filters is also considered and one specific design for this type of processor is presented [Denyer et al. 1983].

Figure 7.13 shows the straightforward, parallel construction of a transversal filter. It has been seen in Section 7.2 that it is not feasible to implement this structure directly using digital techniques, due to the complexity of digital multipliers. However, in the case of analog processors an analog multiplication operation may be realized relatively simply, hence allowing the direct parallel implementation of the transversal filter. Alternatively, the multipliers may be implemented in hybrid form by using multiplying digital-to-analog converters (MDACs). In this instance the multiplications are performed between an analog signal and a digitally stored weight using N MDACs or by multiplexing a single MDAC between filter points if the objective is to save silicon area at the cost of processing speed. Both these approaches (the multiplexed MDAC approach is illustrated in Figure 7.14) to PTF design have been used successfully [Denyer and Mavor 1979(2), Weckler and Walby].

In this section we consider the construction of adaptive filters from the point of view of all-analog realizations of the PTF. One practical implementation of an

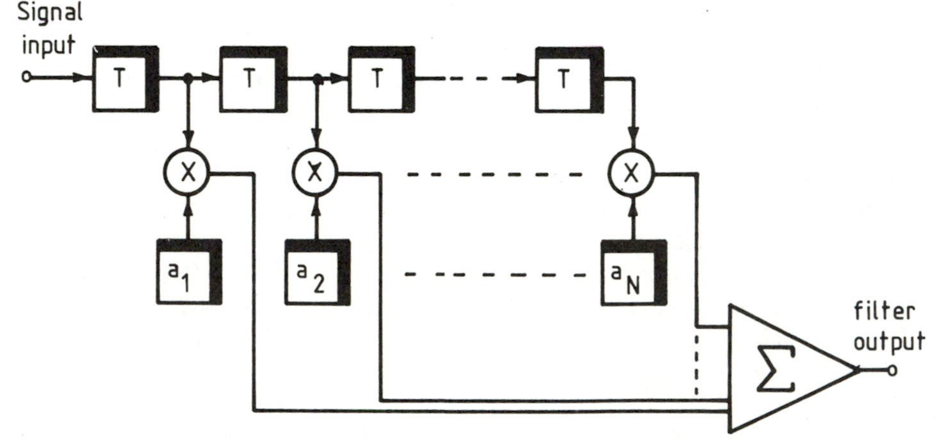

Figure 7.13 Block diagram of a transversal filter in parallel processing format.

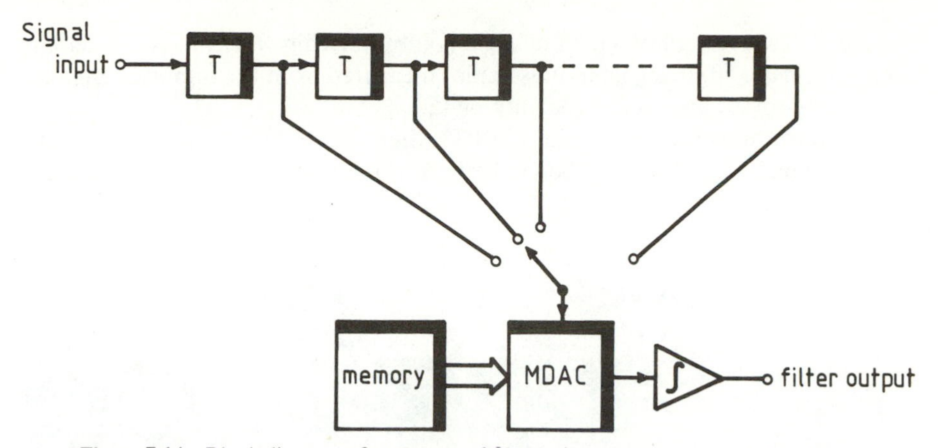

Figure 7.14 Block diagram of a transversal filter using an analog tapped delay line and a single multiplexed multiplying digital-to-analog converter (MDAC).

analog CCD PTF is shown in Figure 7.15. This approach uses a CCD to delay the incoming signal with floating-gate reset taps [Denyer and Mavor 1977] which nondestructively sense the signal along the delay line. The outputs of these floating-gate reset taps are applied to one of the sets of input ports of an array of analog multipliers, which may be implemented using single MOS transistors [MacLennan et al.]. The filter weights are then applied to the other multiplier inputs, the weights being stored on a set of switched capacitor storage sites on the chip. CCD programmable transversal filters based on this architecture have been designed with up to 256 filter points on a single device [Denyer and Mavor 1979(1)].

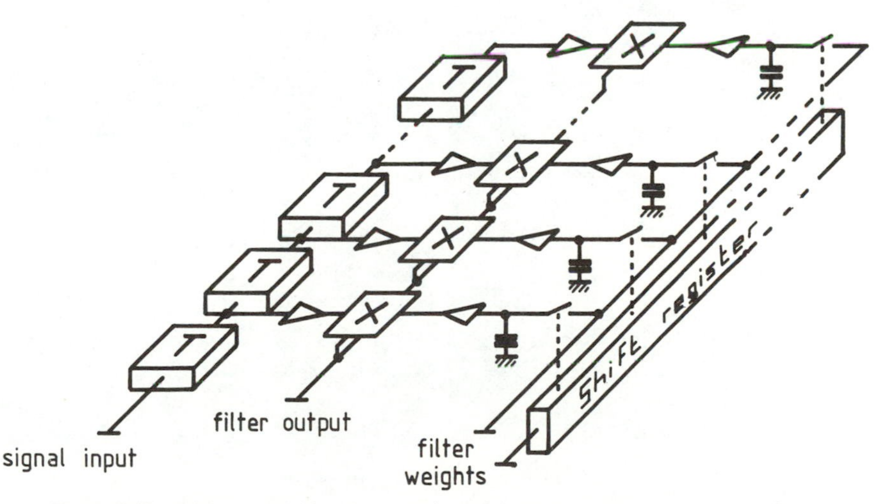

Figure 7.15 Direct parallel implementation of the programmable transversal filter using a CCD approach.

Figure 7.16 shows a photograph of a 256-point filter chip which uses this design. This filter has a 50-dB signal-to-noise ratio (measured from the impulse response) with a real-time bandwidth capability of 150 kHz.

Given the capability to realize a PTF there are then a number of possible ways in which the filter may be configured to operate in an adaptive mode:

Figure 7.16 Photomicrograph of the 256-point CCD programmable transversal filter chip. (Courtesy of Wolfson Microelectronics Institute, Edinburgh University, copyright © 1979 IEE.)

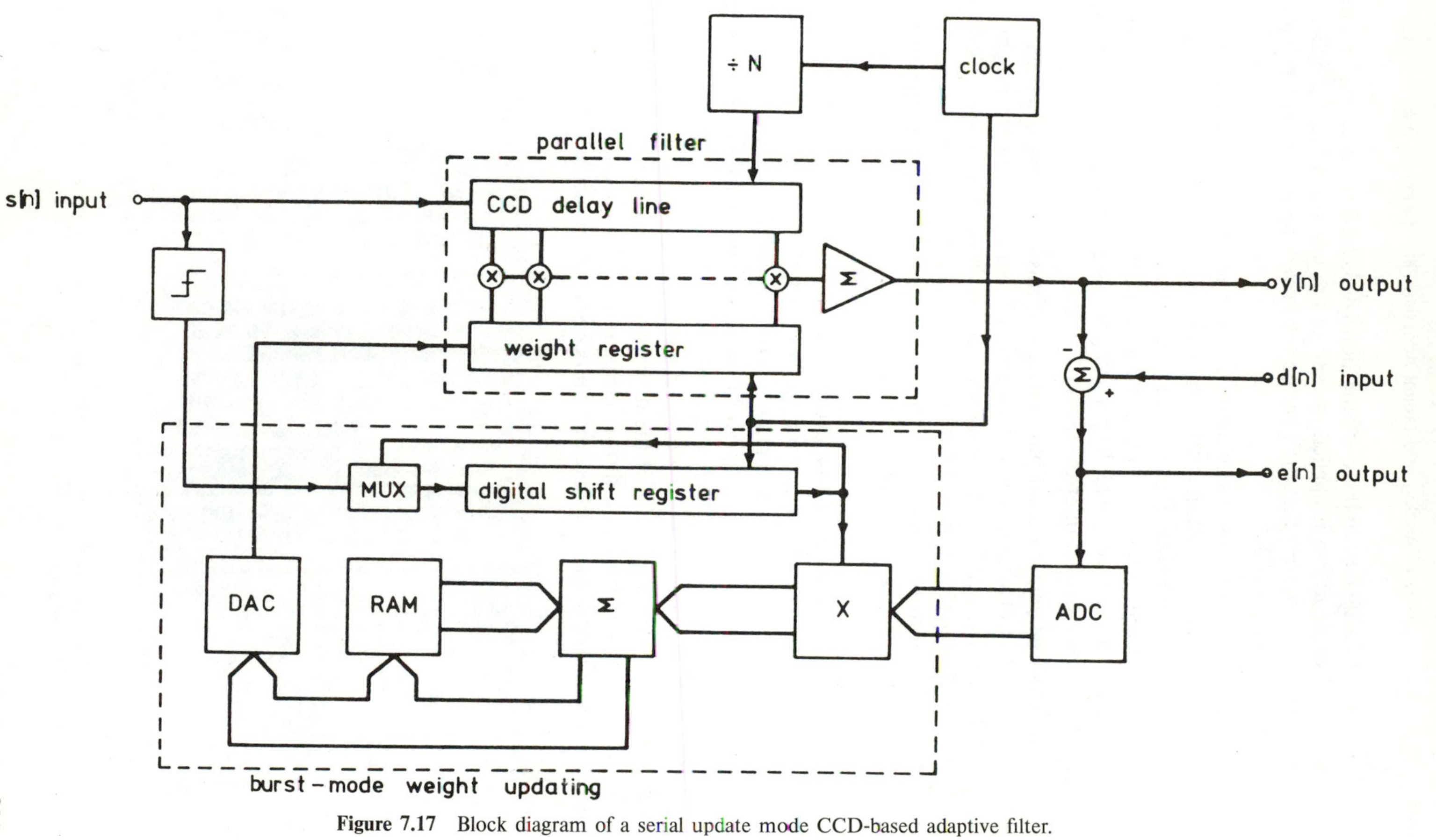

Figure 7.17 Block diagram of a serial update mode CCD-based adaptive filter.

1. *Fully parallel adaption:* in this configuration a dedicated circuit to implement the adaptive algorithm is applied to each of the tap weighting multipliers. This approach requires that the filter be totally redesigned and will be considered in Section 7.3.2.
2. *Burst mode adaption:* in this configuration the adaptive algorithm is implemented as a separate circuit outside the PTF and the individual tap weights are updated in sequence. It is referred to as burst mode adaption because all the tap weight updates are calculated within a single input sample period. Thus the convergence rate of the parallel processor is maintained but at the expense of the overall filter bandwidth.
3. *Serial mode adaption:* the adaptive process in this case is again carried out using a single algorithm computation circuit but only one tap weight is updated during each filter sampling period [Cowan and Mavor 1980, Cowan et al. 1978]. The filter bandwidth is thus maintained but the convergence rate of the filter is greatly reduced, due to the reduced frequency of filter coefficient updates [Cowan and Mavor 1980].

The latter two techniques discussed above provide a potentially powerful approach to adaptive filter implementation, as they use existing CCD PTF designs. The diagram in Figure 7.17 shows a block diagram of a CCD-based system which uses the serial mode adaption process. This is a hybrid processor that mates an analog CCD PTF with digital update circuitry where the filter weights are stored in RAM. The adaption algorithm selected was the clipped LMS algorithm, as it resulted in a greatly simplified implementation of the digital hardware. A picture of a system designed using this architecture is shown in Figure 7.18 [Cowan and Mavor 1980]. This particular system used has as its central processor the 256-point PTF shown in

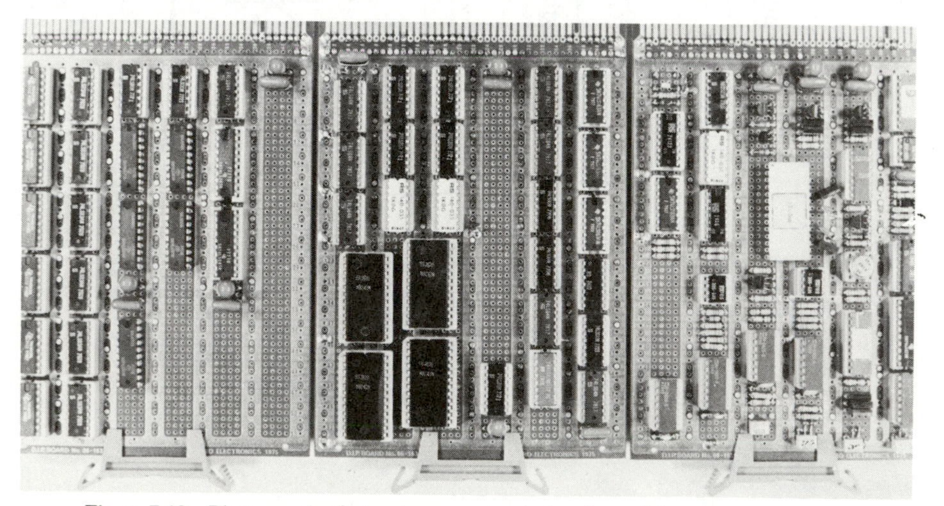

Figure 7.18 Photograph of a hardware system using the serial update architecture of Figure 7.17. (Copyright © 1980 IEE.)

Figure 7.16 and described earlier in this section. The system had an overall power consumption of 7 W with a maximum sampling rate of 300 kHz. Other analog adaptive filter implementations have been implemented with CCD [Sunter et al.] and commercially available BBD [Corl] components.

Some typical experimental results from this system are shown in Figures 7.19 and 7.20. Figure 7.19 shows a graph of convergence time plotted against the system convergence factor. The input signals in this case were pure sinusoids. Some typical results showing oscilloscope traces of the error as convergence proceeds are shown at selected points on the graph. The results shown here correspond well with theoretical predictions of convergence for this algorithm, taking into account the fact that the update rate has been reduced by a factor of 256 (the length of the filter).

The accompanying set of results shown in Figure 7.20 show the cancellation performance of the filter for the same set of convergence factors. The $d(n)$ input signal comprised a pair of equal-amplitude sinusoids; the spectrum of this input is also as shown in Figure 7.20. The higher-frequency sinusoid only was supplied to

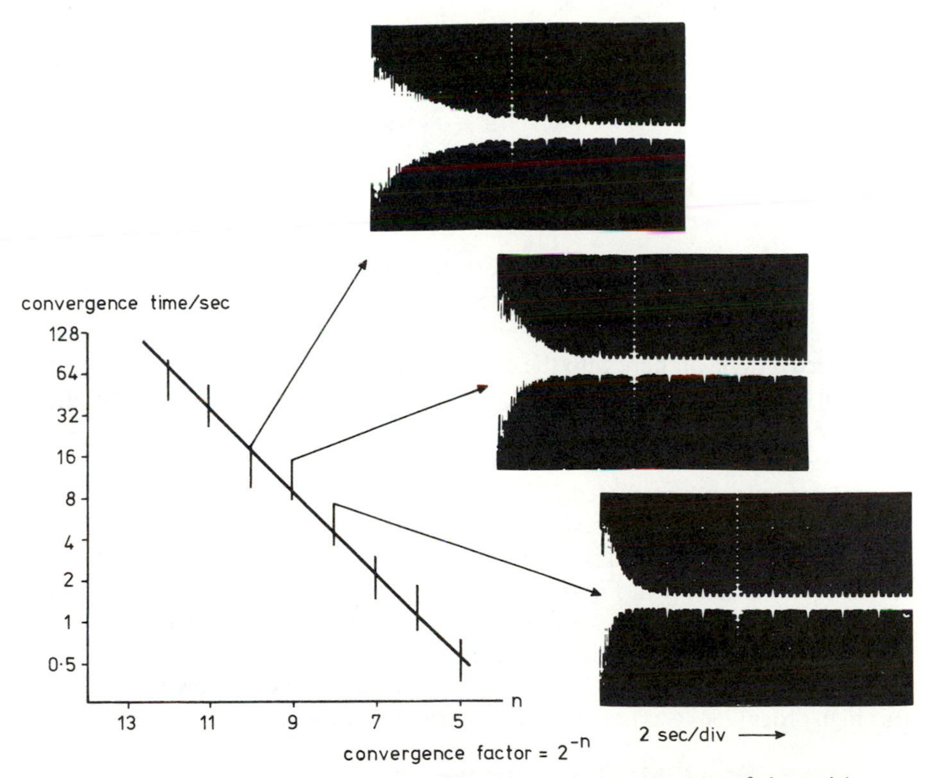

Figure 7.19 Results showing the variation in convergence time of the serial update adaptive filter with varying convergence factor. Horizontal scales 2 s per division, vertical scales linear. (After Cowan and Mavor 1980, copyright © 1980 IEEE.)

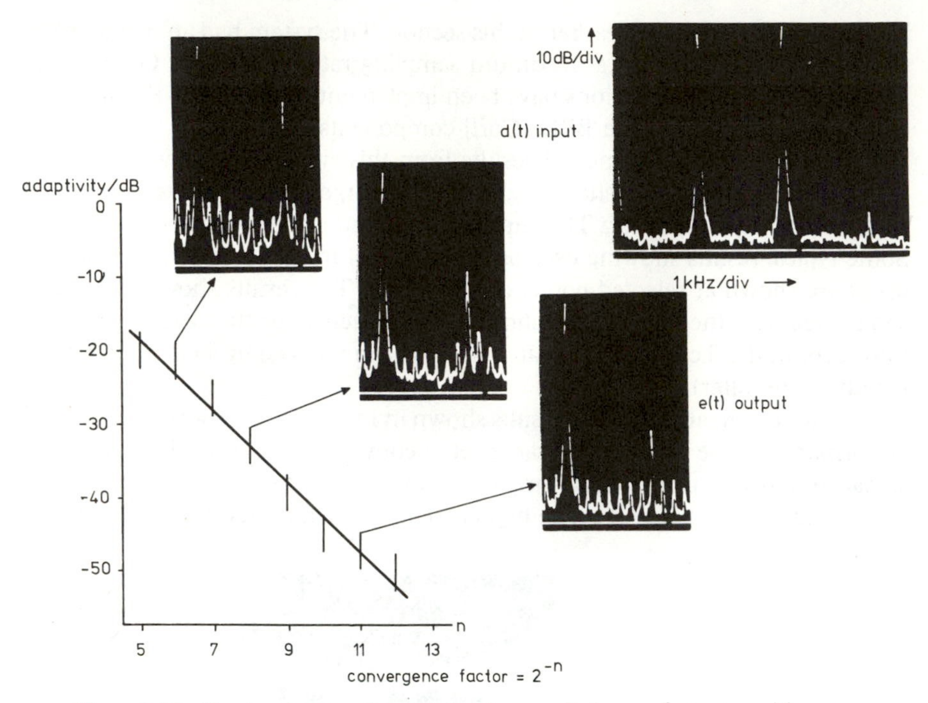

Figure 7.20 Results showing the variation in cancellation performance with convergence factor. Horizontal scales 1 kHz per division, vertical scales 10 dB per division. (After Cowan and Mavor 1980, copyright © 1980 IEEE.)

the $s(n)$ input and the cancellation performance (adaptivity) is measured as the amount in decibels by which the higher-frequency tone is attenuated at the $e(n)$ output relative to the lower-frequency input level. The graph in Figure 7.20 shows the expected trade-off here with the convergence time. Again some typical results showing the spectrum of the $e(n)$ output are shown at selected points on the graph. The best cancellation (about 50 dB) occurs at the lowest convergence factor, which also gives the slowest convergence time.

The final result presented, in Figure 7.21, shows the use of the CCD adaptive filter as an equalizer. In this situation a pseudonoise (PN) sequence was transmitted, at baseband, along a simulated 5 km of telephone cable. The resulting distorted signal is shown in Figure 7.21(a), with the corresponding eye pattern in Figure 7.21(d). After equalization by the CCD filter the signal appears as shown in Figure 7.21(c), with the corresponding eye pattern shown in Figure 7.21(e). It can be seen from this result that good equalization was achieved. However, a filter of this high order (256) was not really necessary to deal with this particular distortion.

7.3.2 Monolithic CCD Adaptive Filter

In addition to hybrid serial update filters the parallel update structure has also been implemented using a CCD with digital update circuitry [Massey et al.]. In

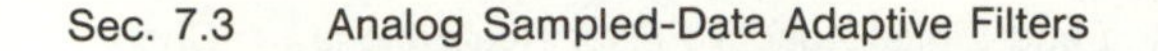

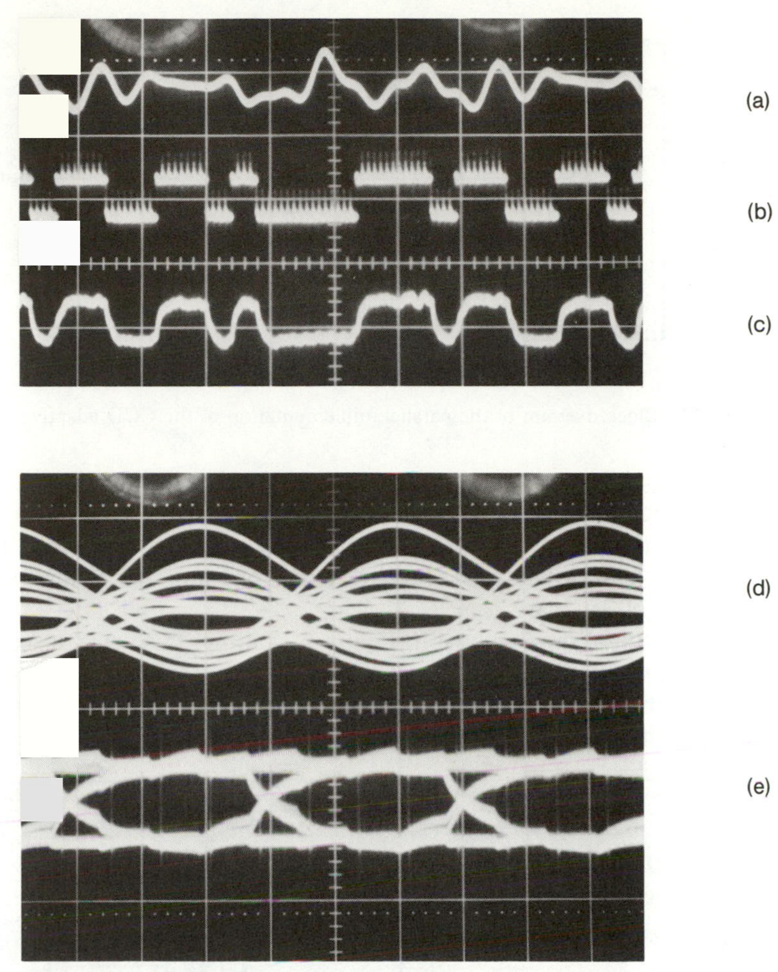

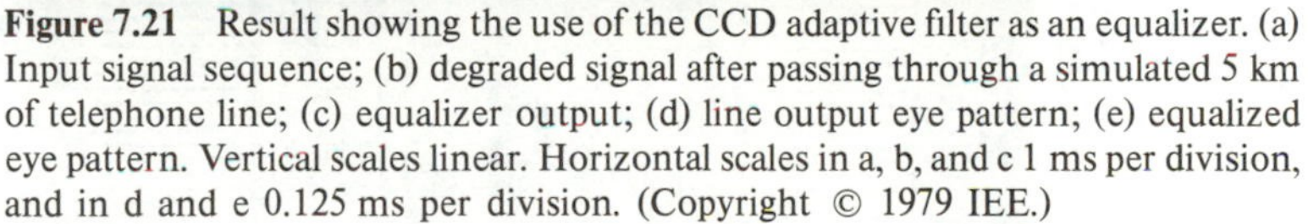

Figure 7.21 Result showing the use of the CCD adaptive filter as an equalizer. (a) Input signal sequence; (b) degraded signal after passing through a simulated 5 km of telephone line; (c) equalizer output; (d) line output eye pattern; (e) equalized eye pattern. Vertical scales linear. Horizontal scales in a, b, and c 1 ms per division, and in d and e 0.125 ms per division. (Copyright © 1979 IEE.)

the remainder of this section we consider a purely monolithic implementation of the fully parallel update architecture. The adaption algorithm used in this case was the linear LMS algorithm. Two separate architectures for this type of monolithic structure have been suggested [Denyer et al. 1983, White et al. 1979]; here we shall consider only the first of these.

A block diagram of the monolithic structure is shown in Figure 7.22. The actual filtering operation is carried out in the same way as that considered in the preceding section. The difference in this chip structure occurs in the existence of individual tap-weight-update circuits at each filter point. This involves the imple-

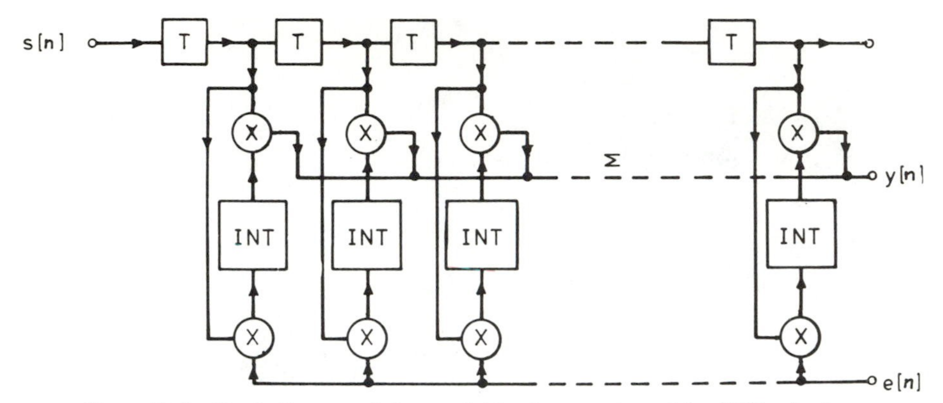

Figure 7.22 Block diagram of the parallel implementation of the CCD adaptive filter.

mentation of a second four-quadrant multiplier and an analog accumulator at each filter point. The structure considered here was implemented as a 65-point filter (shown in Figure 7.23), which used a total chip area of 5 mm × 3.75 mm with a power dissipation of 250 mW.

The experimental result shown in Figure 7.24 shows this device used in a cancellation mode. Again the $d(n)$ input in this case comprises two sinusoidal components and the lower-frequency component is canceled on the $e(n)$ output. It can be seen here that the cancellation achieved was approximately 20 dB. This low

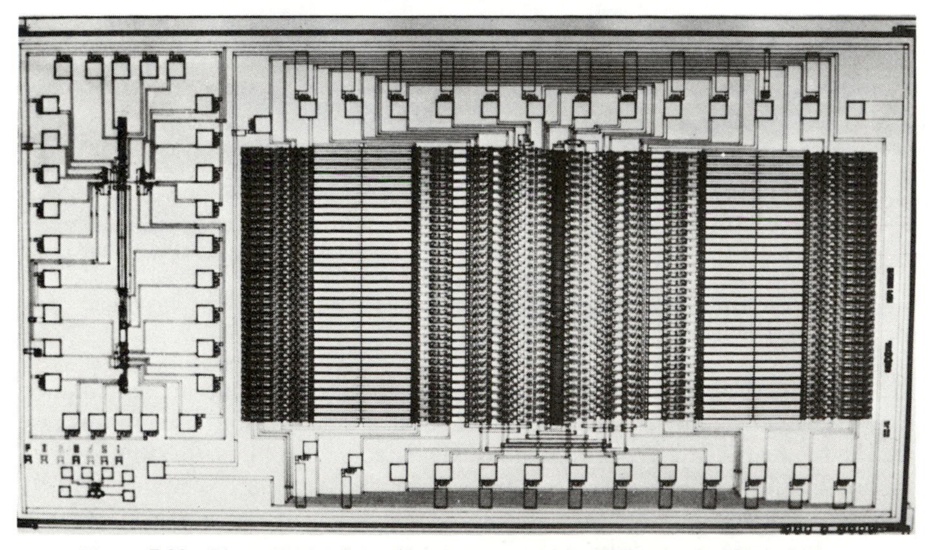

Figure 7.23 Photograph of the 65-point monolithic CCD adaptive filter device. (Courtesy of Wolfson Microelectronics Institute, Edinburgh University; after Denyer et al. 1983, copyright © 1983 IEEE.)

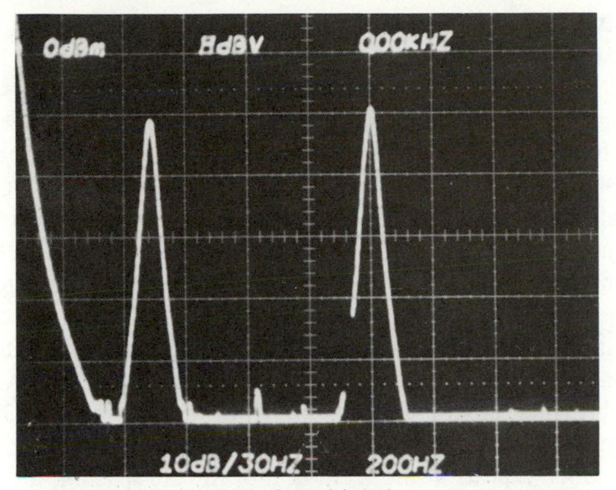

(a) Signal input.

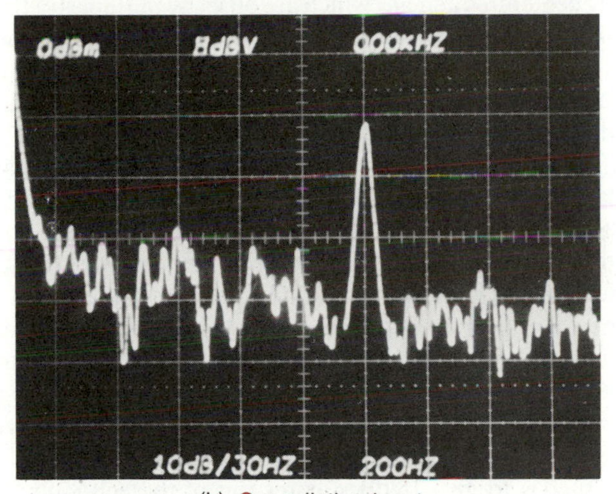

(b) Cancellation input.

Figure 7.24 Experimental result demonstrating the cancellation performance of the CCD adaptive filter. (Copyright © 1982 IEE.) Horizontal scale 200 Hz per division, vertical scale 10 dB per division.

adaptivity was achieved because of the relatively poor noise performance of this prototype device coupled with tap leakage problems.

The maximum sampling rate available using this device was 100 kHz, although sampling rates up to about 2 MHz are feasible using surface channel CCD devices. Still higher bandwidths may be achieved by the use of peristaltic CCDs [Walmsley and Gooding] or surface acoustic wave (SAW) systems [Bowers et al.]. SAW adaptive filters are considered in the following section.

7.4 HIGH-BANDWIDTH ADAPTIVE FILTERS USING SURFACE ACOUSTIC WAVE DEVICES

Other developments in wideband adaptive filters have resulted from the application of surface acoustic wave (SAW) analog FIR filter design techniques. These are intermediate-frequency (IF) bandpass tapped delay lines (TDL) which offer signal bandwidths of 1 to 20 MHz with total delays up to 100 μs. One reported implementation [Masenten] uses a SAW 200-tap delay line to achieve a 2-MHz bandwidth adaptive filter which incorporates the fully parallel tap-weight-update circuitry, as shown in Figure 7.22. As an IF processor the SAW TDL provides, directly, complex processing where 90° hybrids are incorporated with each tap to separate its output into phase (I) and quadrature (Q) weighting networks (Figure 7.25). When realized as hybrid modules these IF weighting networks can achieve 30% fractional bandwidth and 60 dB dynamic range [Masenten]. As this TDL has a sampled transfer function, which repeats at the reciprocal of the tap spacing, the conditioning and input signals can be processed simultaneously in the filter and separated at the output by using center frequencies which are offset by multiples of the sample rate. This 200-tap adaptive filter is a large and expensive processor but its performance is impressive, as it can insert a 20-kHz-wide notch in real time (50 ms) to provide 50-dB suppression of a continuous-wave (CW) jammer anywhere within the 2-MHz filter bandwidth.

Other SAW programmable transversal filter designs that have been employed for adaptive filtering are TDL designs with hybrid field-effect transistor weighting networks [Panasik] and fully integrated storage correlators [Bowers et al.]. The latter all-analog devices incorporate an array of biased diodes to implement the multiplication by the weight values, which are stored as charges on capacitors. They have been used as adaptive filters both for CW interference cancellation [Grant and Kino 1978] and for spurious post-echo suppression [Bowers et al.]. In the latter application both the FIR filtering and tap-weight-update calculation are performed in a single storage correlator with 8 MHz bandwidth and 3 μs interaction region. The device is operated sequentially between filter and tap-weight-update modes to give 15 dB suppression of the echo after 10 iterations, which takes <0.5 ms. Although at present these integrated SAW correlators and hybrid PTFs exist only with small numbers (e.g., <32) of taps, the ability to achieve all the adaptive processing in a single component is especially attractive, compared to the complexity of Figure 7.25.

An alternative realization of wideband adaptive filter uses the frequency-domain approach, where time-domain convolution is implemented by frequency-domain multiplication as described in Chapter 6. SAW techniques permit the design of discrete Fourier transform (DFT) processors using the chirp transform algorithm [Jack et al.], which offers in excess of 10^3 transform points with up to a 60-MHz real-time bandwidth. It has been shown [Grant and Morgul 1982] how a pair of 4-MHz-bandwidth SAW forward and inverse 100-point DFTs can be interconnected via a digital processor to implement the frequency-domain circular con-

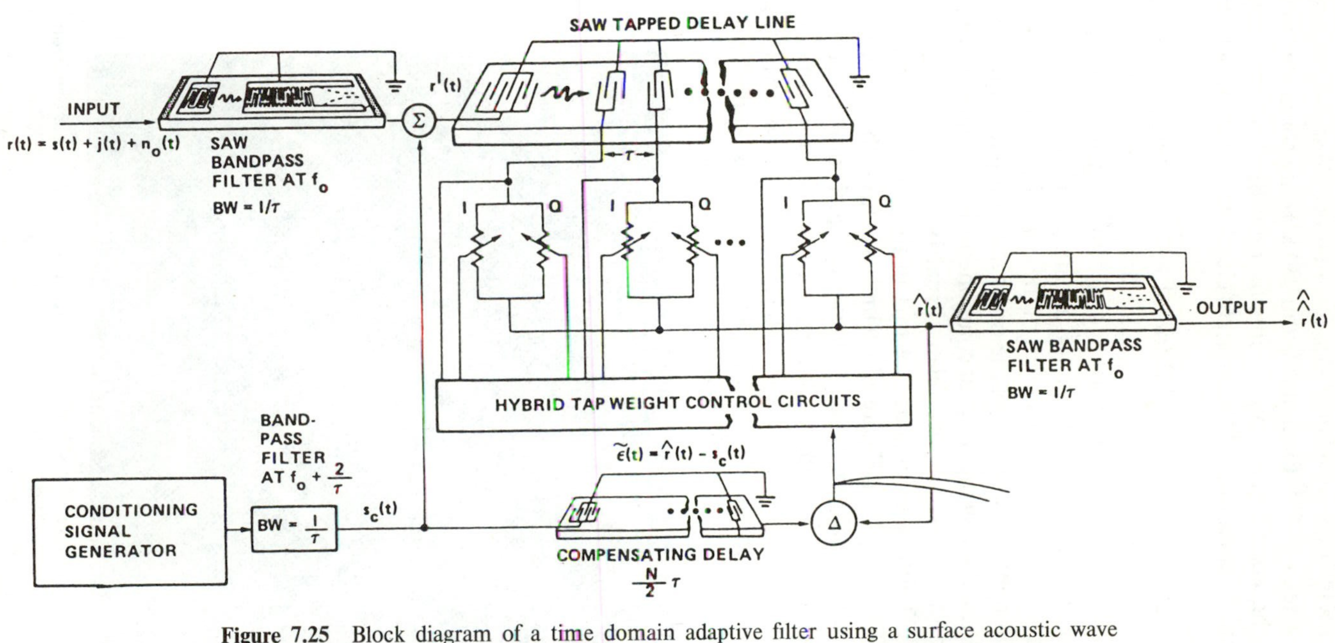

Figure 7.25 Block diagram of a time domain adaptive filter using a surface acoustic wave programmable filter. (Courtesy Hughes Aircraft Company, Fullerton, CA; after Masenten, copyright © 1979 IEE.)

volution adaptive filter described in Figure 6.1. Figure 7.26 shows the operation of this processor when two CW tones at 200 and 930 kHz are connected to the signal input. When the conditioning input contains only the lower-frequency component, the output (lower trace of Figure 7.26) clearly shows the suppression of the 930-kHz signal.

The simulation in Figure 7.27 [Morgul et al.] highlights a key attraction shared by the frequency-domain and lattice adaptive filters. These filters preprocess the input to generate a set of orthogonal samples which make it possible to implement a set of independent loops for updating each of the weighted multipliers. If this is combined with adjustment of the convergence coefficients in each loop, dependent on the power level of the orthogonal outputs, equal convergence rate can be achieved for all input signal components independent of the spread in the eigenvalue in the input autocorrelation matrix.

Figure 7.27 shows a simulation of this in the simplest frequency-domain adaptive filter [Dentino et al.] (Figure 6.1), which is input with the two CW tones as described previously, except that the conditioning signal comprises the higher-frequency tone in this case. The upper simulation shows the performance with a fixed (constant) convergence coefficient throughout the processor, which gives performance equivalent to an adaptive FIR filter. In comparison, the lower trace

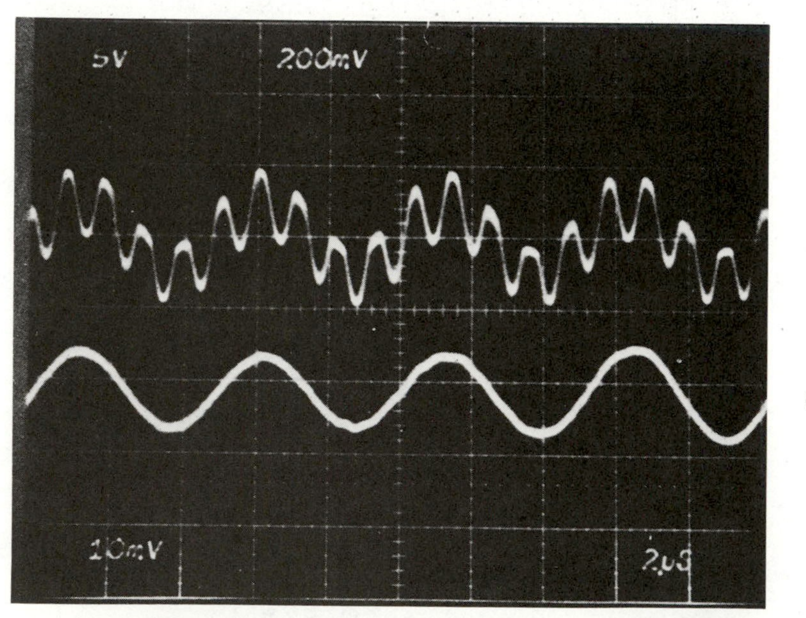

Figure 7.26 Experimental result showing the operation of a frequency-domain SAW adaptive filter. (a) Input consisting of sinusoids at 200 and 930 kHz; (b) filter output at 200 kHz. Horizontal scale 2 μs per division, vertical scale linear. (After Grant and Morgul 1982, copyright © 1982 IEEE.)

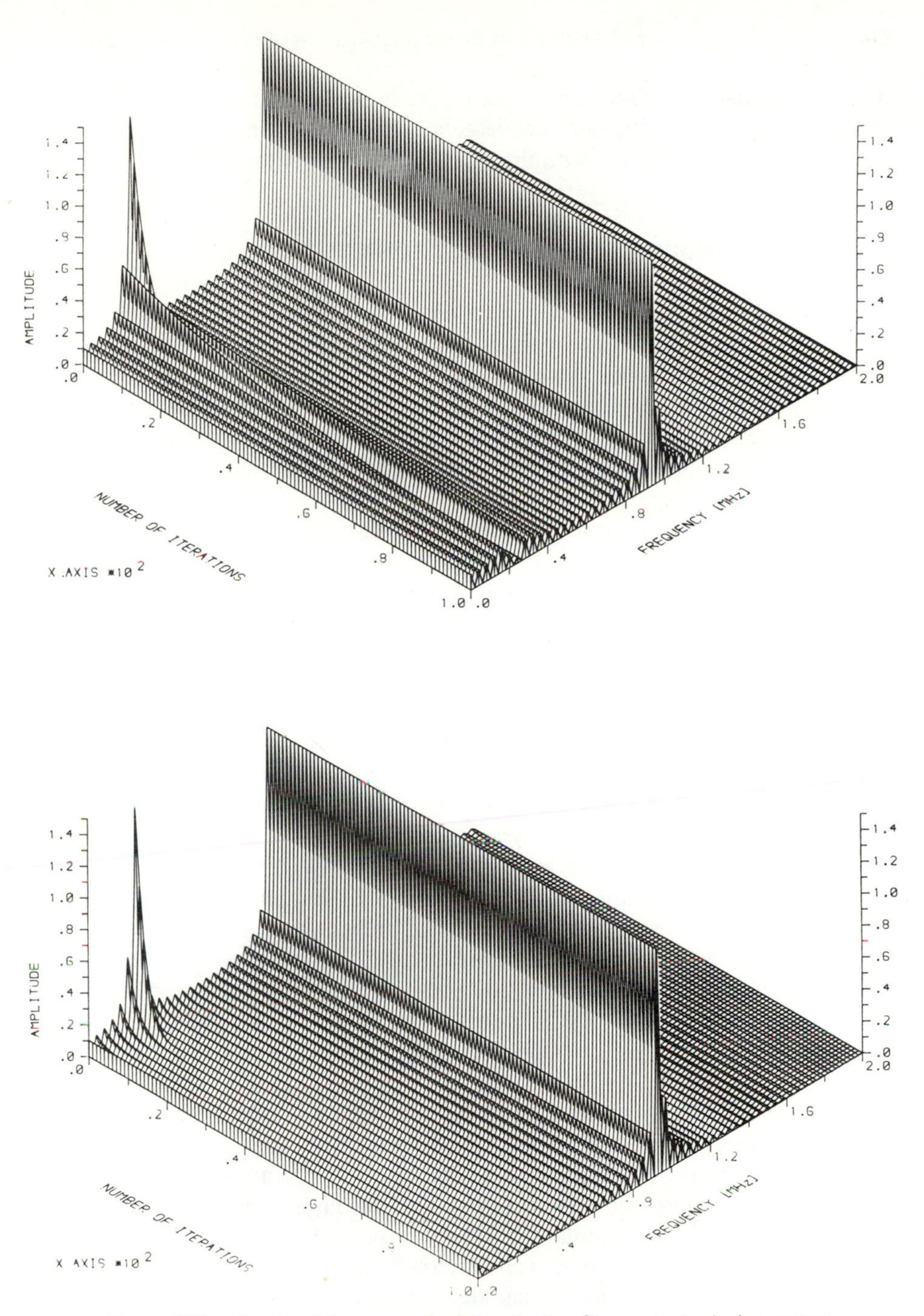

Figure 7.27 Simulated frequency-domain adaptive filter outputs during convergence. (a) Unnormalized LMS adaptive algorithm; (b) normalized convergence factor LMS adaptive algorithm. (After Morgul et al., copyright © 1984 IEEE.)

shows how independent selection of convergence coefficient for each transformed signal component gives the same convergence rate for all signal components, a feature that cannot be achieved with the single global error of the FIR adaptive filter.

7.5 FUTURE DESIGNS USING VLSI TECHNOLOGY

The preceding three sections have been concerned with a survey of techniques used in the implementation of real-time adaptive filters using currently available technologies. In this section we consider the possible alternatives that are made possible with the introduction of VLSI processing. This applies to silicon, silicon on sapphire (SOS), gallium arsenide (GaAs), and other semiconductor processes. Although silicon processing is widely used, a great deal of research is being performed on GaAs processing techniques.

The advantages of VLSI processing are really apparent only in digital techniques since analog structures such as CCDs derive no real advantage from the feature-size shrinkage involved. However, it is projected that in digital technology VLSI will result in an increase in circuit speed by an order of magnitude with a twentyfold increase in chip complexity for the same chip size as that currently used in LSI processing. This would make available systems having 10^6 active devices per chip with gate delays reduced to something in the order of 10 ps. This would allow us, for instance, to design ten 12-bit digital multipliers on a single chip with individual multiply delays in the subnanosecond region.

From the sort of figures quoted above it is obvious that it will be possible to design complex signal processing systems as single chips with realistic sampling rates which enable them to act as real-time processors. However, the very complexity involved in designing VLSI chips acts as a fundamental limitation to their cost-effectiveness, unless a practical automated design and test facility is available. As an example, the estimated design time involved in producing the recent 32-bit microprocessors is in the region of 130 person-years. With extensions to VLSI complexity and given the highly ordered structures afforded by systolic array approaches [Kung and Leiserson 1980], the design time rises to something like 700 person-years per device.

With this level of investment it is obviously advisable to keep the number of individual designs to a minimum, with maximum usage for each design. It is proposed that this should be done by the increasing use of programmable general-purpose signal processing devices which would be readily interfaced to microcomputer controllers. The particular structure considered for implementing linear filters of the type considered in this chapter is shown schematically in Figure 7.28. Basically, this system consists of a parallel multiplier in which the inputs are serviced by two-port RAMs. The output of the multiplier may be routed to an accumulator, and the additional facility is a RAM or ROM used to store address and control sequences for the processor. This type of structure is clearly nonoptimum for adaptive filter operation, though the throughput speed would certainly allow

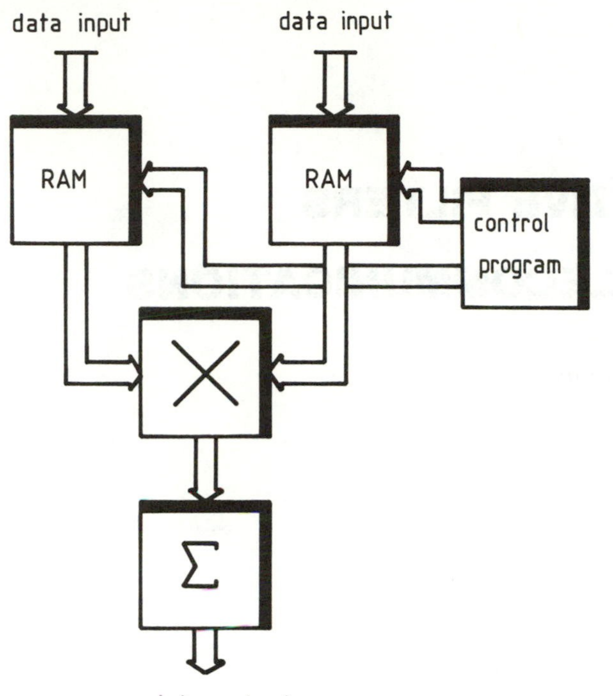

Figure 7.28 Block diagram of a general signal processing module for linear filtering applications.

operation at better than voice-bandwidth sample rates. Whether or not this type of processor could be applied at video and sonar bandwidths for any appreciable filter length remains to be seen.

The only feasible alternative to these general-purpose processors, using VLSI, is the custom-design procedure. This depends solely on the availability of automated layout design aids [Denyer et al. 1983]. These computer-aided design (CAD) tools allow design at the systems level through the use of predesigned and tested system blocks. The designer may then simulate the design in software before the integrated circuit mask patterns are automatically generated by the CAD software. One such system [Denyer et al. 1982], based on bit-serial arithmetic architectures [Lyon], has been suggested which is claimed to potentially cut major chip design times down to a few person-days. This is the sort of performance that must be achieved if custom-design techniques are to be a feasible alternative to mass-produced programmable components.

8

ADAPTIVE FILTERS IN TELECOMMUNICATIONS

Peter F. Adams

8.1 INTRODUCTION

Telecommunications is a growing and changing industry which has proved to be fertile ground for the application of adaptive filters. The reasons for this are threefold: rapid advances in silicon technology, especially the advent of large-scale integrated (LSI) and very large scale integrated (VLSI) circuits, have made possible the implementation of adaptive filters at commercially acceptable costs; second, a rapid growth in data communications has created a need for adaptive filtering to overcome impairments inherent in existing telephony networks; and third, a desire to provide improved speech communications where echoes cause subjective impairment or instability. The two broad areas of data transmission and speech communications provide a natural division for the material in this chapter. However, it is instructive to remember first the two different roles that adaptive filters play, namely as equalizers and cancelers (see Section 1.2).

For equalization the adaptive filter is cascaded with an unknown linear channel $C(f)$ and its purpose is to approximate the inverse of $C(f)$. In the cancellation role the adaptive filter is in parallel with the unknown linear channel and is required to approximate $C(f)$. Adaptive filters are used in both these roles in telecommunications applications.

This distinction between roles is important because it results in different constraints on the operation of the adaptive filter: for example, interfering signals or noise at the output of the unknown channel have a different effect on the adaptive filter in each case; also for equalization, but not for cancellation, the channel characteristics can affect the rate of convergence of the adaptive filter.

In other chapters of this book various adaptive filter structures and adaption algorithms are described. Here we shall be concerned almost exclusively with transversal filters adjusted using the stochastic gradient least-mean-squares (LMS) algorithm and its variants. Although other structures and algorithms have been investigated for telecommunications applications, they are of less practical importance. This is a testimony to the simplicity of the transversal structure and the robustness of the stochastic gradient LMS algorithm. Other structures and algorithms are mentioned where appropriate.

The bulk of the material that follows is concerned with digital data transmission over telephony channels and metallic pair cables. This is a reflection of the vast amount of research and development that has been expended in this field and its importance in providing the means of digital communication over a network dominated by the needs of speech communication. The remainder of the chapter is concerned with applications where adaptive filters are required to suppress echoes in speech communications. Alternative methods of achieving the same results are already used but adaptive filters provide a subjectively more acceptable performance.

8.2 DATA TRANSMISSION

Although data transmission in the form of telegraphy predates telephony, speech communication came to dominate the evolution of telecommunications networks. Developed countries, therefore, have telephony networks that are unrivaled in their ubiquity and offer worldwide communication. When the growth in computer usage created a need for data communications it was not surprising that telephony networks initially offered the best medium for this communication. Unfortunately, transmission systems in telephony networks were optimized for analog speech waveforms and introduce various impairments that impede data communications. The most serious of these impairments are linear distortions, and linear filters could be used to equalize or cancel the distortion. However, such distortions vary widely between different network connections, so it became necessary to use adaptive filters.

Today, adaptive filters are widely used to provide equalization in data modems which transmit data at rates of 2400 bits/s up to 16,000 bits/s over speech-band channels (nominally, 300 to 3400 Hz). Although it is theoretically possible to achieve even higher rates, it is practically difficult to obtain a satisfactory error-rate performance without recourse to wider bandwidths. Higher-speed data modems (48,000 to 72,000 bits/s) are commercially available for operation over wider-bandwidth (60 to 108 kHz) channels, and some of these use adaptive equalization [Parkinson and Harvie].

Recently, there has been a growing interest in duplex data transmission over speech-band circuits, which has resulted in adaptive filters being investigated for use as echo cancelers. As yet, very few modems using echo cancelers are commercially available, but that situation may well change in the next few years. Both these applications are described in this section, but first an outline of the types of linear distortion encountered in telephony channels is necessary.

8.2.1 Linear Distortions in Telephony Networks

Linear distortions arise in many different ways in telephony networks, but three distinct types can be identified: amplitude distortion, group-delay distortion, and echoes. Figure 8.1 illustrates how these arise in a typical telephony network connection. A subscriber is usually connected to his or her local switch by metallic pair cable; within the speech band this introduces amplitude slope, as shown in Figure 8.2(a). Between the local switch and other switches there may be loaded junction cable which introduces group-delay distortion at the top end of the speech band, as shown in Figure 8.2(b). Between switches four-wire circuits are used to enable signal amplification and multichannel transmission systems to be employed. Multichannel transmission systems use band-limiting filters which introduce both group-delay and amplitude distortion, as shown in Figure 8.2(c) for frequency-division multiplex (FDM) carrier system filters. Hybrid transformers are used to separate the go and return paths of the four-wire circuit and should ideally introduce infinite attenuation between the two paths. In practice the attenuation is finite, allowing signals to circulate around the four-wire loop, creating echoes. Those appearing back at the transmitter are referred to as talker echoes, while those arriving at the receiver are called listener echoes. Impedance mismatches in the network are a further source of echoes. Listener echoes give rise to ripples in the frequency response of the channel, the amplitude of the ripples being proportional to the echo-to-signal ratio and the frequency of the ripple being proportional to the echo delay.

Real network connections are often more complicated [Duffy and Thatcher, Ridout and Rolfe 1970] than this simple model and are becoming more so as modern pulse-code modulation (PCM) transmission systems and digital switches are introduced [Duerdoth]. However, the three basic impairments remain and identifying them separately helps us to understand what the adaptive filters used to combat linear distortion are required to do and how they behave.

Effects of modulation and demodulation. Because the telephony channel is bandpass and generally passes through multichannel transmission equipment which introduces small frequency offsets, data transmission systems use modulation to place the signal spectrum in the usable bandwidth and demodulation to recover the data and remove offsets. A simple system is shown in Figure 8.3: a stream of binary data (symbols) at rate $1/T$ is band-limited to $\frac{1}{2}(1+\alpha)/T$ Hz where α is the roll-off factor $(0 < \alpha \leq 1)$, and modulated onto a carrier of frequency f_c; if f_c is chosen to be near the center of the speech band, the full signal spectrum can be received at the far end provided that $f_c - \frac{1}{2}(1 + \alpha)/T > 300$ and $f_c + \frac{1}{2}(1 + \alpha)/T < 3400$. Demodulation by a carrier of the correct phase

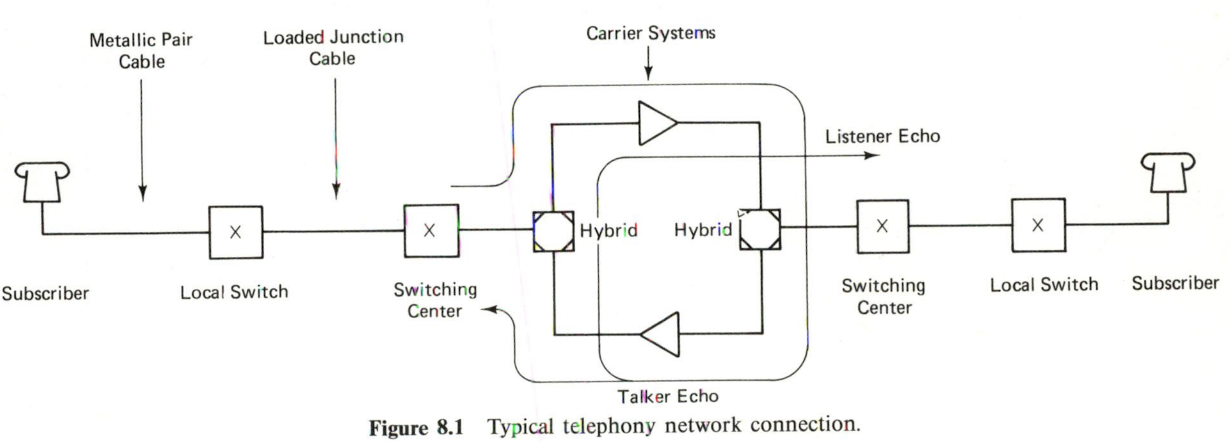

Figure 8.1 Typical telephony network connection.

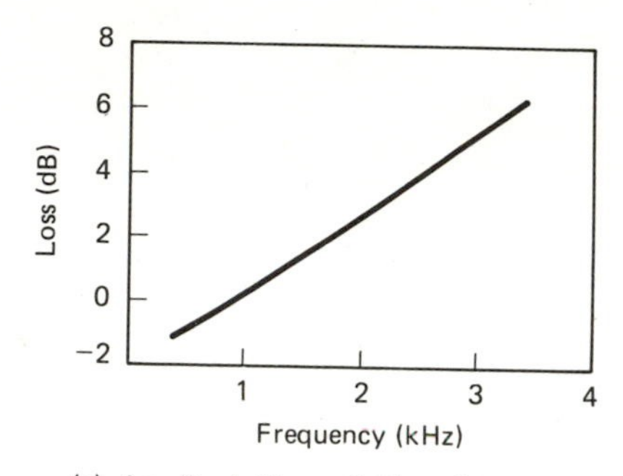

(a) Amplitude Slope of 4 km of Local Cable

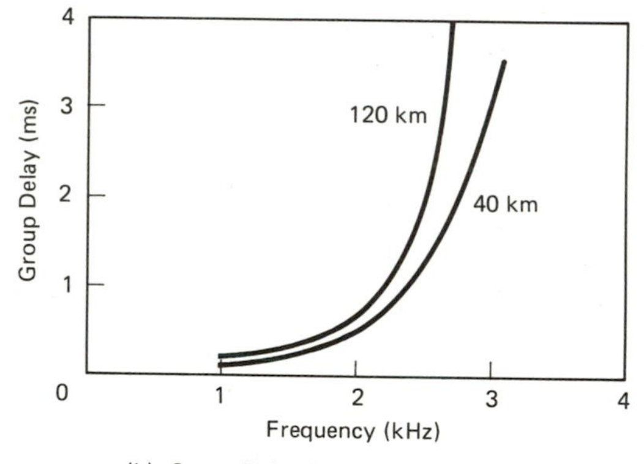

(b) Group Delay Distortion of Loaded Cable

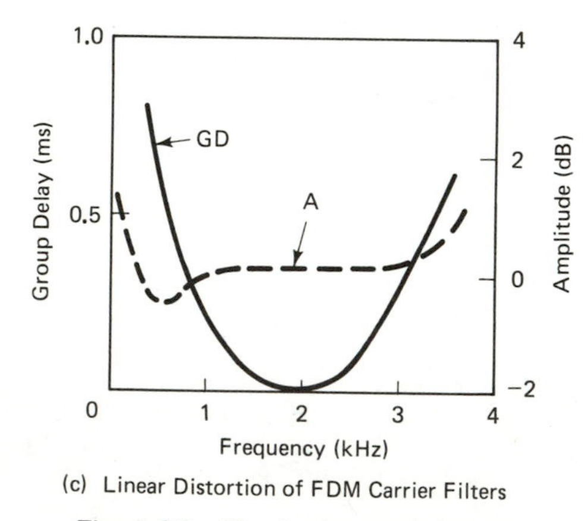

(c) Linear Distortion of FDM Carrier Filters

Figure 8.2 Circuit characteristics.

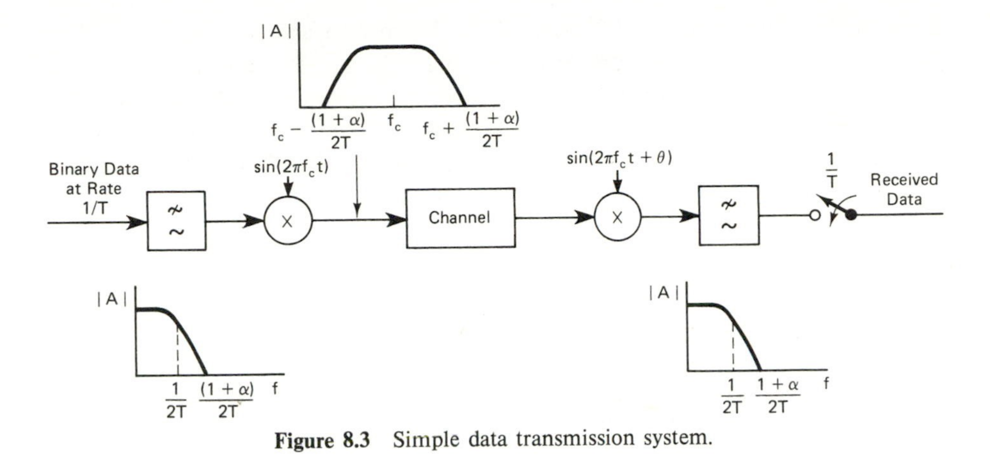

Figure 8.3 Simple data transmission system.

recovers the baseband signal, which is sampled every T seconds at the appropriate instant to detect the data with a low probability of error. If the channel is nondistorting and the two band-limiting filters are correctly designed, then, at the sampling instant, data symbols do not interfere with each other. If the channel introduces linear distortion, intersymbol interference (ISI) is caused, which degrades the performance of the system. If a single impulse (data symbol) is applied to such a system, then at the output of the receiver low-pass filter an impulse response is obtained which is the equivalent baseband impulse response of the channel. From a knowledge of the three basic linear channel impairments, f_c, the demodulating carrier phase θ, and the band-limiting filter responses, the equivalent baseband impulse response may be calculated. The ISI caused by the linear distortion is governed by the impulse response sampled at T spaced intervals.

Simple double-sideband amplitude modulation (DSBAM) as shown in Figure 8.3 uses twice the bandwidth required for the baseband signal; practical data transmission systems use more efficient modulation methods [Lucky et al. 1968], of which one is of particular interest: quadrature amplitude modulation (QAM). A QAM signal is formed by summing two DSBAM signals where the two carriers are of identical frequency but are 90° out of phase, as shown in Figure 8.4. At the output of a nondistorting channel, the two DSBAM signals are separated by demodu-

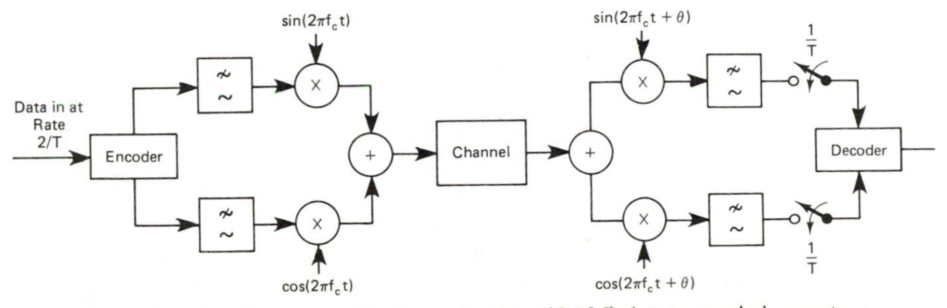

Figure 8.4 Quadrature amplitude modulation (QAM) data transmission system.

lating with two carriers again 90° out of phase. Distortion, as well as introducing ISI into each of the two equivalent baseband channels, causes interference between the two channels. The equivalent baseband channel can be drawn as the cross-connected networks shown in Figure 8.5. A very convenient way of representing this is to regard the two data inputs (and outputs) as real and imaginary and then the equivalent baseband response may be represented by a complex impulse response.

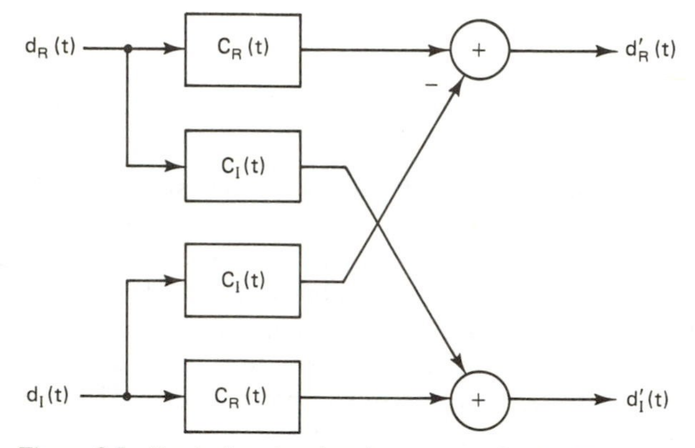

Figure 8.5 Equivalent baseband response of a QAM system.

The concept of a complex channel response is very useful for the design of QAM modems. The imaginary part of the channel impulse response is simply the Hilbert transform of the real channel impulse response. To equalize or cancel a complex channel response, a complex adaptive filter is required.

Complex adaptive filters. In earlier chapters the adaptive filters considered have been operating on real signals; the extension to handle complex signals is straightforward and is presented here without proof. Figure 8.6 shows a complex transversal filter updated using the complex stochastic gradient LMS algorithm. Indicating complex quantities by an asterisk superscript, the output of the filter is given by

$$\hat{y}^*(n) = \mathbf{S}^{T*}(n)\mathbf{H}^*(n) \tag{8.1}$$

Adaption by the stochastic gradient algorithm with a fixed-gain constant μ is by the recursion

$$\mathbf{H}^*(n+1) = \mathbf{H}^*(n) + \mu[\mathbf{S}^*(n)]'[y^*(n) - \hat{y}^*(n)] \tag{8.2}$$

where the prime indicates conjugation of the complex quantities in the vector. The analysis of this algorithm parallels the real case discussed in Chapter 3 and leads to similar convergence properties and residual error. A number of simplifications of the basic algorithm are commonly encountered in data transmission. These are detailed in the following sections where appropriate.

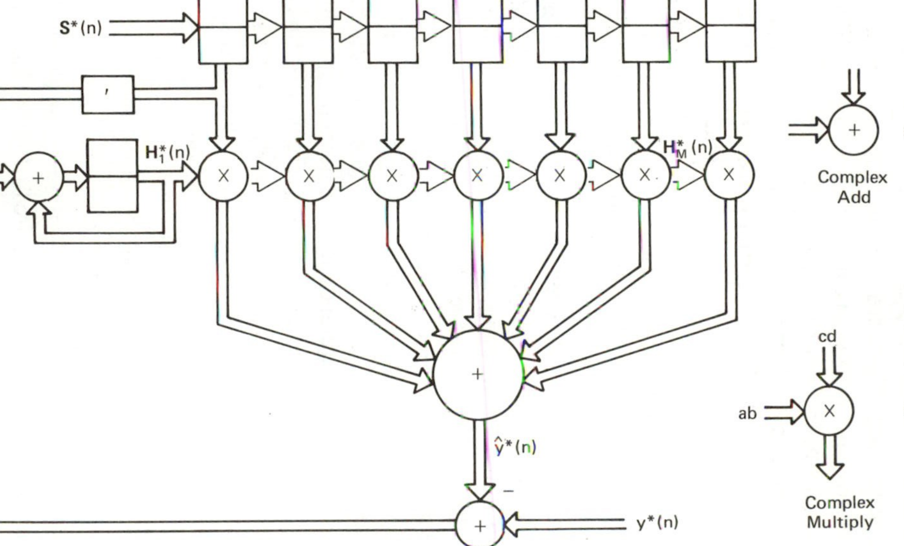

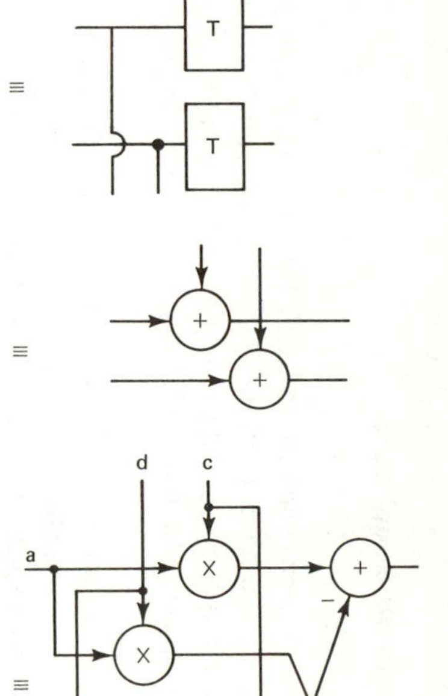

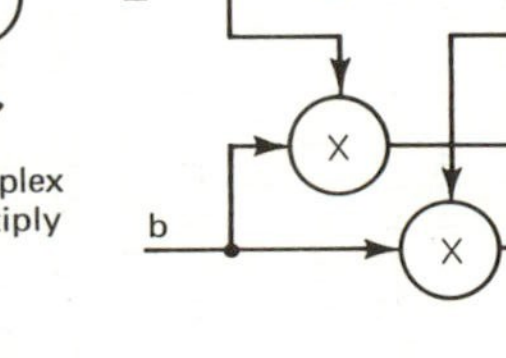

Figure 8.6 Complex adaptive filter structure.

8.2.2 Speech-Band Equalizers

Most speech-band modems conform to CCITT-recommended modulation formats [Folts], which, for the high-speed modems needing equalizers, involve either pure phase modulation or combined phase and amplitude modulation. Both types of modulation can be viewed as forms of QAM and so require the use of complex adaptive equalizers.

As well as linear distortion, speech-band channels generally introduce frequency offset and phase jitter onto the data signal. The modem receiver therefore has to use some form of carrier-phase tracking circuitry to remove frequency offset and reduce phase jitter. There are two common modem structures for combining complex adaptive equalizers with carrier phase tracking; Figure 8.7 shows them in block diagram form. In the first, equalization is performed on the complex baseband signal after demodulation using quadrature carriers obtained via a digital phase-locked-loop (DPLL) from carrier-phase error estimates generated at the quantizer. The quantizer is the device that decides which of the two-dimensional signal states is being received at the time of sampling and outputs the complex number corresponding to the signal state. The error signal for adapting the equalizer is the difference between the input and output of the quantizer. In the second structure the order of demodulation and equalization is reversed. As the equalization is performed on the modulated data signal the error signal for the equalizer has to be modulated using the recovered carrier phase information; although, involving more signal processing, the second structure is often preferred because it does not introduce delay via the filters and equalizer in the phase tracking loop, so that rapid phase jitter is more easily tracked. The DPLL and the equalizer both derive tracking information from the quantizer, and as the equalizer taps are a function of the demodulation carrier phase (a change of phase of $+\phi$ will rotate each complex tap by $-\phi$), careful design of a modem is required to ensure that the two loops do not interact adversely. One way of ensuring this is to use a joint gradient algorithm [Falconer 1976(1)] or, alternatively, make one loop much slower to respond to changes than the other. In telephony networks the linear distortion characteristics do not usually vary significantly rapidly for this to be a disadvantage. The convergence behavior and residual error of the two equalizer arrangements, known as baseband and passband equalization, respectively, are equivalent.

Types of equalizer. As well as the way in which the equalizer is combined with carrier-phase tracking, there are further variations on the way the adaptive transversal filter is used as an equalizer. For simplicity we will describe these variations in terms of their action on a real equivalent baseband channel; they can be applied to complex baseband channels and to the passband structure as well. There are three important types of equalizer used in speech-band data communications:

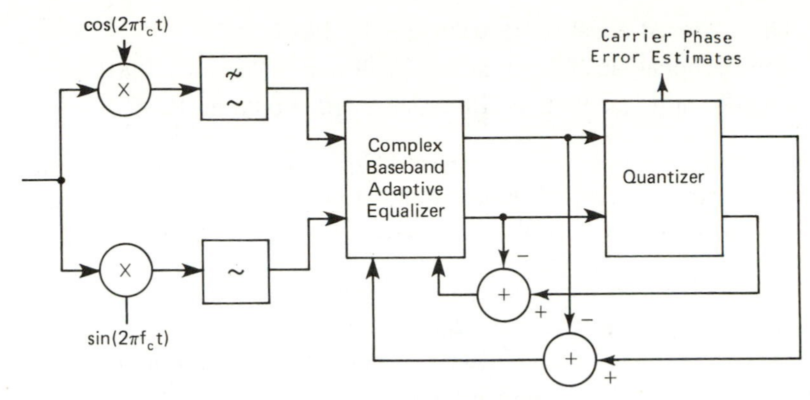

(a) Baseband Equalization

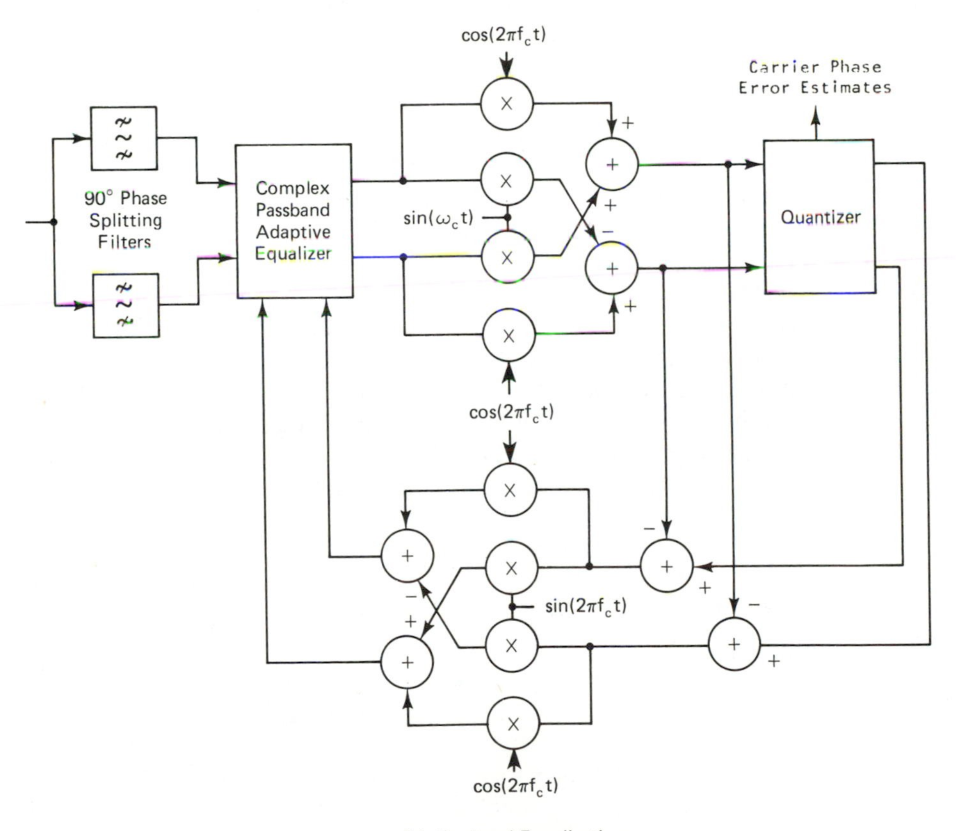

(b) Passband Equalization

Figure 8.7 Modem structures.

1. The T-spaced equalizer [Proakis, Lucky 1966]
2. The fractional tap (FT) equalizer [Gitlin and Weinstein 1981]
3. The decision feedback (DFB) equalizer [Belfiore and Park]

To understand the reasons for using these different types of equalizer we must first elaborate on the concept of equalization presented in the introduction. The impulse response of the equivalent baseband channel sampled at rate $1/T$ may be represented by the z-transform $C(z)$. The frequency response of $C(z)$ is defined completely by the response in the bandwidth 0 to $0.5/T$ Hz. The coefficients of $C(z)$, and therefore the frequency response $C(f)$, are a function of the sampling phase. The unsampled baseband data signal occupies a bandwidth of $\frac{1}{2}(1 + \alpha)/T$ and the sampling process causes spectral components above $0.5/T$ to fold over and add to components below $0.5/T$. For a distortionless channel, properly designed band-limiting filters, and a correct choice of sampling phase the fold-over process results in the sampled channel frequency response having flat amplitude and linear-phase from 0 to $0.5/T$ Hz. Distortion in the baseband channel in the region up to $\frac{1}{2}(1 + \alpha)/T$ will cause the sampled channel frequency response to deviate from this ideal; so $C(f) \neq 1$ and the job of the equalizer is to restore as far as possible a flat amplitude and linear-phase response. A transversal filter with taps spaced at T intervals and sample rate $1/T$ can do this and is known as a T-spaced equalizer. In situations where the group-delay distortion is changing relatively slowly in the region $\frac{1}{2}(1 - \alpha)/T$ to $\frac{1}{2}(1 + \alpha)/T$, the T-spaced equalizer operates very well and has been widely used. However, when the group-delay distortion is more severe, the summation of components about $0.5/T$ can lead to deep nulls in the sampled channel amplitude/frequency response, especially if the timing phase chosen for the sampling is inaccurate. As the equalizer attempts to remove these nulls it can amplify channel noise by an unacceptable amount. This disadvantage may be avoided by using a fractional tap (FT) spaced equalizer.

An FT equalizer is shown in Figure 8.8. The equalizer is now an adaptive filter with taps spaced at nT/m, where n and m are integers $(n < m)$, and sam-

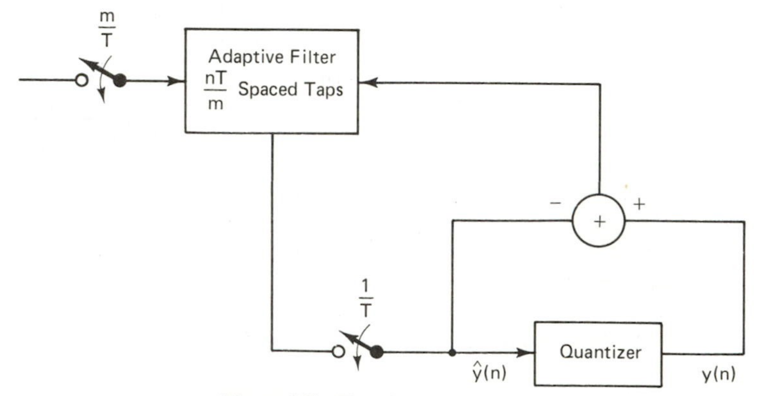

Figure 8.8 Fractional tap equalizer.

ple rate m/T. Sampling at rate $1/T$ takes place after the equalizer. Such an adaptive filter can correct the channel response up to $0.5m/(nT)$ hertz. If $\frac{1}{2}(1 + \alpha)/T < 0.5m/(nT)$, any adverse fold-over problems can be avoided because the channel is corrected before T-spaced sampling. It should be noted that it is the T-spaced sampled channel spectrum that is equalized; the frequency response before T-spaced sampling of the combined equalizer and channel $L(f) = H(f)C(f)$ is, ideally, flat amplitude and linear phase, that is, $\exp(-2\pi jf t_0)$, where t_0 is an arbitrary delay, from 0 to $\frac{1}{2}(1 - \alpha)T$ Hz, but from $\frac{1}{2}(1 - \alpha)/T$ to $\frac{1}{2}(1 + \alpha)/T$ it is such $L[(1/2T) - f] + L[(1/2T) + f] = \exp(-2\pi jf t_0)$.

FT equalization is so effective that the residual error performance of the equalizer is virtually independent of timing phase [Ungerboeck, 1976]. In addition, the FT equalizer can provide more optimum filtering of the received data signal, giving a better signal-to-noise ratio at the data detector. The penalty paid for this improved performance is that the number of taps for a given equalizer time span is increased by the factor m/n and the number of delay elements by m. However, for representative telephony channels with only amplitude and group-delay distortion for a fixed number of taps, the FT equalizer gives a better performance than does the T-spaced equalizer [Qureshi 1982]. With listener echo present, however, the T-spaced equalizer with its greater time span may be preferable.

For equivalent baseband channels with severe amplitude distortion both the T-spaced and FT equalizers enhance channel noise because they introduce gain to combat the amplitude losses. Another alternative in this case is the DFB equalizer [Belfiore and Park], shown in Figure 8.9. A pure DFB equalizer is shown on the right-hand side of the illustration: the detected data symbols are used as the input to a transversal filter whose output is subtracted from the received signal. If the main (largest) sample of the channel impulse response is the first, the ISI samples that follow are removed by the transversal filter, whose taps are equal to the ISI samples. Thus the pure DFB equalizer is, by the definition given in the introduction, operating as an ISI canceler, not as an equalizer. However, common usage has sanctioned the term "DFB equalizer."

Because the filter operates on noiseless data (post decision) the channel noise is not enhanced and equivalent baseband channels with severe amplitude distortion can be equalized more effectively. However, for pure amplitude distortion the impulse response is symmetrical about a peak and the DFB equalizer cannot cancel the prepeak ISI. Therefore, the DFB equalizer is usually preceded by a T-spaced equalizer which has the job of equalizing the prepeak ISI. Comparisons [Falconer 1976(2)] of T-spaced and DFB equalizers suggest that there is a performance advantage to be gained from the use of the DFB equalizer especially when the data transmission system bandwidth is such that severe distortion is being experienced at the edges of the data signal spectrum. The DFB equalizer is also very good for removing listener echo with no noise enhancement (echoes cause pronounced ripples in the amplitude frequency response). Linear equalization and cancellation can also be combined in other ways to give improved performance. It has been shown [Gersho and Lim], for example, that using tentative decisions

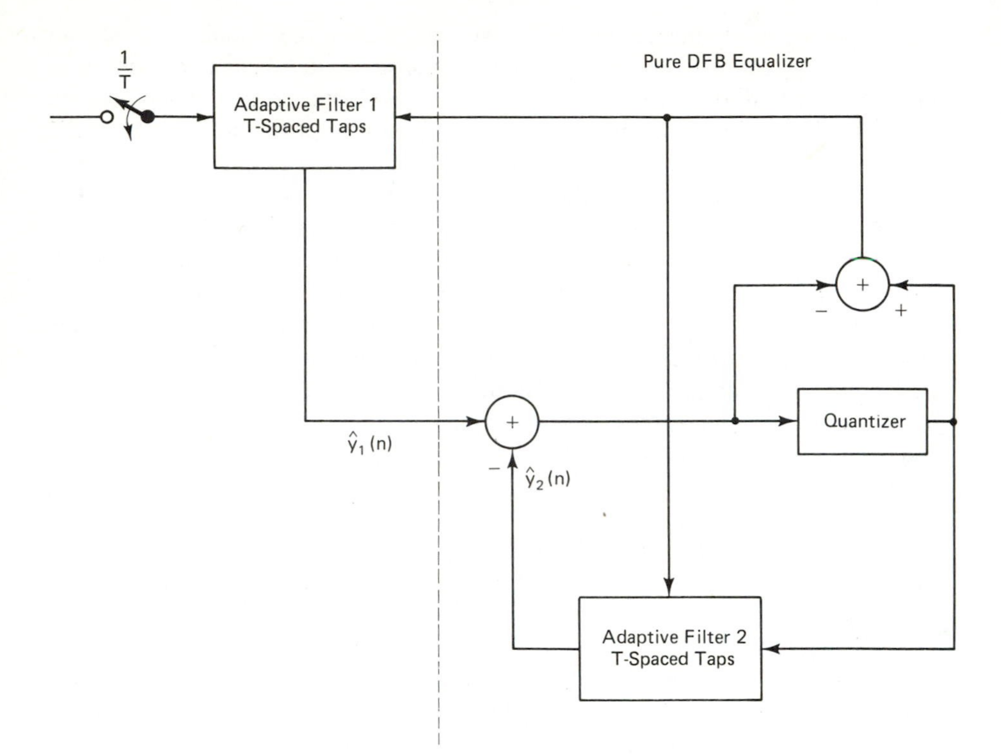

Figure 8.9 Decision feedback (DFB) equalizer.

obtained after linear equalization as inputs to an ISI canceler can give improved performance.

Equalizer adaption. The job of an equalizer is to approximate the inverse channel response with a finite number of taps. There are various ways of doing this, but the most robust is to adjust the equalizer taps so that the sum of the mean square residual ISI and noise is minimized, that is, if the combined channel and equalizer response is

$$L(z) = z^{-M} + \sum_{\substack{-\infty \\ i \neq M}}^{\infty} p_i z^{-i} \tag{8.3}$$

where M is the delay introduced by $L(z)$ to the main sample, then the LMS equalizer minimizes

$$E\left\{ \sum_{\substack{-\infty \\ i \neq M}}^{\infty} |d_i p_i|^2 + \sigma^2 \right\} \tag{8.4}$$

where σ^2 is the variance of the noise at the output of the equalizer and d_i are the transmitted complex data elements. The algorithm for achieving adaption to this state is the complex version of the stochastic gradient LMS algorithm given by (8.2).

An immediate problem that arises is that of the reference signal $y^*(n)$ required to form the error signal for the equalizer. By the very nature of data communications the input to the channel is separated from the receiver. There are two ways in which a reference signal may be obtained. The first is to have a stored reference; the second is to use the output from the decision circuit in the modem, as indicated in Figure 8.7. The stored reference, which has to be synchronized with the transmitted sequence, is used to train the equalizer initially; but for tracking during transmission of data the decision-directed technique must be used. Decision-directed training without any stored reference is possible, but high error rates before convergence can lead to false convergence of the equalizer to nonglobal quasi-stable minima [Mazo].

The convergence properties of the LMS algorithm are described in Chapter 3, where the rate of convergence is shown to be a function of the number of equalizer taps, the gain constant μ of the update loop and the power spectrum of the input signal to the filter. Generally, the number of taps will be chosen to meet the equalization requirement and the value of μ is governed by stability constraints and the amount of tap jitter that can be tolerated.

The power spectrum of the input to the filter is determined by the equivalent sampled baseband channel amplitude/frequency characteristic and the power spectrum of the transmitted data sequence. It is usual to ensure that the power spectrum of the data sequence is white by employing data scramblers and descramblers [Ridout and Harvie 1982] at the transmitter and receiver, respectively. The convergence of the filter is then a function of the equivalent baseband channel amplitude/frequency characteristics only.

Fortunately, it has been shown [Ungerboeck 1972] that the channel characteristics only weakly affect the convergence rate and as, for most applications, the convergence rate of the equalizer is not particularly critical, the standard complex stochastic gradient LMS algorithm is adequate. There is one application, however, where the rate of convergence is crucial. In some data communication networks a central modem polls each of a number of out-station modems all connected to a multipoint circuit. To receive a reply from each of the modems, the central modem has to train its equalizer in turn for the channel between each out-station and itself. Often, the messages returned from the out-stations are short, so that the train-up time must also be short if it is not to be a significant proportion of the transmission time.

Various schemes have been proposed for achieving fast equalizer convergence, including frequency-domain equalization [Walzman and Schwartz], matrix inversion algorithms [Butler and Cantoni], Kalman filter techniques [Falconer and Ljung 1978], orthogonalization techniques [Qureshi 1977], and cyclic equalization [Mueller and Spaulding 1975]. Some of these are dealt with in earlier chapters of this book.

It is desirable from an implementation point of view to try to minimize the complexity of the adaptive filter. To maintain linearity there is not much that can be done to reduce the accuracy requirements of the filter itself. However, the variables in the stochastic gradient algorithm can be modified drastically without destroying its ability to converge, albeit at a slower rate [Duttweiler 1982]. In the later section on implementation the predominance of digital realizations of adaptive speech-band equalizers is stressed. Digital signal processing can be much simplified if for multiplications the multiplier and/or the multiplicand are reduced in accuracy. The gradient algorithm variables that can be treated in this way are μ, the error signal, and the signal inputs to the correlation multipliers. As μ is a fixed quantity, setting it to 2^{-i}, where i is an integer, results in a simple shift in the complex error signal words. This simplification (Chapter 7) is often used in adaptive filters for data modems.

A further simplification is to employ only the sign of the signal inputs to the multiplier, the clipped algorithm in Table 7.1. At first sight this seems a dangerous thing to do because it destroys the nondivergence property of the LMS adaption algorithm [Claasen and Mecklenbrauker]. However, it has been found [Moschner, Macleod et al., Duttweiler 1982] that the effect is simply to slow convergence by a small factor depending on the amplitude probability density function of the input signal. This is usually an acceptable penalty to pay for the simplification of half the multiplications in the adaptive filter to adds or subtracts.

Operations on the error signal also slow convergence. We shall return to this topic in Section 8.3, where simplifications in processing are even more important and drastic simplification of the error signal with a significant slowing of convergence can be made acceptable.

Equalizer complexity. The number of taps required in the complex transversal equalizer is governed by the severity and type of the distortions the modem is expected to work over, the carrier frequency f_c, the roll-off factor α, the signaling rate $1/T$, and the required performance. The distortions encountered by the data transmission system depend on its application. Many high-speed modems are required to work over dedicated conditioned circuits. A conditioned circuit is one which, within the network, has been equalized using fixed filters to within a recommended (e.g., CCITT Recommendation M1020) maximum group-delay and amplitude variation. The adaptive equalizer then has the job of removing any residual distortion and so needs very few taps. On the other hand, modems required to function over switched network channels encounter far more severe distortions and require much longer equalizers, especially if they must deal with long-delayed listener echoes.

The carrier frequency affects the amount of significant ISI by virtue of where it places the spectrum of the signal. As the carrier frequency is increased, for example, the upper frequencies of the signal spectrum will experience more and more distortion as the edge of the speech band is approached. Similarly, increasing the signaling rate or α will widen the signal bandwidth and thus affect how much band-edge distortion the data signal encounters.

The performance requirement of the equalizer can be expressed in a number of ways, but the most useful one from an equalizer design point of view is the mean-square error at its output (i.e., the mean-square residual ISI plus the variance of any noise). The target value of this quantity depends on the number of signal states in the modem line signal (e.g., signal phases for PSK) and the tolerable error rate.

Implementations. The first consideration in devising an implementation of an adaptive equalizer is the number of taps required; typical choices range from about 8 taps for a 4800-bit/s polling modem up to 64 or more for a 9600-bit/s modem intended for switched network applications. The second consideration is the sampling rate of the equalizer. Typically, these range from 600 samples per second up to 4800 samples per second or more. Although analog realizations of speech-band equalizers are possible, commercially competitive modems now almost exclusively use digital signal processing (DSP) realizations which are both cheaper and give better performance.

An important consideration in a DSP realization concerns the word lengths required for each part of the equalizer structure. These will vary with the performance requirements of the equalizer, the number of taps, and the gain constant. Two types of DSP realization can be used, each of which affects the word-length consideration in a different way. The DSP hardware can be realized either as a dedicated, or semidedicated, custom-designed LSI or VLSI circuit, or with a more general microprocessor architecture. The former is usually more efficient, but the latter allows for much greater flexibility in design so that one design of IC can, by reprogramming, be used for different modems, or even different signal processing applications entirely. Both approaches are used, although now that more complex VLSI implementations are possible, efficiency is not so important and there is a tendency to opt for the more flexible approach, allowing the development costs to be amortized over a greater number of products.

In the custom-designed approach the word lengths of each part of the equalizer can usually be specified independently and so are minimized to reduce the circuit complexity. In the microprocessor architecture there is usually a global word length which must obviously be greater than the maximum word length required by the adaptive filter (and any other DSP functions required in the modem if these are also implemented on the same device). The precision requirements for adaptive equalizers have been studied theoretically [Gitlin and Weinstein 1979], but usually simulation studies are used to determine the necessary word lengths. Generally, the tap coefficients require greater word lengths than the signal samples. Typical values range from 12 to 20 bits for the coefficients and 6 to 10 bits for the signal samples, depending on the number of taps, the gain constants used, and the desired performance (Chapter 7).

An example of custom-designed LSI circuits [Brownlie et al.] implementing a complex adaptive filter for speech-band modem applications is shown in Figure 8.10. It consists of three different LSI circuits: an adaptive filter processing IC, a shift register IC, and an IC that performs the data detection and error signal generation as well as various other modem functions. Two of each of the processing and storage chips are combined with one of the data detection chips to form a 2400-sample per second 72-tap complex adaptive equalizer mounted on a hybrid circuit substrate. The circuits are realized in 5-μm NMOS technology working at about 2 MHz clock rate and so by modern standards not particularly complex or fast. Other LSI implementations are described by [Guidoux and Le Riche, Murano et al.]. Speech-band adaptive equalizers have also been realized in standard bit-slice microprocessors [Watanabe et al.], but competitive products are now usually based on LSI and VLSI circuits.

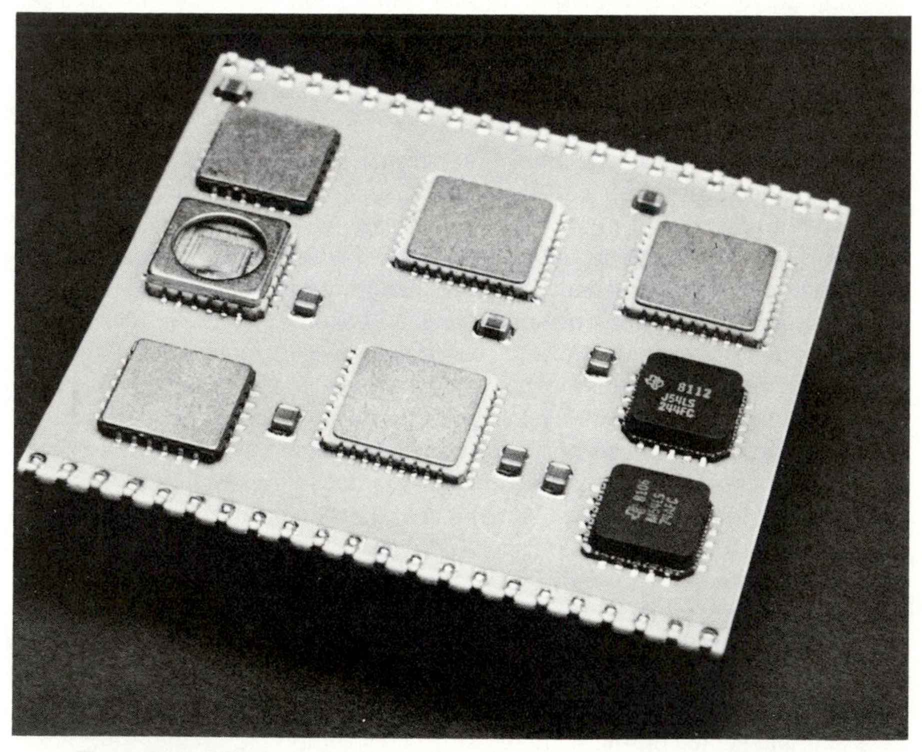

Figure 8.10 LSI circuit implementation of an adaptive equalizer. (Courtesy of British Telecom Research Laboratories)

8.2.3 Echo Cancellation for Speech-Band Data Transmission

There is a growing need in data communications for duplex transmission over two-wire switched circuits. Where the data rate required is 2400 bits/s or less, frequency-division techniques are employed so that fixed filters can be used to separate a received signal from talker echoes. Above 4800 bits/s the limited bandwidth available in the speech channel precludes the use of frequency division because the data signals would require too many states for reliable detection without sophisticated and expensive processing. Cancellation of the talker echo using an adaptive filter is the only way of achieving two-wire duplex data transmission at the higher rates. In fact, echo cancellation has also been applied at 2400 bits/s in a commercially available modem [Stein] as an alternative to the frequency-division approach and higher rate designs are emerging.

Figure 8.11 shows how an adaptive filter can be used to cancel talker echoes. As illustrated, the adaptive filter has a single input and output and is a wholly real

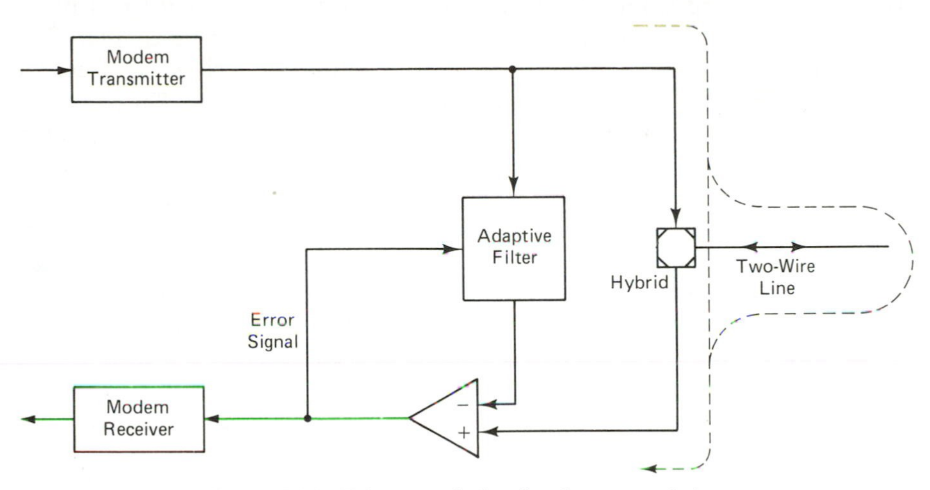

Figure 8.11 Echo cancellation for data transmission.

filter. To cancel all the echo frequency components in the bandwidth of the received data signal, the sample rate of the adaptive filter must be at least twice the highest frequency present in the data signal spectrum. Therefore, if the sample rate is f_s, then

$$f_s > \frac{1 + \alpha}{T} + 2f_c \tag{8.5}$$

The direct application of a real adaptive filter has two disadvantages:

1. The adaptive filter is driven by analog samples or, if a digital implementation is used, digitally encoded analog samples. As we shall see later, the dynamic range of the filter is usually required to be large (e.g., >60 dB), so the filter delay line is required to store samples very accurately. Also, each of the multiplications in the filter is between two accurately represented quantities.
2. The input signal has sample-to-sample correlation imposed on it by the filters in the modem transmitter, which tends to slow the convergence of the filter.

Both these disadvantages are circumvented by using data-driven echo-canceler structures.

Data-driven echo cancelers. A data-driven echo canceler [Weinstein 1977(2)] is shown in Figure 8.12. A complex adaptive filter is driven by the transmit data after it has been encoded into its complex form prior to modulation. Because the canceler is canceling the line signal, a modulator operating at the transmitter carrier frequency follows the adaptive filter. Note that the sample rate of the filter is an integer multiple m of the modem signaling rate $1/T$ and still obeys (8.5). To generate the complex error signal for the filter, a complex line signal is formed with a Hibert transformer and then demodulated to adapt the baseband filter. The adaptive filter is required to model the equivalent baseband echo response convolved with the response of the spectrum shaping filters in the modem transmitter. The interpolation filter restores the real line signal from the T/m spaced samples to a continuous waveform ready for resampling by the modem receiver. This is necessary because the sample timing in the receiver is not necessarily of exactly the same frequency as the transmit timing. At first sight this structure looks far more complicated than the use of a real adaptive filter. However, under certain conditions often pertaining in data transmission systems, it allows the amount of signal processing to be significantly reduced. There are also modifications to the basic structure, giving further savings in processing. [Baudoux and Macchi, US Patent 4162378 (1978)]

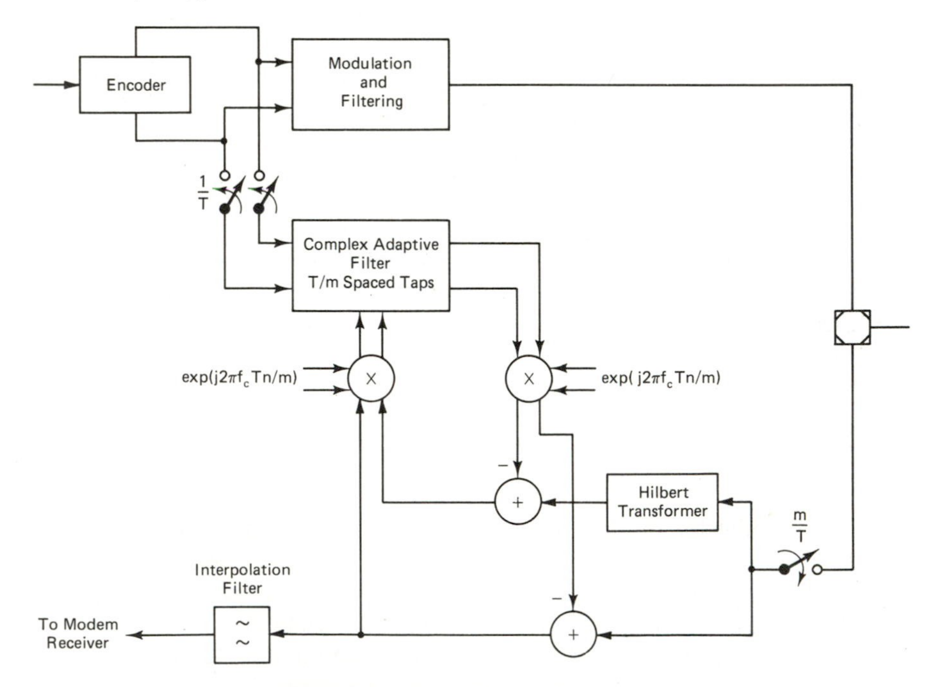

Figure 8.12 Data-driven echo canceler.

For many QAM signal formats the data elements after encoding consist of a few discrete levels. If a digital delay line is used in the adaptive filter, very few bits of storage are required for each delay element. This also means that in a digital realization one input to the tap and correlation multipliers has very few bits; the multipliers are, therefore, very simple to implement. The multiplications in the modulator remain as complicated operations. However, if the data system is such that its carrier frequency f_c and signaling rate $1/T$ are related so that $2\pi f_c T/m$ is a multiple of $\pi/2$, the multiplications by $\sin(2\pi f_c Tn/m)$ and $\cos(2\pi f_c Tn/m)$ become multiplications by 0 or ± 1 or, by scaling by $\sqrt{2}$ and shifting by $\pi/4$, just ± 1. This condition is met in a number of modulation formats.

Another useful structure is obtained by reversing the order of modulation and adaptive filtering as shown in Figure 8.13. Provided that in this case the carrier

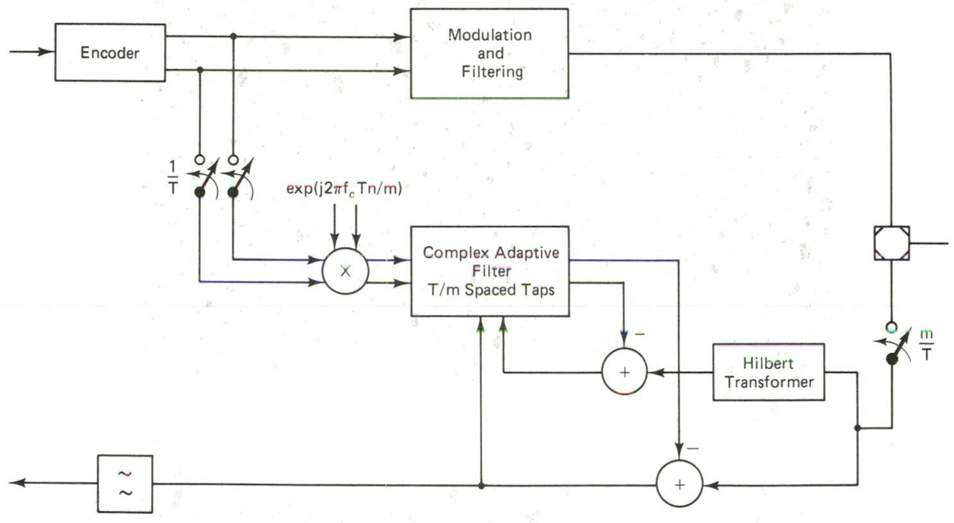

Figure 8.13 Modulated data-driven echo canceler.

frequency and signaling rate are such that $2\pi f_c T$ is a multiple of $\pi/2$, the data entering the adaptive filter are again very simple. In addition, the error signal does not need demodulating. This structure has an additional advantage when it is required to cancel the real line signal only as shown in Figure 8.14. As the adaptive filter is required to produce only the real output, half the processing (that which produces the imaginary output) disappears. The error signal is now purely real, so the tap updating is simpler, but the penalty for this is that the mean convergence rate of the filter is approximately halved.

Adaptive operation. In the data-driven structures the adaptive filters are driven by a succession of data symbols at T intervals with $m - 1$ zero values between them. This means that the adaptive filter operates as m independent adaptive filters, each producing an output every T seconds, the outputs being multi-

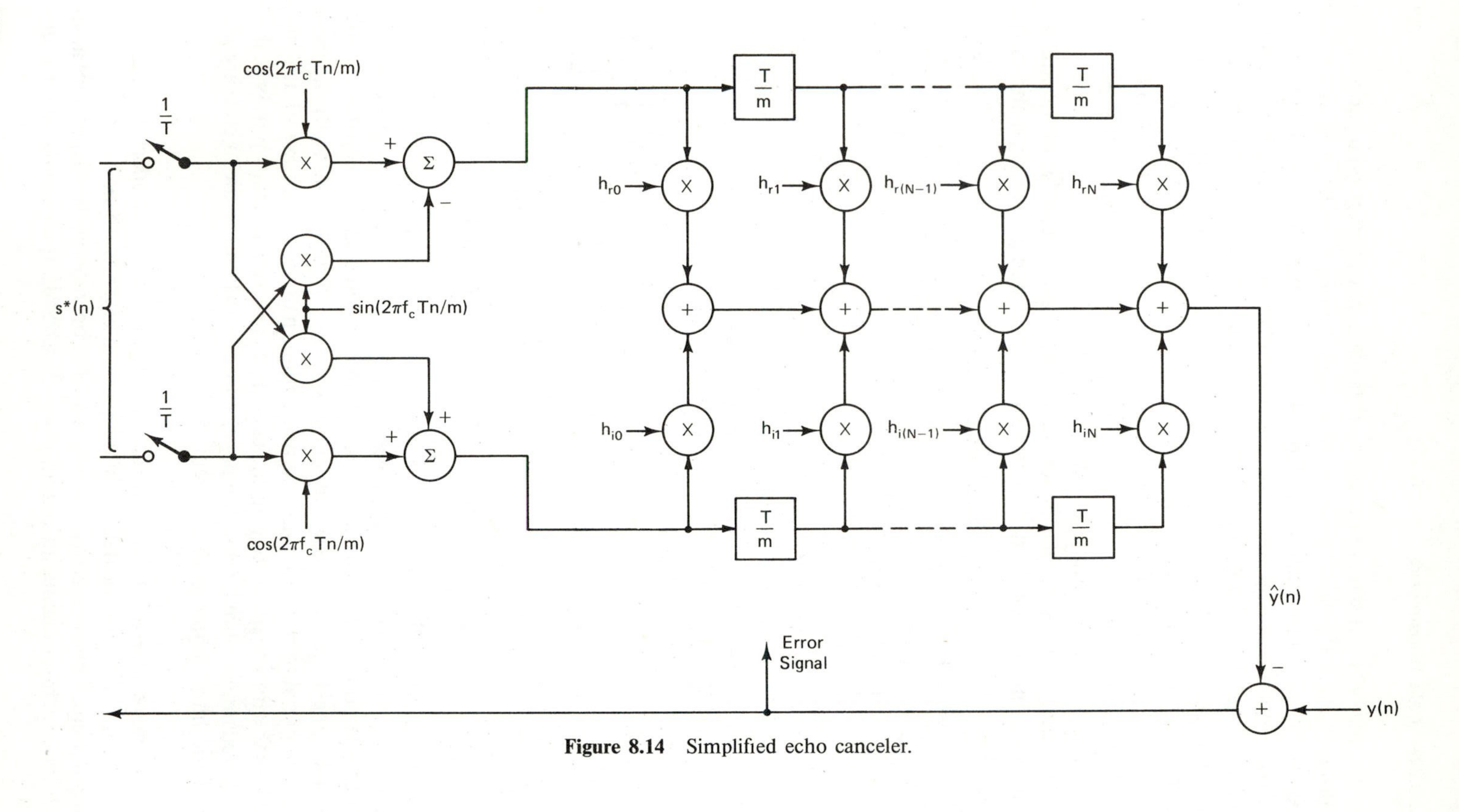

Figure 8.14 Simplified echo canceler.

plexed in time. We can, therefore, examine the convergence of a single filter of sample rate $1/T$.

As the echo canceler is driven by the data (or modulated data) and the data are normally scrambled before encoding, the input signal to the adaptive filter is spectrally white. Therefore, the echo canceler convergence is dependent only on the number of taps of the adaptive filter, the amplitude probability density function of the data symbols, and the value of μ. Analysis [Weinstein 1977(2)] of the evolution of the mean-square tap misadjustment power (MSTMP) gives a recursion formula.

$$E\{|\xi(n+1)|^2\} = [1 - 4\chi\mu A + \chi\mu^2(B + 4(N-1)A^2)]E\{|\xi(n)|^2\} + 2\chi\mu^2 NAE\{|w|^2\} \tag{8.6}$$

where N is the number of T spaced taps, $2A$ is the average value of the square of the modulus of the complex data elements, B is the average value of the fourth power of the modulus of the complex data elements, χ is 1 for complex error signals and $\frac{1}{2}$ for real error signals, and $E\{|w|^2\}$ is the expectation of the uncancelable component of the received signal. This includes echo components outside the span of the echo canceler, noise, and most important, the wanted data signal from the far end. The analysis assumes that $w(n)$ is uncorrelated with the input to the adaptive filter. To ensure that the wanted data signal is uncorrelated with the transmit signal, different scramblers and descramblers are usually employed for each direction of transmission. The residual MSTMP after convergence is obtained by iterating (8.6) to give

$$E\{|\xi(\infty)|^2\} = \frac{2\mu NAE\{|w|^2\}}{4A - \mu(B + 4(N-1)A^2)} \tag{8.7}$$

The fastest convergence is obtained by using the optimum gain constant.

$$\mu_{\text{opt}} = \frac{2A}{B + 4(N-1)A^2} \tag{8.8}$$

Unfortunately, for μ_{opt} and large enough N the residual MSTMP is, from (8.7), approximately $\frac{1}{2}E\{|w|^2\}/A$. The residual tap misadjustment gives rise to an uncanceled residual echo of power $2AE\{|\xi(\infty)|^2\} = E\{|w|^2\}$. As in a well-designed system the desired data signal from the distant transmitter is the dominant component of the received signal, then for μ_{opt} the residual echo is as large as the wanted data signal! Therefore, the gain constant must be reduced to μ_t, which is the value of μ giving an acceptably small ratio $E\{|\xi(\infty)|^2\}/E\{|w|^2\}$. The trade-off between convergence rate and residual echo is illustrated in Figure 8.15. As μ_t is small, the adaptive filter can track only very slowly time-varying echo responses.

Echo characteristics of speech-band circuits. Echo cancelers in general have to model two types of echo signal. One is the result of leakage across the hybrid in the modem itself. Although, theoretically a hybrid can be balanced to

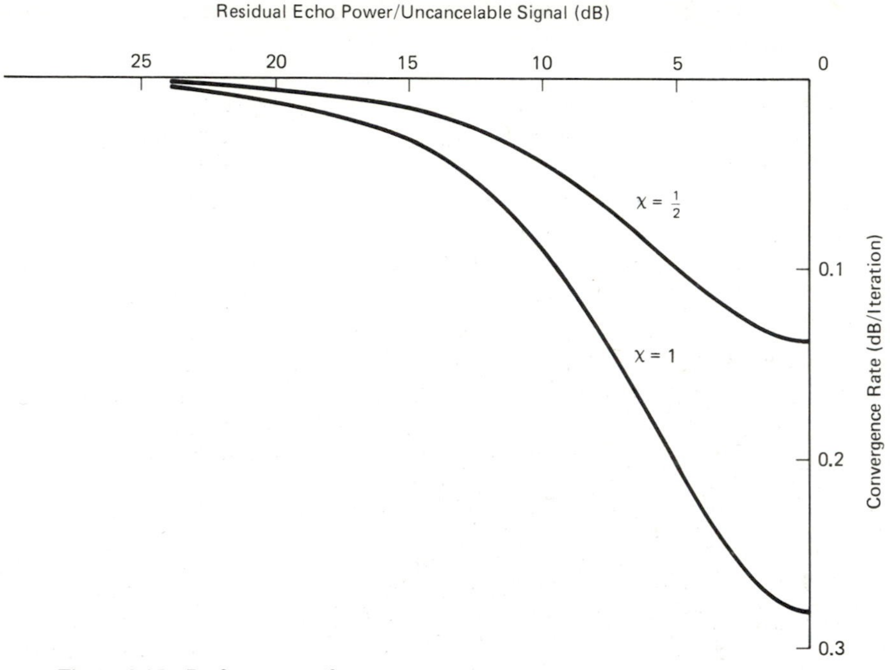

Figure 8.15 Performance of an echo canceler shown in terms of residual echo versus convergence rate. ($N = 16, A = 1/2, B = 1$)

prevent any leakage between the transmitter and the receiver, in practice this is very difficult, especially when a modem is required to work on any line, and the amount of trans-hybrid loss can be as low as 8 dB. Thus there is a large talker echo component with very little delay associated with it. The second type of echo results from the network itself, as shown in Figure 8.1. Generally, more than one discrete echo can occur. These echoes are usually smaller than the trans-hybrid echo (e.g., 20 dB down) and can have a delay varying from a few milliseconds to hundreds of milliseconds if a satellite circuit is involved. Also, because the four-wire portion of the telephony circuit is nominally zero loss, there is only a weak correlation between echo amplitude and delay. A further problem is that the network echoes may be offset in frequency due to the modulation and demodulation taking place in four-wire carrier systems. Although the offsets are usually quite small (e.g., <1 Hz), the resulting time variation of the echo response is difficult for an echo canceler to track.

Implementation considerations. Four fundamental parameters control the complexity of the echo canceler: the maximum echo signal level, the minimum received signal level, the required signal-to-uncanceled echo ratio, and the number of echo canceler taps. Earlier it was stated that the trans-hybrid loss could be as little as 8 dB; as the hybrid echo is the dominant one, this gives a maximum echo level of −8 dB relative to the transmit level. The maximum loss over a full switched network connection is ~48 dB. Typically, a received signal-to-uncanceled echo ratio of better than 20 dB is required. This means the echo canceler must suppress echoes by more than 60 dB. Together with the fact that the echo canceler may have hundreds of taps, achieving this level of performance requires a digital implementation as far as possible; an analog implementation would introduce too much spurious noise. As the line signal is analog, the echo canceler must therefore contain an analog-to-digital converter (ADC) and a digital-to-analog converter (DAC).

Three structures are possible depending on whether the subtraction of the echo-canceler output is done digitally, by an analog sampled-data subtraction, or by subtraction of continuous analog signals. These three methods are illustrated in Figure 8.16 for the case of cancellation of the real line signal. Structure 1 requires an ADC and a DAC of such an accuracy that the quantization noise introduced onto the received wanted signal is insignificant. Structure 2 avoids having an ADC and DAC in the signal path, but the interpolation filter creates delay in the tap update loop, and because out-of-band components appear as noise in the signal path, the interpolation filter requires a high stop-band attenuation. Structure 3 avoids these problems: The loop delay is minimized, there is no ADC or DAC in the signal path, and the interpolation filter has a much less severe out-of-band attenuation requirement. The DAC still needs to be of sufficient accuracy that the quantization noise of the echo-canceler output is small compared to the received signal. However, the ADC does not need to be of such accuracy. In fact, it is possible to reduce it down to a single (sign) bit [Claasen and Mecklenbrauker] with the penalty of very much slower convergence.

It is important in all these structures to maintain good linearity in the transmitter, the hybrid, the ADC, and the DAC when they are in the signal and/or echo paths. The reason for this is that nonlinear distortion cannot be modeled using a linear adaptive filter. If the echo-cancellation requirement is for 60-dB suppression, any nonlinear distortion components must be more than 60 dB below the transmit signal level, as must any extraneous noise in the analog circuitry. These requirements place severe, although not impractical constraints on the circuit design.

Depending on the exact application, ADCs and DACs in the signal or echo paths have word lengths of 10 to 12 bits or more. The tap coefficient word lengths depend additionally on the number of taps and the value of μ; typically 20 to 32 bits is required. These very long word lengths can be reduced by using averaged gradients [Gitlin and Weinstein 1978] for the updating, but the number of bits of storage per tap is still large because of the need to store the tap update during averaging. Clearly, speech-band echo cancelers require a great deal more storage for implementation than do equalizers, although using the data-driven structures, the processing per tap is usually simpler.

If the echo canceler is required to deal with network echoes that are offset in frequency, the complexity is even greater. One technique [Weinstein 1977(2)] for

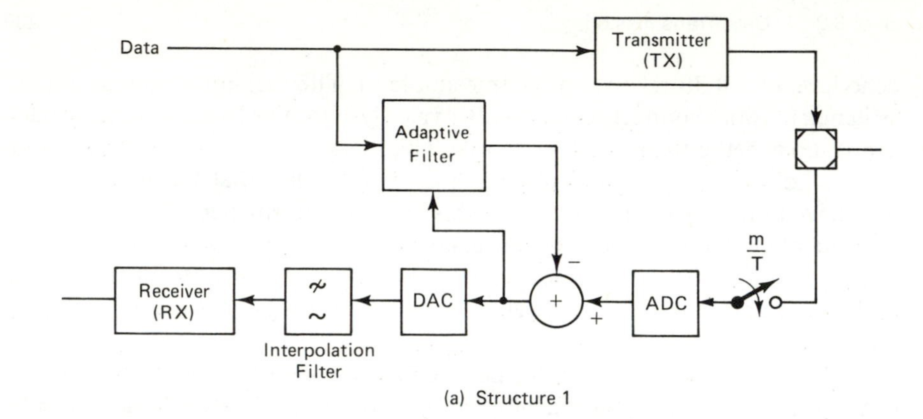

(a) Structure 1

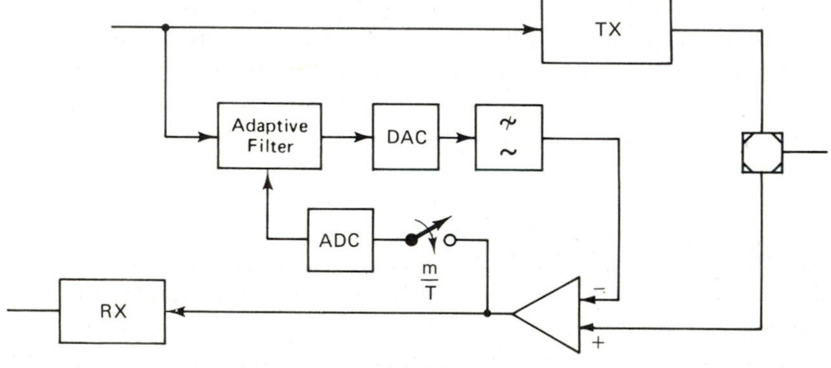

(b) Structure 2

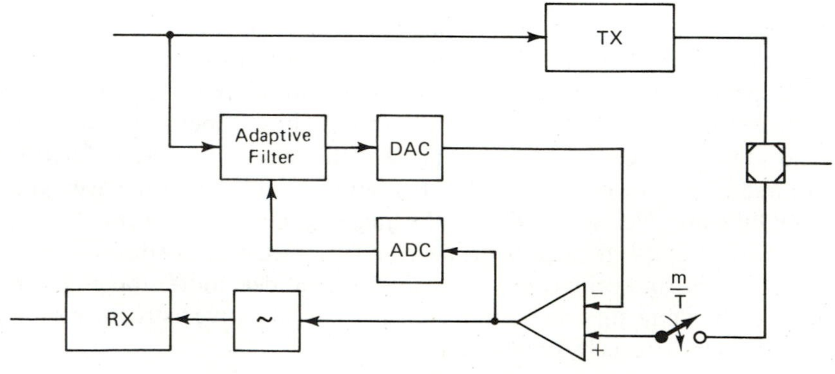

(c) Structure 3

Figure 8.16 Implementation of various echo-canceler structures. (Adapted from Verhoeckx et al. Copyright © 1979 IEEE.)

dealing with this case is to use a second adaptive transversal filter combined with a phase-tracking circuit to remove the frequency offset. Clearly, this is a much more complicated implementation.

Design example: A 9600-bit/s echo-canceling modem. With the exception of the modem mentioned earlier, echo-canceling modems have yet to make a major commercial impact, although this situation is likely to change over the next few years. The following was a typical experimental design constructed for tests in the U.K.

The design used an experimental 4 × 4 QAM modem with a carrier frequency of 1800 Hz and signaling rate of 2400 baud, and was aimed at providing a large coverage of the U.K. national network. The echo canceler was of the type shown in Figure 8.14 and structure 1 of Figure 8.16. Its time span was 26 ms (256, $T/4$ spaced complex taps) and it suppressed echoes to 60 dB below the transmit signal level. An efficient parallel processing hardware structure allowed the large amount of storage (approximately 15 kilobits) and processing (approximately 15 million add/subtract operations per second) to be implemented by standard MSI TTL and MOS integrated circuits on a single board of approximately 12 in. × 5 in. A very much more compact version using VLSI circuits should be possible in 3-μm technology.

Experimental echo-canceling modems were used [Adams 1980(1)] to establish the feasibility of the echo-canceling technique for 9600-bit/s duplex transmission over the U.K. switched network. Tests on a wide variety of network connections showed [Adams and Elliott 1983] that the technique did work and in particular that there is normally no frequency offset on U.K. network echoes. Shown in Figure 8.17 are oscillograms of typical echo-canceler tap values obtained by reading out the digital tap values (real and imaginary interleaved) through a DAC. These clearly illustrate the observations made earlier about the characteristics of echoes.

The experimental evidence is, therefore, that echo cancellation is a feasible technique for duplex data transmission over switched networks that contain no frequency offset echoes, have sufficient linearity, and do not have very long delayed echoes. Although echo cancelers can be devised to cope with all these problems, the implementation costs may well render them commercially unattractive.

8.3 DIGITAL TRANSMISSION OVER LOCAL NETWORKS

Telephony networks are making increasing use of digital transmission and switching techniques to provide considerable cost savings in network implementation and maintenance compared to the currently dominant analog equipment. The scale and pace of the introduction of digital equipment varies from country to country, but the time can be foreseen when most local switches are interconnected by digital channels. These channels would still be aimed primarily at telephony but

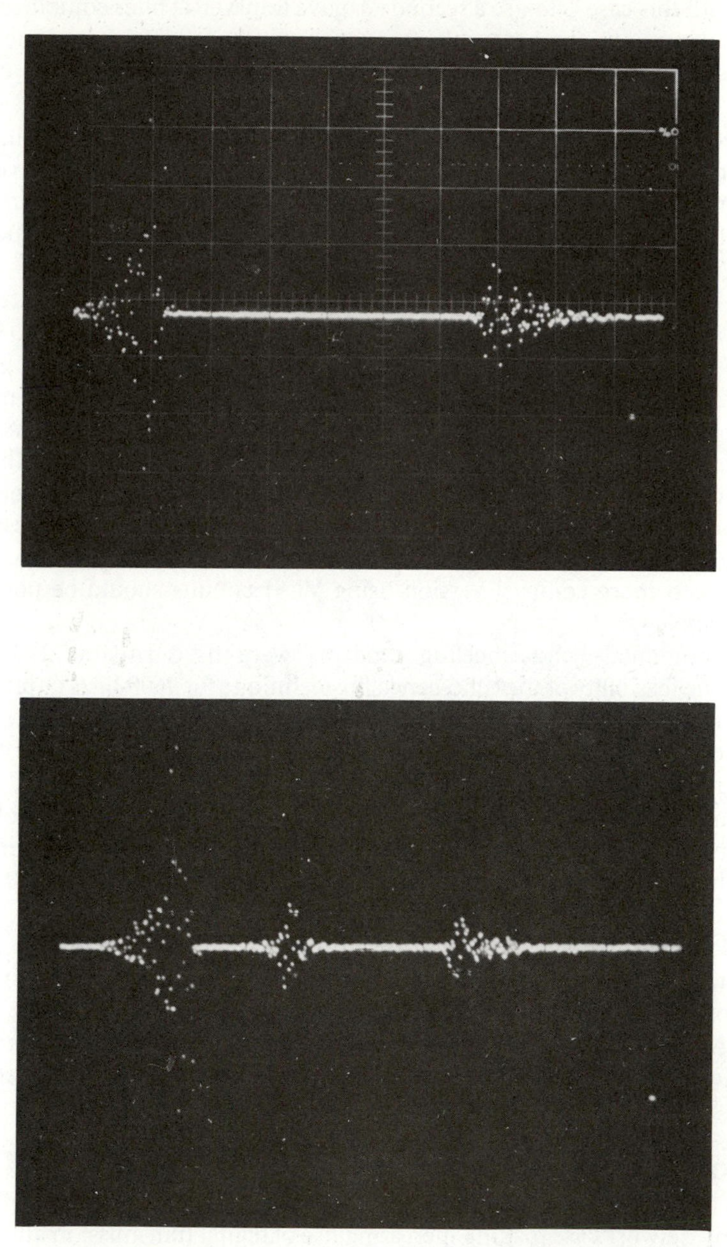

(b)

Figure 8.17 Echo-canceler impulse response (converged tap-weight values) for connections between (a) Aberystwyth and Dumfries (approximately 300 miles) and (b) Inverness and Ipswich (approximately 600 miles). The horizontal scale is 2.6 ms per large division, and the vertical scale is linear. (Courtesy of British Telecom Research Laboratories)

would also be capable of providing data communication an order of magnitude faster than speech-band modems (e.g., in Europe the standard CCITT recommended bit rate of 64 kilobits/s is used). The next step in the spread of digital communication is to connect each subscriber to his or her local switch by a digital transmission system. Such systems must be very cheap, must operate over existing local line plant, and because most subscribers can have only a single two-wire circuit, provide two-wire duplex operation. The last requirement has stimulated much interest in the use of adaptive filters as echo cancelers.

Subscribers' loops consist of twisted-pair metallic cables with a variety of diameters, several of which may occur on any one connection, and lengths varying from a few meters to several kilometers; consequently, the transmission characteristics of the connections vary widely. Figure 8.18 shows the insertion loss and phase characteristics for a length of a typical local network cable. Also shown is a plot of crosstalk attenuation. These characteristics illustrate two conflicting requirements. For a system to work over the longest possible line lengths, the power spectrum of the digital signal should be confined to as low a frequency band as possible to avoid crosstalk interference. However, by using modulation or line codes which have their spectrum at the higher frequencies (where the linear distortion is less), no equalization is necessary. The reach of systems using such techniques is limited by crosstalk interference. This is not too significant a problem, as the majority of subscribers, especially in cities, are close to their local switch. Generally, therefore, two types of subscriber loop transmission system may be identified: those that are essentially self-equalizing (without an adaptive equalizer) but of limited reach, and those that to obtain a longer reach use baseband transmission

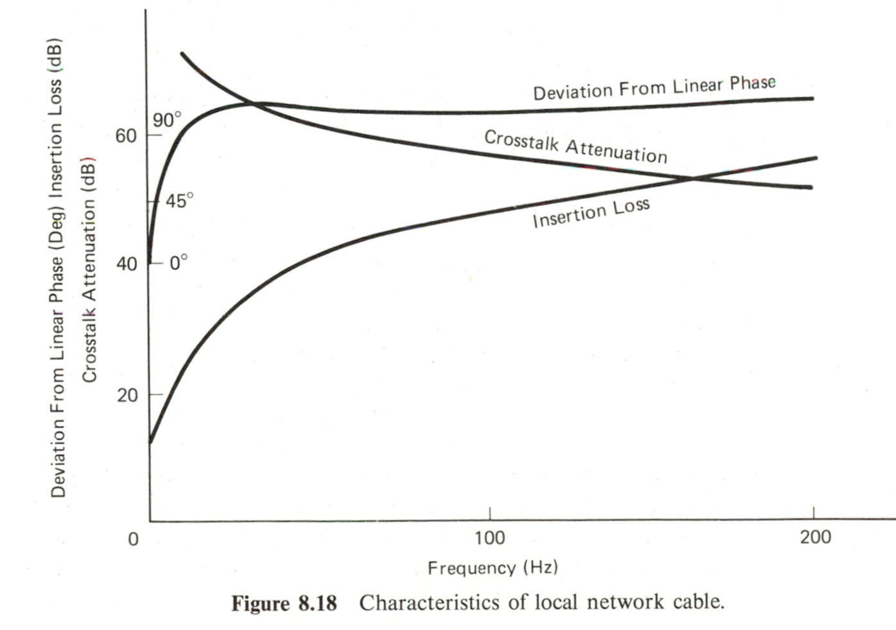

Figure 8.18 Characteristics of local network cable.

with adaptive equalization to counter the linear distortion at low frequencies. Both types have been the subject of research and development and systems are now becoming commercially available.

The precise system bit rates used vary depending on the application but are generally in the region of 100 kilobits/s. The adaptive filters are thus required to work up to sample rates of several hundred samples per second—an order of magnitude or more up on the adaptive filters used in the speech-band modem applications. Essentially all the techniques described previously for equalization and cancellation may be applied to subscriber loop transmission. However, the higher sample rate for the adaptive filters makes different demands on implementation technology. Fortunately, the adaptive filters often need fewer taps, which, combined with the absence of the complication of carrier phase tracking, allows some diverse adaptive filter implementations to be used [Van Gerwen and Verhoeckx, Agazzi et al. 1982(1), Vry and Van Gerwen 1981, Adams et al. 1981(1), 1984]. The following examples are chosen to illustrate this diversity.

8.3.1 Echo Cancellation for WAL2 Transmission

A good example of a self-equalizing line code is the WAL2 line code [Boulter], so called because its signaling waveforms shown in Figure 8.19(a) resemble the second Walsh function. It has the power spectrum shown in Figure 8.19(b). The smaller spectral lobes above $2/T$ hertz are usually suppressed by a line filter.

To design an echo canceler for this system we need to cancel all the significant energy in the echo signal, which, because it occupies the frequency band 0 to $2/T$, requires an adaptive filter working at a sample rate of $4/T$. The length of echo canceler depends on the degree of echo suppression required, which in turn is a function of the loss experienced by the signal received. Generally, however, the number of taps required is fairly small because the WAL2 code also tends partially to equalize the echo response. In the two examples that follow the number of $T/4$ taps are 12 and 24, respectively, for systems with worst-case received signal losses of 30 and 40 dB, respectively. The examples are chosen because they illustrate the interplay between technology and adaptive filter implementation and show how currently available components can be used to make compact and inexpensive adaptive filters.

Analog realization. One particular application [Adams et al. 1981(1)] of WAL2 transmission called for a system with a bit rate of 80 kilobits/s to operate over cables with up to 30-dB loss at 80 kHz. Experiments showed that a 12-tap echo canceler was sufficient to give adequate echo suppression. To realize the echo canceler cheaply without recourse to LSI circuit technology, the analog circuit shown diagrammatically in Figure 8.20 was developed which was both very compact and inexpensive. It can be seen that the implementation corresponds to structure 2 of

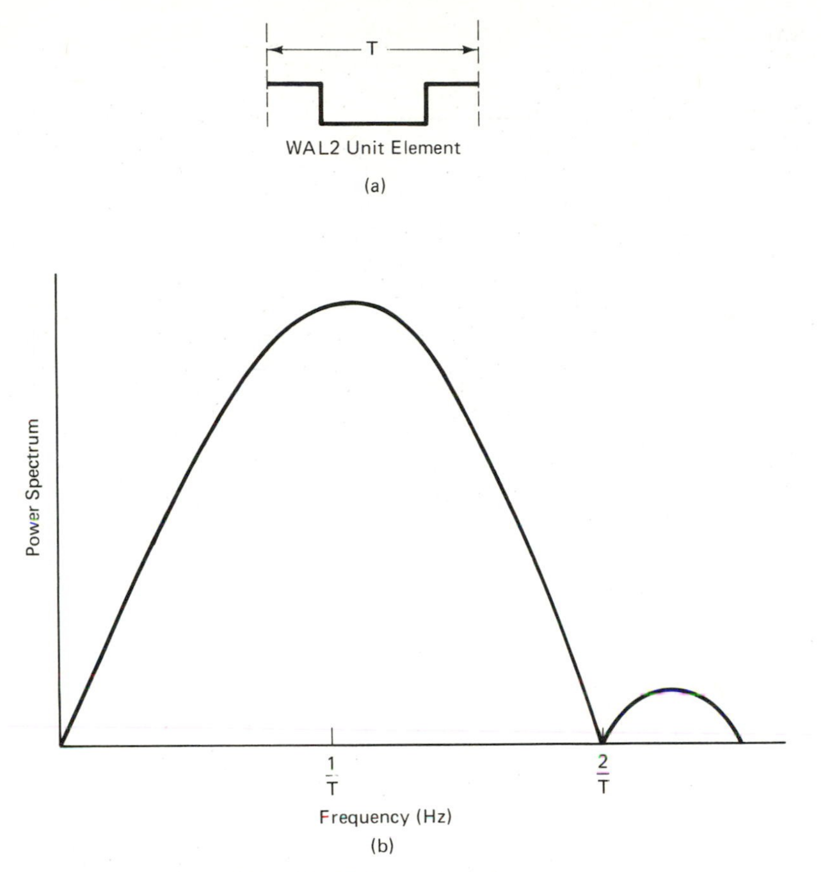

Figure 8.19 WAL2 line code and power spectrum.

Figure 8.16, but with the adaptive filter, apart from the delay line, implemented by analog circuits. The reason for choosing this particular structure was that the use of an interpolation filter, where shown, helps to isolate switching transients and noise from the received signal path. The circuit implementation used extremely simple switching multipliers [Glen] which exploit cheap bidirectional analog switches. The automatic gain control (AGC) for the system was included in the tap-update loop to help overcome direct-current offset problems in the filter implementation and was decoupled from the loop by virtue of its fast adaption time compared to the filter. The filter suppressed echoes sufficiently to give very good performance over lines with a 30-dB loss at 80 kHz and converged on the worst-case line in <70 ms. Figure 8.21(a) shows a photograph of the adaptive filter (excluding the AGC and filter) which consisted of only 12 ICs and consumed <150 mW. Experiments showed that the limitations of this design were due to nonideal performance of the analog circuit elements, not to the limited number of taps. Therefore, an analog design of this type has a limited field of application.

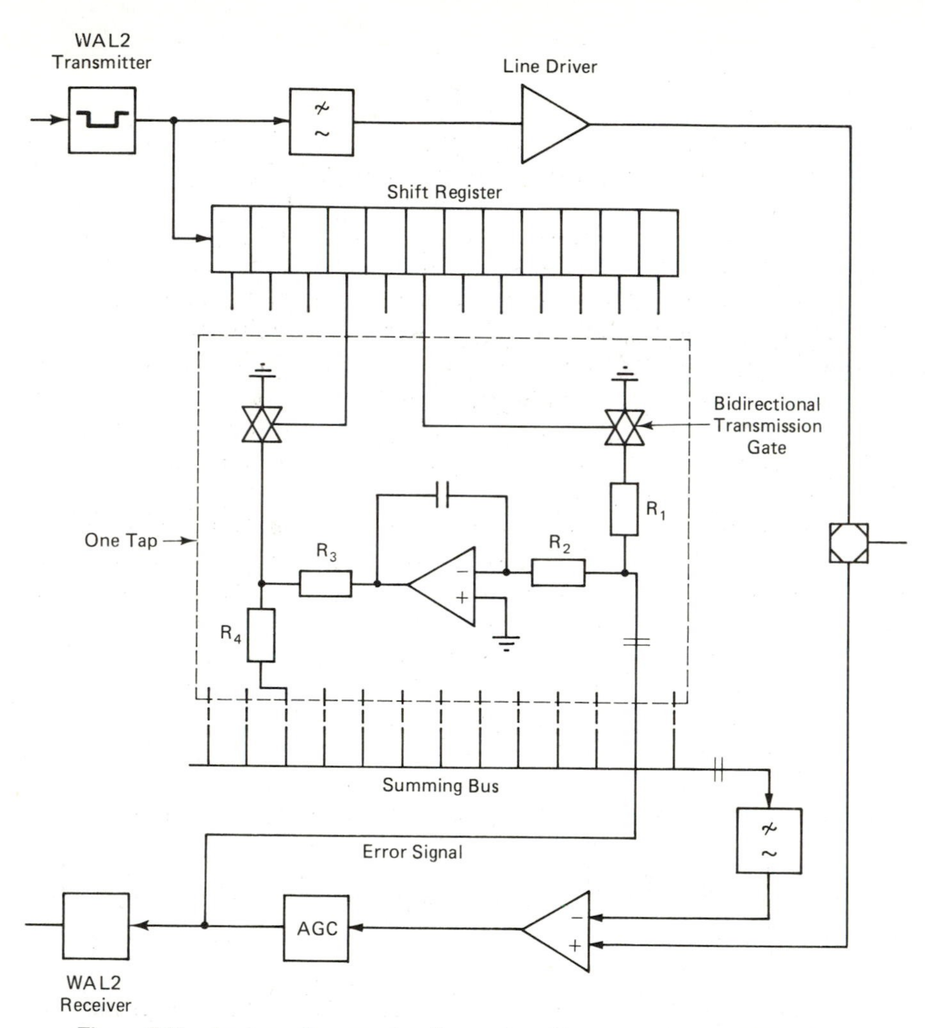

Figure 8.20 Analog echo canceler. (Inset after Glen, copyright © 1982 IEEE.)

Digital realization. An 80-kilobit/s WAL2 system required to operate over a 40-dB loss at 80 kHz needed a digital echo canceler with 24 taps. A direct digital realization of the preferred structure 3 in Figure 8.16 is rather complex unless LSI circuit implementation is used. A realization that allows easy implementation is the lookup-table approach [Holt and Stueflotten], which is a special case of the distributed arithmetic adaptive filter outlined in Chapter 7. The idea is to look up the required cancellation signal stored in a random access memory (RAM) by using the transmitted data as the address for the RAM. However, it is not necessary to address the RAM with 24 bits corresponding to the 24 taps. As stated in Section 8.2.3, a data-driven echo canceler operating at m times the data signaling rate can be viewed as m independent echo cancelers. Thus the number of

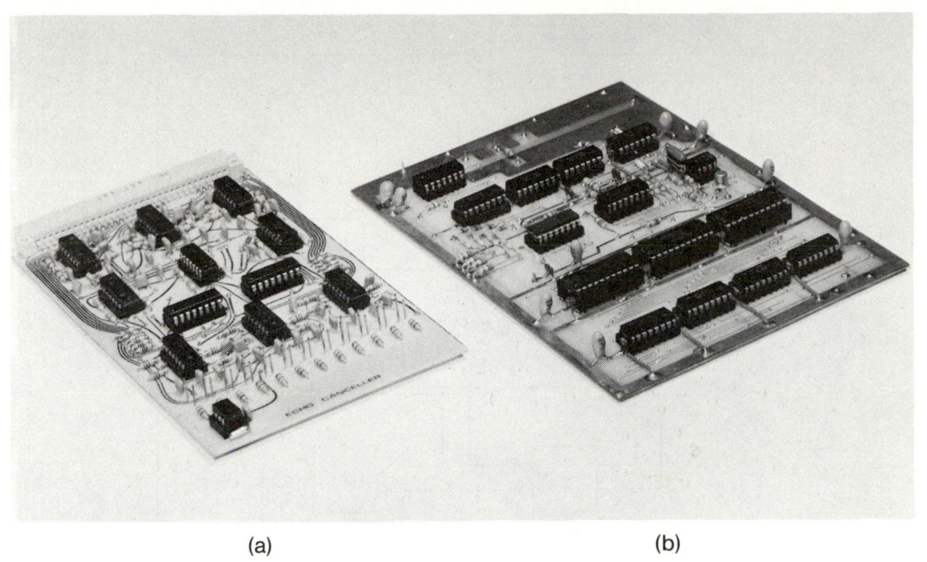

(a) (b)

Figure 8.21 Echo-canceler hardware: (a) analog design and (b) digital lookup-table design. (Courtesy of British Telecom Research Laboratories)

address lines for this example is required to be six to give a $6T$ time span, plus two to divide the RAM into four parts corresponding to the four independent echo cancelers. Figure 8.22 shows a block diagram of the echo canceler. The adaptive algorithm for the echo canceler is very simple: As each echo location is addressed, the RAM contents are converted to an analog signal by the DAC; the signal is subtracted from the received signal, and the error signal, after analog-to-digital conversion, is simply added to the current RAM location word to form a better estimate of the echo, which is then read back into the RAM. At each sampling instant only one location in the RAM is updated, so the convergence rate of this realization is much slower than the normal one where every tap is updated. In local network applications, however, this is not a problem; the stability and invariability of the echo on a given connection means that power-down storage may be used to hold the echo samples when the system is not being used. When it is required to be used, any minor adjustment to the RAM contents would take place very rapidly. A further simplification is to replace the ADC with a sign detector; convergence can still be ensured, although it is slower. Figure 8.21(b) shows a photograph of an experimental version of such an echo canceler which used CMOS technology and consumed <300 mW. The worst-case convergence time using just a sign update error signal was found to be <500 ms from the condition when the RAM contents all start at zero.

Another property of this adaptive filter realization is that the table lookup places no constraints on the type of echo path response, other than its time duration. Therefore, any nonlinearity in the transmitter, line interface, or echo-canceler DAC is also modeled.

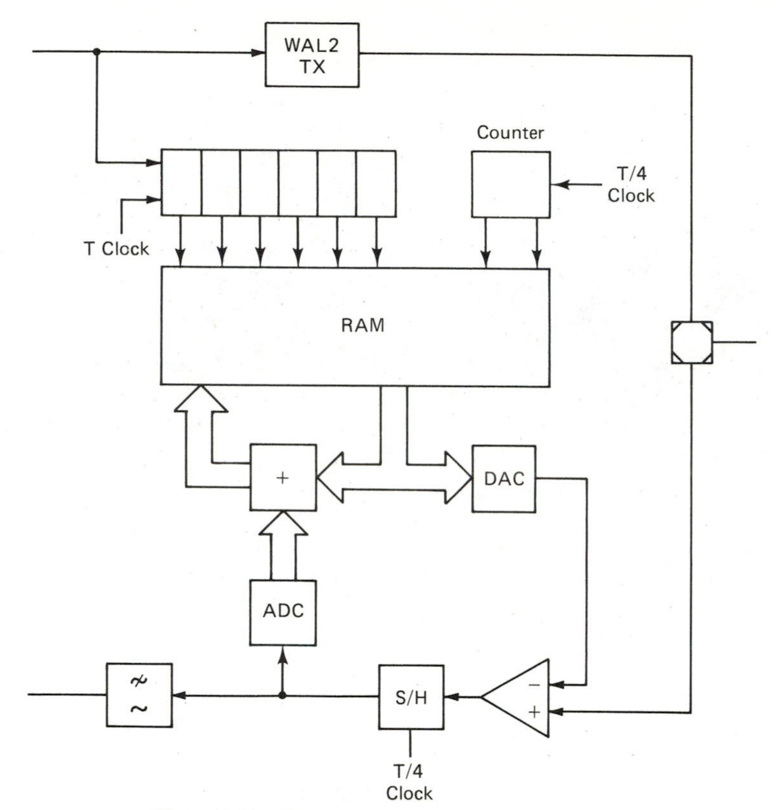

Figure 8.22 Digital lookup-table echo canceler.

8.3.2 Baseband Transmission

To obtain longer reach, baseband transmission systems are essential. However, these experience much longer echo responses and severe ISI, as shown in Figure 8.23. The straightforward way of dealing with this is to use a receiver with a $T/2$ echo canceler and an equalizer–the baseband equivalent of the echo-canceling modems of Section 8.2.3. In many local network transmission systems the timing clock used for the customer to local switch direction is locked to the clock derived from the received signal from the local switch. This synchronization of the clocks introduces some problems for the echo canceler but also allows another structure to be used, as we shall see later.

The pulse responses of local network cables, as exemplified in Figure 8.23, exhibit two common features: the leading edge rises rapidly to a maximum and the response eventually decays away. On the shorter lengths the maximum of the pulse response is only slightly more than one T interval after the start of the pulse; on longer cables the pulse has risen to over half its maximum after one T. Any of the equalizers described in Section 8.2.2 may be used, but with a suitable choice of

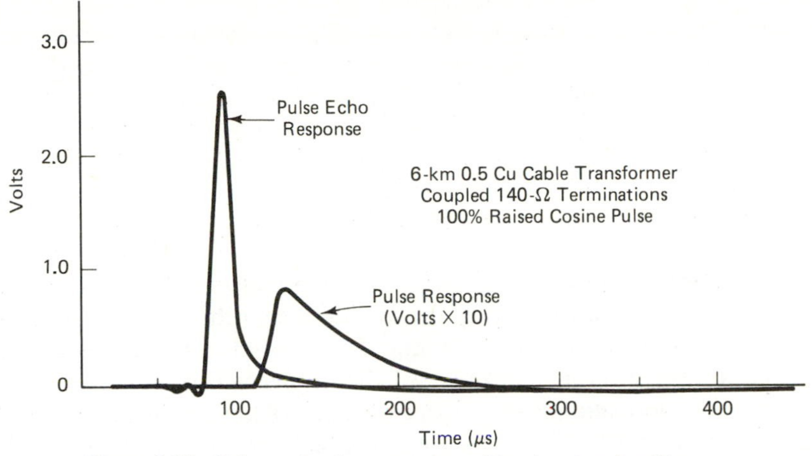

Figure 8.23 Pulse and echo responses of local network cables.

sampling phase, the DFB equalizer stands out as being eminently suitable. The decaying tail of the response suggests that some simple fixed linear equalization may be used to reduce the length of the tail and the size of the DFB equalizer taps—an important point for the reduction in error extension effects, that is, the tendency of the DFB equalizer to generate additional errors from a single decision error because an error results in uncanceled ISI. We will return later to the subject of equalization; first we consider the design of an echo canceler for structure 3 of Figure 8.16.

Simplified echo canceler for baseband transmission. Although the echo canceler is required to have only $T/2$ tap spacing compared to $T/4$ in the wider-bandwidth WAL2 system, the longer echo pulse responses require the number of taps to be substantially more (e.g., as great as a $12T$ or larger time span). The analog realization is not capable of expansion to this number of taps because of implementation problems. For a binary baseband system the lookup-table approach would require a memory of 2^{13} words or more of at least 16 bits accuracy. Although RAM storage is becoming very cheap and compact, this is still a not inconsiderable amount and it does not take many more taps before it becomes prohibitive. In such cases a digitally implemented data-driven linear adaptive filter is the more acceptable alternative. If mild nonlinear distortion of the echo is a problem, there are techniques available [Agazzi et al. 1982(2), Cowan and Adams 1984] for modifying the linear filter to account for them. Structure 3 of Figure 8.16 is the best one, for the reasons given earlier, but one of the main items of expense in the structure is the ADC in the error signal loop. As mentioned in Section 8.2.2, shortening the word length of the error signal can simplify the correlation multiplications; it can also simplify the ADC. The greatest simplification comes when the error signal is reduced to a sign-only representation. In the absence of a receive signal the echo canceler will converge. With a received signal present, convergence is assured only if the received signal is sometimes smaller than the error signal from the canceler. If it is not, the sign of the received signal controls the updating and the filter will not converge. In a system with synchronous timing for the two directions of transmission there is no guarantee that this will be true. This problem is usually avoided by adding a dither signal to the echo-canceler update loop. The same problem is found with the lookup-table echo canceler of Section 8.3.1, and it can be dealt with in the same way. An echo canceler implemented in this fashion has been integrated for a digital 1 + 1 carrier system using WAL2 transmission [Vry 1982].

***T*-spaced echo cancellation and DFB equalization.** DFB equalization may be combined with any of the fractional-tap echo cancelers described in the preceding sections, but the synchronous clocking of many local network transmission systems suggests that provided that a means can be found to recover timing-phase information from T-spaced samples of the line signal, the echo canceler need only have T-spaced taps, with a consequent saving in processing. Fortunately, the shape of the leading edge of the pulse response of local network cables enables such a timing recovery scheme to be implemented [Ehrenbard and Tompsett, Adams et al. 1984]. Therefore, the combined T-spaced echo canceler and DFB equalizer shown in Figure 8.24 may be used. To generate an error signal the output of the decision device is fed to an automatic reference control (ARC) to take account of the loss in the cable. Such a configuration is preferred to an AGC because it gives a better joint convergence characteristic [Falconer and Mueller 1979] for the echo canceler and DFB equalizer. It can be shown [Mueller 1979, Adams et al. 1984] that in the absence of decision errors the convergence of the two

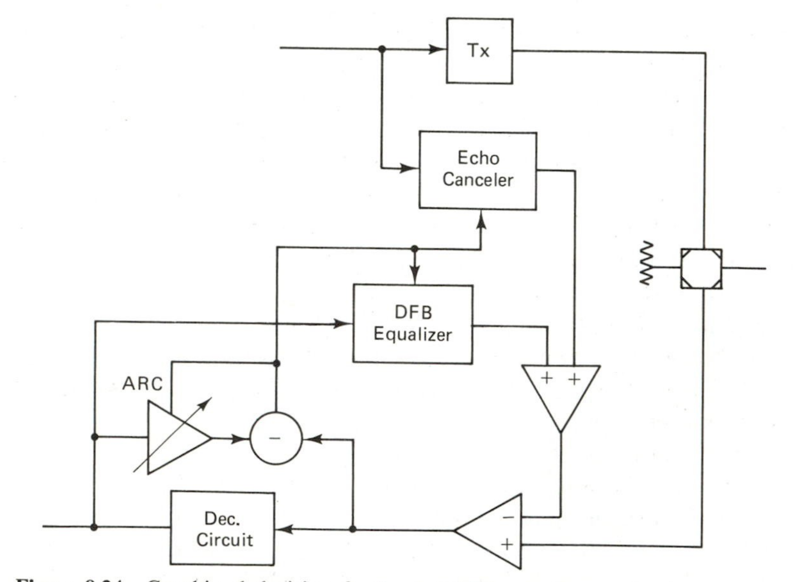

Figure 8.24 Combined decision feedback (DFB) equalizer and echo canceler incorporating automatic reference control (ARC).

adaptive filters is identical to the convergence of a single adaptive filter with $N + M + 1$ taps, where N is number of echo canceler taps and M the number of DFB equalizer taps. In this structure the received signal is no longer present on the error signal, so the gain constant can be near its optimum value, giving the fastest possible convergence (and an increase of approximately 3 dB in residual noise). A computer simulation of the structure's convergence showed that for binary signals,

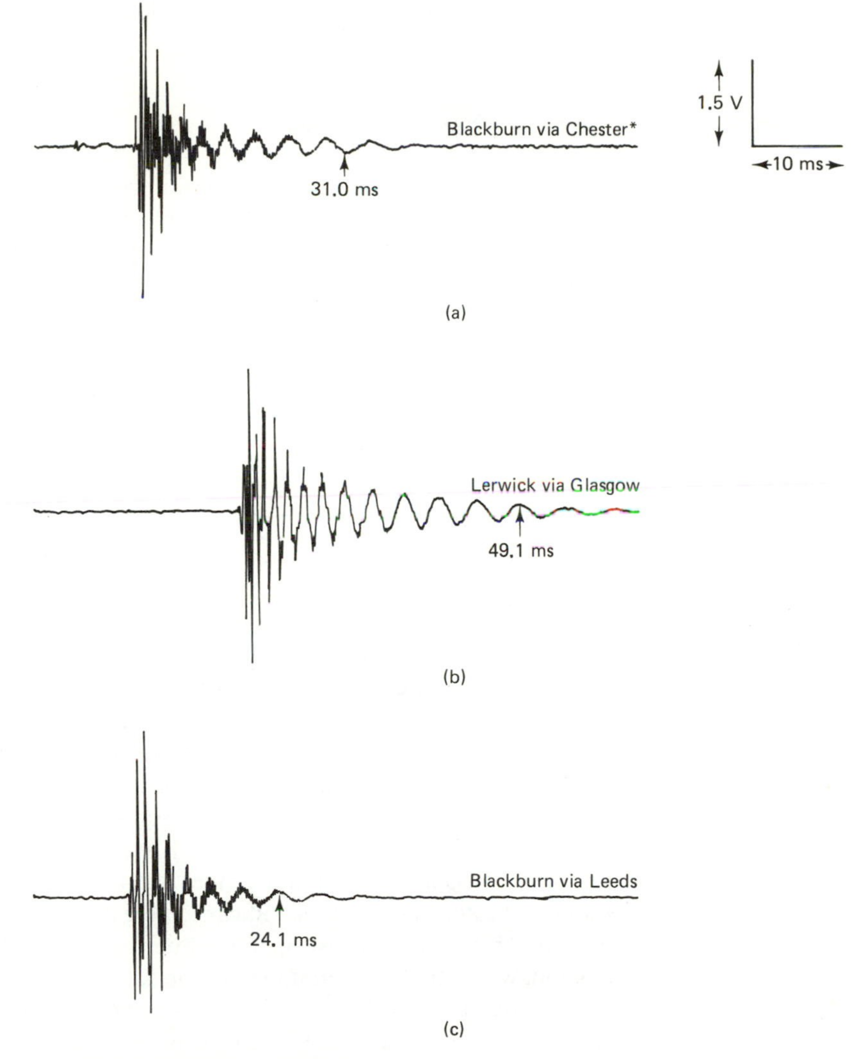

Figure 8.25 Typical echo impulse responses of various telephone network connections from a U.K. international gateway. (After Hoppitt, copyright © British Telecommunications plc (1978).)

$N = 20$ and $M = 13$ and, assuming that the decision device generates no errors, convergence to -70 dB takes <6 ms for a 100-kilobaud system (Section 8.3.1). In practice, errors due to the high initial interference from uncanceled echo and ISI slow down convergence, but the use of a suitable training routine can overcome this.

To implement this structure cheaply requires the use of VLSI circuit techniques. Recent studies [Adams et al. 1984] have shown that incorporating all the processing on a single chip for systems operating at about 100 kilobits/s over long cables is possible.

8.4 ECHO CANCELLATION FOR TELEPHONY

Echoes in speech-band circuits affect speech communications in two ways: subjective degradation and circuit stability. Subjective degradation of speech arises in long-haul circuits where echoes delayed by more than about 30 ms can upset the normal conversational process. Stability problems can arise in applications such as loud-speaking telephones and audio teleconferencing, where high-gain amplification can cause circuits to "howl." The traditional solution to both these problems is to use voice-activated attenuation of the return channel in the four-wire part of the circuit (Figure 8.1). However, under certain conditions, voice-activated attenuation can cause subjective degradations such as speech clipping and circuit deadness. Echo cancellation provides a better solution. Two applications are now described to highlight the design features of adaptive filters for speech cancellation.

8.4.1 Network Echo Cancelers

Network echo cancelers are located at a convenient point in the four-wire part of the telephone circuit. Their purpose is to cancel the echoes of speech on the go path which appear on the return. A circuit requires two echo cancelers, which will probably not be co-located, as the nearer the ends of the four-wire circuit they are, the shorter the delay of the echoes to be canceled. To design the echo canceler a knowledge of the echo delay and duration is required with reference to the position of the echo canceler in the network. Figure 8.25 shows some typical echo responses, measured at an international gateway in the United Kingdom, obtained from the U.K. national network [Hoppitt 1978]. Such echoes can have delays as long as 20 ms and very long oscillatory tails with significant energy up to 30 ms beyond the start of the echo. This is in contrast to the echoes experienced in speech-band data transmission, where the band limitation of the data signal causes the echo tail to die away more quickly (see Figure 8.17). Most echoes, however, have shorter delays and the amount of the tail that has to be canceled is dependent on the degree of echo suppression required. Published designs of echo cancelers aim at canceling echo components up to 16 ms in duration to give a balance return loss enhancement of 25 to 30 dB. Further reduction of the residual echo is achieved by using techniques such as center clipping [Horna 1977, Weinstein 1977(1)]. With the usual 8000 samples per second used to encode a speech channel, the adaptive filter is required to have 128 taps. If realized digitally, the input samples for a filter would need to have an accuracy of 12 bits to account for the wide dynamic range of telephony speech. With the usual long word lengths required for the tap values, a speech echo canceler has to perform a large number of complicated multiply operations and to store long word-length values. Clearly, speech echo cancelers even in this minimal form are more difficult to implement than are their data-driven counterparts. If such problems as time-varying echoes are to be overcome, the degree of complexity can become very high.

Problems of adaption. In data transmission the input to the echo canceler is well behaved in terms of signal power level and spectrum. The speech input for network echo cancelers is not so obliging; in particular, the following properties create additional serious problems:

1. The speech varies dramatically in power from talker to talker and between different syllables for a single talker. Consequently, if a fixed μ is used, it must be set very low, to avoid instability. This is often overcome by dynamically normalizing μ to the instantaneous power of the input samples in the filter delay line [Duttweiler 1978].
2. Some speech syllables (e.g., voiced sounds) are highly correlated, which can also result in instability if μ is set too high.
3. Conversations are essentially half-duplex in nature, and therefore the input to the adaptive filter disappears from time to time, making tracking of any time-varying echoes difficult.
4. If a conversation were truly half-duplex, the appearance of speech from the far end would not matter because the lack of an input signal would prevent the tap jitter due to a noisy error signal. However, conversations contain periods of double talking during which speech is present in both directions at once. To prevent misadjustment of the echo canceler, most designs incorporate double-talk detectors [Weinstein 1977(1)], which inhibit the tap updating.

Implementations. A number of experimental network echo cancelers have been built and as the costs of implementation fall, the commercial exploitation of such designs is beginning to take place. The simplest designs ignore frequency offset problems, use double-talk detectors to prevent misadjustment, and in some cases use a center clipper to improve the overall performance. A typical block diagram of such a design is shown in Figure 8.26. To reduce the hardware complexity, pseudologarithmic coding [Horna 1977] of the speech samples and the tap values is employed. Designs of this type have been integrated onto a single VLSI 5-μm NMOS circuit [Duttweiler and Chen 1980] and, exploiting the fact that a number of echo cancelers may well be co-located, realized in a 12-channel multiplexed MSI TTL hardware structure [Duttweiler 1978].

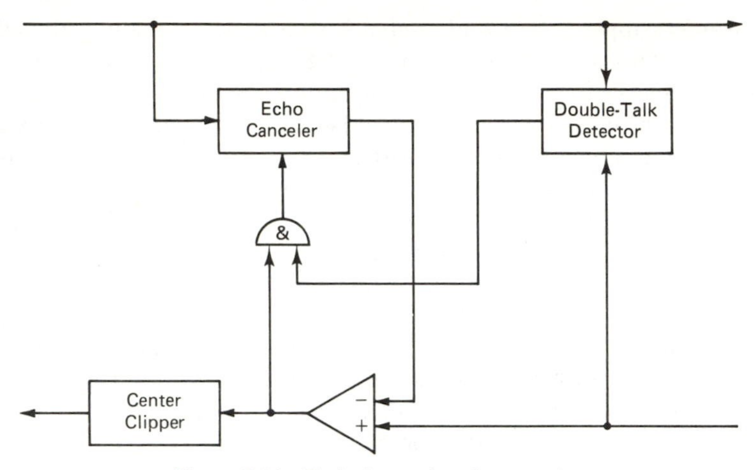

Figure 8.26 Typical speech echo canceler.

Experimental designs have also been investigated which attempt to deal with the time-varying echoes resulting from frequency offset. These are generally much more complicated, using such techniques as delaying the return path while fast tap estimation is performed [Demytko and Mackechnie], and using phase-adaptive structures [Weinstein 1977(1)].

8.4.2 Terminal Echo Cancelers

Speech terminals such as loud-speaking telephones (LSTs) and audio teleconferencing facilities involve the use of amplifiers to drive loudspeakers. Acoustic isolation of the microphone and loudspeaker is difficult and instability can result. Echo cancelers can be used to reduce the effect of acoustic coupling, and proposals have been made to apply echo cancelers to both audio teleconferencing [Horna 1982, Ceruti and Pira] and to LSTs [South et al.]. The echo canceler requirements in both these applications are different from those of network echo cancelers, due partly to their position in the overall connection and partly to the fact that their function is different. The LST application is described next as an illustrative example of a terminal speech echo canceler.

Loud-speaking telephone. The use of adaptive filters in LSTs is illustrated in Figure 8.27. Acoustic coupling between the loudspeaker and the microphone and leakage across the hybrid circuit combine to create feedback resulting in instability. By using adaptive filters to model the acoustic coupling and the hybrid leakage, the amount of feedback can be reduced sufficiently to render the telephone stable when combined with some shallow voice switching.

The acoustic path can be considered as a multireflection medium with an impulse response duration of several hundreds of milliseconds. The hybrid leak-

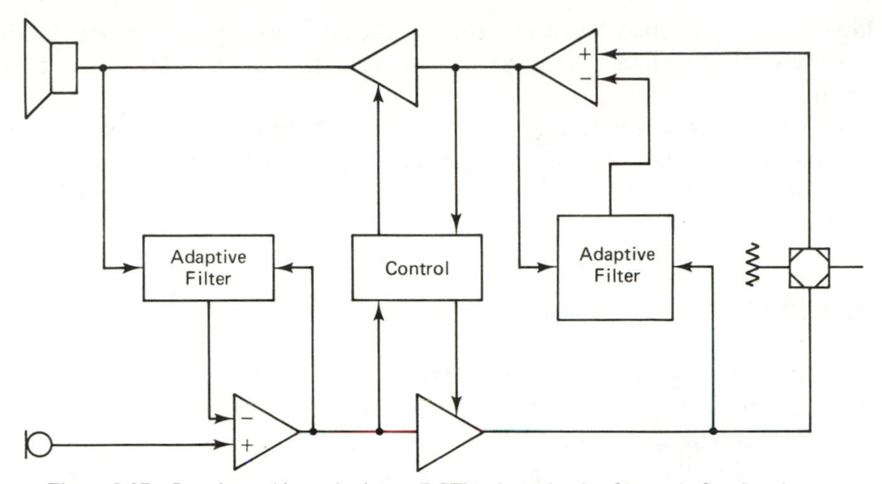

Figure 8.27 Loud-speaking telephone (LST) using adaptive filters. (After South et al., copyright © 1979 IEEE.)

age, as in the data transmission application, has two components: the direct leakage across the hybrid and the echoes, which will have the characteristic outlined in Section 8.2 (but with longer decays), obtained from the network. However, the aim of the echo cancelers is to produce stability, so that the longer delayed network echoes and acoustic reflections which are of smaller magnitude can be ignored. An experimental design [Hoppitt 1982] using adaptive filters with a limited time span was found to give good results. A foreground/background method [Ochiai et al.] of adaption was used to overcome the problem of double talking. The method, which is claimed to be more reliable than double-talk detection, uses two echo-path models, foreground and background. The foreground model is the best estimate of the echo path so far and is used to calculate the echo replica for cancellation. The background model is continuously updated by the stochastic gradient LMS algorithm until, by comparison of the error signals from the foreground and background models, it is deemed to be a better model, at which time it replaces the foreground model.

8.5 OTHER TELECOMMUNICATIONS-RELATED APPLICATIONS

The applications covered in this chapter have been concerned with the use of the telephony network for data and speech communications. There are other application areas which are also in the realm of telecommunications. For example, adaptive equalizers are used in modems operating over high frequency and microwave radio channels [Monsen 1980, Foschini and Salz, Dinn] where the channel characteristics can change rapidily (see Section 9.2.1). In the area of speech coding, where low-bit-rate digital speech may make more efficient use of expensive long-distance channels, adaptive filters are used in linear predictive coding [Markel and Gray 1976].

As the costs of processing come down, it is probable that adaptive filters will find increasing application. Similarly, as the time–bandwidth product of the filters increases, new applications will emerge. It is certain that adaptive filters have an assured future in telecommunications.

ACKNOWLEDGMENTS

The author thanks various colleagues at British Telecom Research Laboratories for their contribution to his understanding of adaptive filter applications in telecommunications. Thanks are also due to Dr. P. Cochrane for his encouragement and help in editing the manuscript. Acknowledgment is made to the Director of Research of British Telecom for permission to make use of information contained in this chapter.

9

OTHER ADAPTIVE FILTER APPLICATIONS

P. M. Grant

9.1 INTRODUCTION

This text has reviewed in Chapters 3 to 6 the theoretical development of several distinct techniques for the design of adaptive filters, studied their implementation in Chapter 7, and in Chapter 8 investigated their application as equalizers and echo cancelers in data communication systems. This chapter reviews other applications for these adaptive filters, which are predominantly in the areas of adaptive estimation [Assefi] and spectral estimation [Childers, Haykin 1979, Durrani, Kay and Marple, Papoulis].

These applications fall into two general categories which are dependent on the speed of the processor. Those that require real-time processing have been covered partly by the equalization and echo-cancellation examples in Chapters 7 and 8, and other examples, such as adaptive multipath compensation and speech processing, will be discussed further here. The other category is off-line, batch, or non-real-time processing, which is applied, for example, to seismic data processing [Wait] and biomedical data analysis [Linkens].

Seismic survey data are normally held on magnetic tape and the information is preprocessed to enhance the required information in a computer program that models the desired adaptive processor before applying image reconstruction algorithms. Here the preprocessor complexity influences the length of the program and hence the speed of computation, but in general, speed is of less importance in off-line processing. However, the analysis of multisensor data often requires sophisticated distributed array processors to achieve realistic processing rates for the sophisticated algorithms that are adopted.

One of the key advances over recent years has occurred in real-time adaptive filter applications, where rapid developments in technology are reducing the cost and increasing the sophistication of adaptive filters. Several designs of analog adaptive filters [Cowan 1978–1981(1), Denyer et al. 1983], digital adaptive transversal filters [Denyer et al. 1982, Duttweiler 1980], and adaptive lattice filters [Ahmed et al. 1981(2), Rutter et al. 1983(1) and (2)] now exist, which can process signals at sample rates exceeding 100 kHz (see Chapters 5, 7, and 8).

In addition, general-purpose signal processing circuits, such as programmable signal processors [Chapman] and fast, high-accuracy (32-bit) microprocessors [McDonough et al.], also make it possible to realize adaptive filters for the sample rates used in speech processing and data modems for local area networks. These developments are forcing systems designers to consider seriously increasing the use of adaptive processors to improve the performance of next-generation systems.

This final chapter first covers other applications of adaptive filters in adaptive estimation, over and above those reported in Chapters 7 and 8, then reviews their use in spectral estimation before briefly concluding with other application areas such as spatial nulling and bearing estimation in adaptive arrays.

9.2 ADAPTIVE ESTIMATION

Adaptive estimation [Assefi] covers the use of an adaptive processor or filter to measure and identify the key parameters that define a signal or are present in an unknown system. Chapter 1 explained how the adaptive filter could be employed to perform two basic functions: the modeling of the direct or inverse impulse response of a system [Figure 1.2(a) and (b)]. Both these techniques are now considered further here for adaptive estimation.

9.2.1 Inverse System Modeling

The adaptive equalization examples discussed in Chapter 8 for speech-band data modems and local network digital transmission provide two examples of the use of inverse system modeling adaptive estimation techniques. Similar processing techniques are also being applied to multipath compensation in high-frequency (HF), troposcatter, and digital microwave radio communication systems, in addition to spread-spectrum transmissions in urban digital radio, where severe multipath arises from reflections off buildings [Turin 1980].

When analog radio systems are subject to multipath, the transmission bandwidth is restricted to less than the reciprocal of the multipath delay by the dispersion in the propagation medium. In a frequency modulation system the dispersion also introduces degrading intermodulation products. Hence it was common to apply some signal, code, or diversity reception to alleviate the problem [Brayer]. However, with digital transmission, adaptive processing can be used to measure a multipath which is slowly fading with respect to the data rate [Anderson et al. 1979, Monsen 1974] and use it as a form of implicit diversity to improve the overall system performance. Thus the capacity of digital troposcatter and other systems are not restricted to the same extent by the multipath returns. In many systems two- or fourfold path diversity reception techniques are incorporated and the receiver combines the separate processed returns from each channel before making the decision as to the polarity of the received data. Combining can be implemented at RF, IF, or baseband.

Typical values for multipath decorrelation spreading factors in time and frequency for three systems are as follows. In the 2- to 30-MHz long-range (>100-mile) HF radio systems, the fading rate or Doppler spread is of the order of 0.1 Hz and the multipath delay spread is approximately 1 ms [Monsen 1980]. In the 0.4- to 5-GHz troposcatter links these factors become 1 Hz and 100 ns, respectively, while in the shorter-range (<50-mile) microwave line-of-sight radio systems they are typically 0.01 Hz and 10 ns. The multipath delay spreads of these three systems are related to the length of the propagation path.

One of the earliest processors proposed for overcoming multipath degradations in low-intersymbol-interference systems was the RAKE filter [Price and Green, Bitzer et al.], which bears a very strong resemblance to the FIR adaptive filter (Figure 7.22). It operates by raking together all the separate multipath components and adjusting their amplitude and phase in weighting multipliers to achieve a coherent summation. An alternative structure is the correlation filter [Sussman], which uses a single-stage IIR filter (Figure 9.1) where the previous information bits are used as a coherent reference to implement a matched filter [Turin 1976] to each data bit [Unkauf]. Maximum-likelihood-sequence estimation techniques [Forney, Ungerboeck 1974] have also been proposed for this application, but their complexity is much higher than that of simpler adaptive FIR filters such as the RAKE.

In line-of-sight microwave radio links the fades are slow with respect to the transmitted bit rates, with it taking from a few seconds to almost a minute for a multipath notch (Figure 9.2) to move across the 30- to 75-MHz-wide bandwidth of these equipments. The primary requirement in the new digital microwave radio systems is that they must accommodate the same number of 3.3-kHz bandwidth telephone channels as the earlier analog systems. This is providing the thrust behind the development of bandwidth-efficient modulation techniques such as 16- or 64-state quadrature amplitude modulation (QAM) and other approaches [Chamberlain et al.] as well as the use of dual (orthogonal) antenna polarization techniques for frequency reuse, which all give rise to increased intersymbol interference. When this is combined with the 20- to 30-dB deep fades experienced in these equipments [Rummler], effective equalization schemes are required before low-bit-error-rate communications can be established. Figure 9.2 shows the typical frequency and group delay response for a three-ray multipath fade with a notch at 160 MHz in a system with a 140-MHz intermediate frequency. Figure 9.3(a) shows a simulation of the degraded output eye diagrams for the 0.3 raised cosine channel when subject to the multipath fade shown in Figure 9.2.

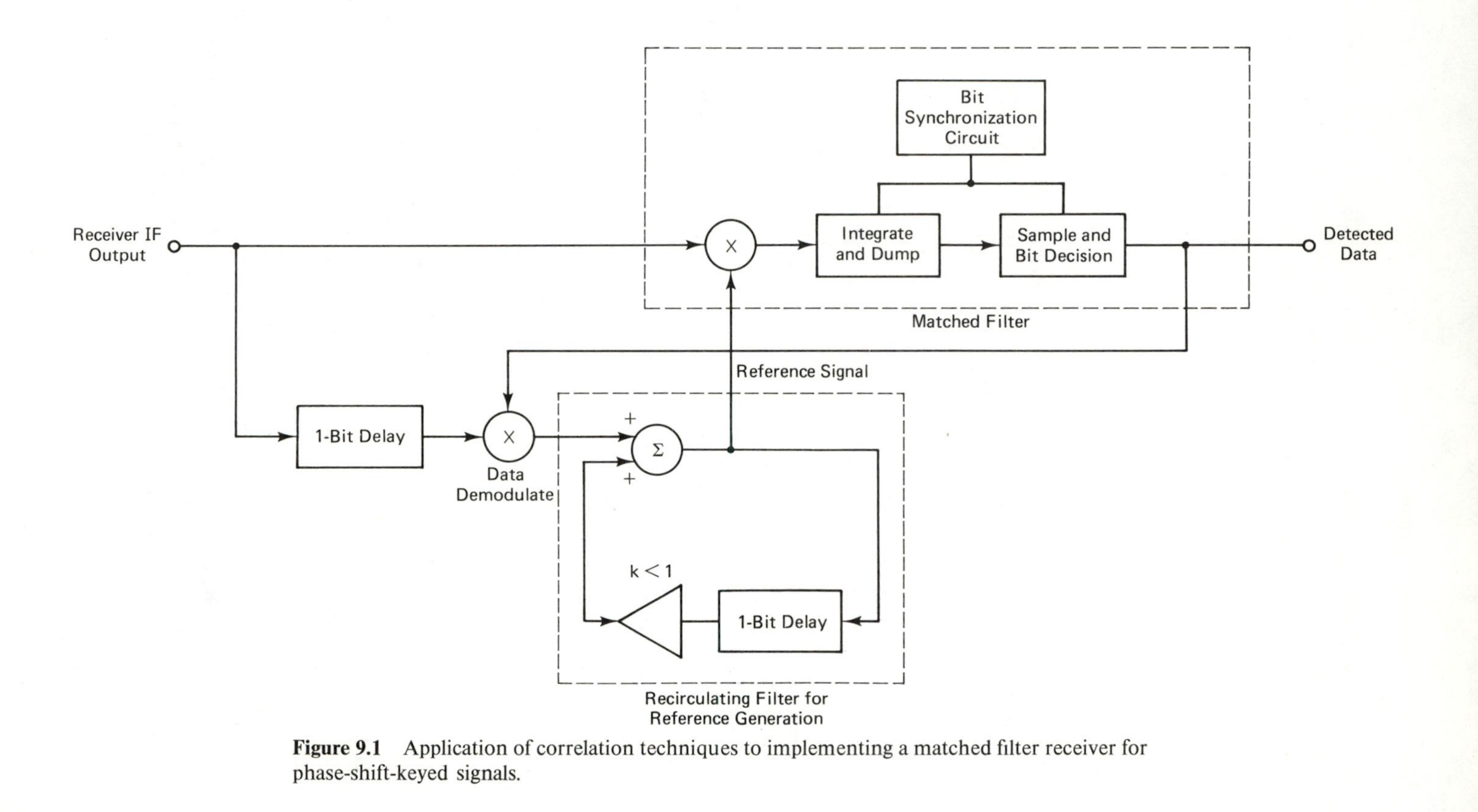

Figure 9.1 Application of correlation techniques to implementing a matched filter receiver for phase-shift-keyed signals.

Current approaches for handling multipath interference in microwave radio have used diversity reception techniques, but this only compensates for flat rather than frequency selective fades. Equalization has been based on frequency-domain approaches [Hartmann and Bynam], where the spectral response of the receiver is altered with adjustable filters to compensate for the frequency-selective distortion by attempting to level the energy in the received spectrum. This approach can be implemented with a derivative of the simple correlation filter [Sussman], but the receiver must be manually preset in advance [Murphy et al.] to handle either a minimum or a nonminimum phase fade [Rummler, Chap. 4]. There is thus thrust toward the FIR adaptive filter techniques, which provide programmable phase compensation to accommodate both fade types and achieve superior performance. Figure 9.3(b) shows the improvement in eye diagram that can be obtained by incorporating a five-tap adaptive FIR filter to compensate for the multipath fade causing the distortion of Figure 9.2.

Decision feedback adaptive equalizers (Figure 8.10) are also being incorporated in reduced-bandwidth quarternary-phase-shift-keyed 6- and 11-GHz digital systems. These equalizers [Dudek and Robinson 1980 and 1981], which operate at the European 140-megabit/s standard rate, have clearly demonstrated that they can compensate for the intersymbol interference which is introduced by the narrow-bandwidth (60 MHz) transmitter filters. These are incorporated to provide the reduced bandwidth capability of 0.8 times symbol rate and to achieve the consequent high transmission rate of 4 to 5 bits/s per hertz.

The use of a pure decision feedback technique without the linear equalizer of Figure 8.10 greatly simplifies the overall design of the equalizer, as the feedback data comprise a regenerated bit stream which requires only bipolar shift registers, compared to the multilevel registers in the adaptive FIR filters reported in Chapter 7. The tap-weight adjustment can be obtained with a variable current source [Dudek and Robinson 1981] which is summed with the incoming data in a common load at the input of the quantizer. Such a design can be realized with 6- to 8-ns settling delay, which permits operation at 74 megabaud, corresponding to the 140-megabit/s transmission rate (Figure 9.4). These equalizers also permit the use of orthogonal polarization in the antenna, which doubles the traffic to 280 megabits/s in the same bandwidth allocation and provides equivalent number of subscribers to the previous analog FDM/FM transmission systems. Error rate calculations for decision feedback equalizers under frequency-selective fades are provided by [Bogusch et al.].

The two-tap complex equalizers reported in Figure 9.4 have also shown that they can compensate for 15 dB of two-ray multipath fade with an echo delay of one-tenth of the symbol period, without the need for diversity reception. If combined with the normally applied height diversity reception techniques, much more severe multipath is tolerable. There are problems with pure DFB equalizers when a change of fade type occurs, as the symbol timing circuitry must always track the strongest signal. This problem is not so severe in adaptive FIR filters, where the change in fade type requires only a time reversal of the adaptive filter weights. Five

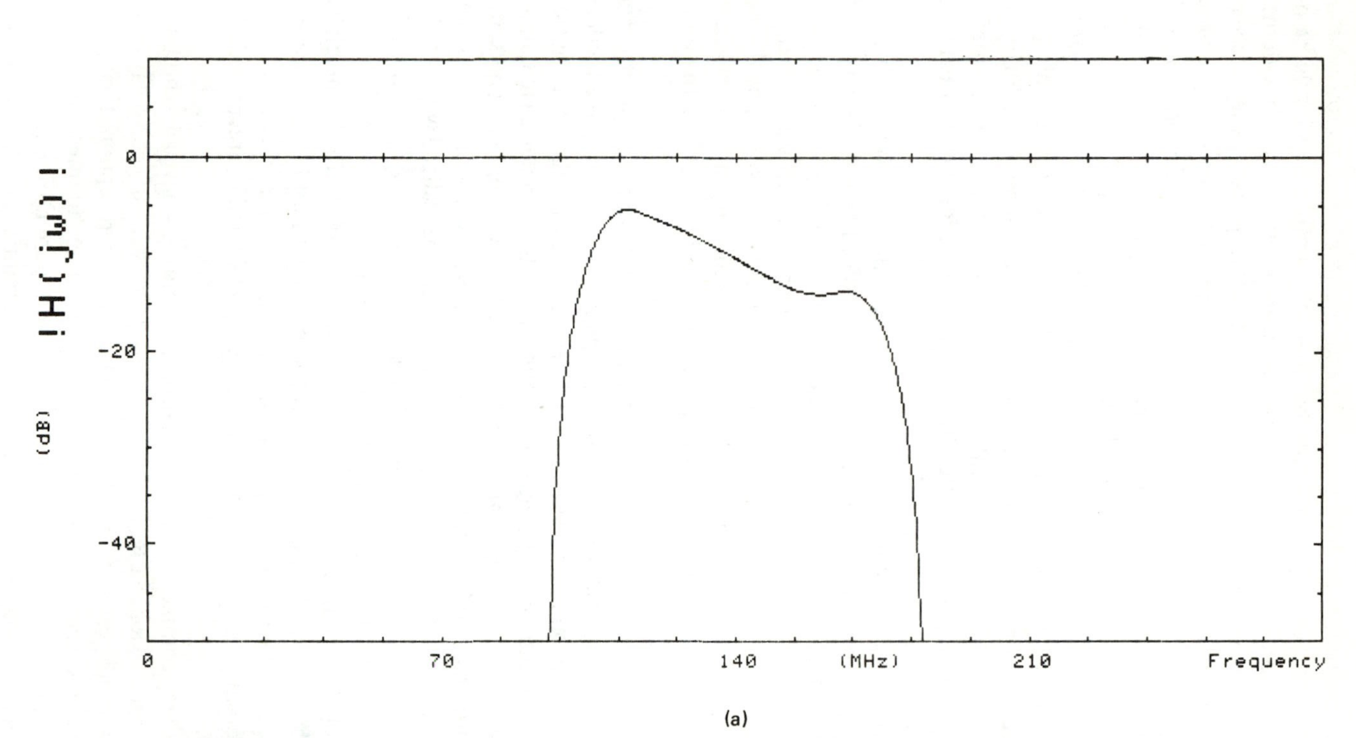

(a)

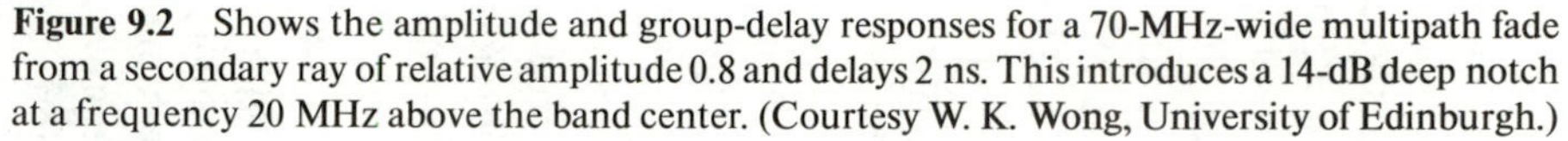

Figure 9.2 Shows the amplitude and group-delay responses for a 70-MHz-wide multipath fade from a secondary ray of relative amplitude 0.8 and delays 2 ns. This introduces a 14-dB deep notch at a frequency 20 MHz above the band center. (Courtesy W. K. Wong, University of Edinburgh.)

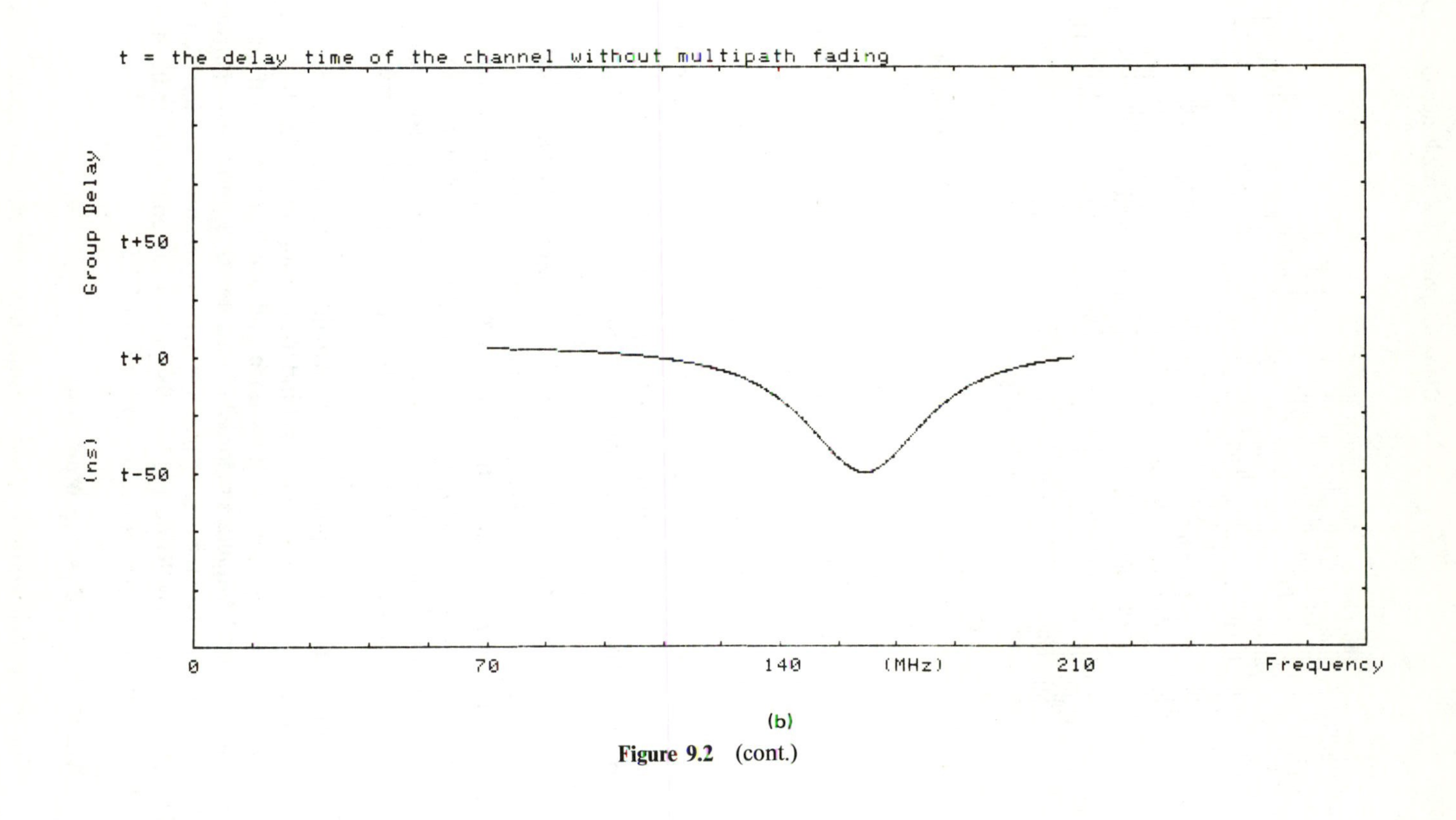

(b)

Figure 9.2 (cont.)

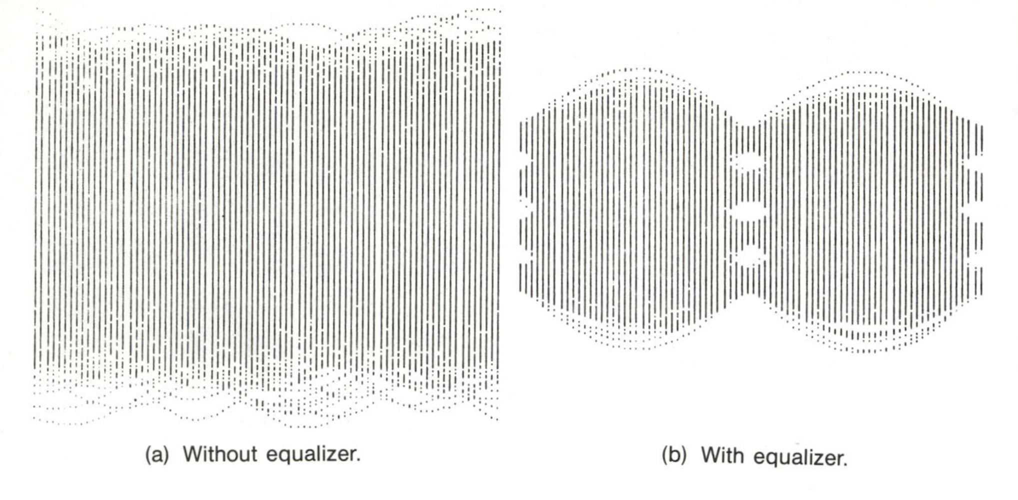

Figure 9.3 Simulations of receiver eye diagrams with and without equalization for the multipath fade of Figure 9.2. Equalization was simulated with a five-tap complex adaptive finite impulse response filter. (Courtesy of W. K. Wong, University of Edinburgh.)

to nine-tap equalizers [Amitay and Greenstein] appear to adequately compensate for expected fades.

The problems of multipath cancellation become slightly more difficult in troposcatter links as the data rate reduces to 2 megabits/s due to the increase in fading rate in these longer-transmission-path systems. These problems become even more severe in HF links. First, the lower 3- to 30-MHz carrier frequency reduces the available channel bandwidth to the 3-kHz separation of analog radio systems. Thus although the fade rate reduces in absolute terms compared to a troposcatter link, in this narrowband channel the relative fade rate, with respect to the channel baud rate, is increased by a factor of approximately 100. As a result the received signal can no longer be considered as stationary, and adaptive processor designs, other than the gradient search algorithms, must be applied to provide the faster tracking rates.

Short-duration block processing is normally used to overcome these difficulties, and this provides another thrust behind the development of fast-tracking adaptive filters such as the Kalman [Lawrence and Kaufman, Godard] and lattice [Griffiths 1977 and 1978] filters reported in Chapters 2, 3, and 5. Alternatively, the adaptive filter weights can be calculated by open-loop matrix inversion techniques [Kretschmer and Lewis, Kung and Hu 1983] which are computationally demanding but will be realizable with developments in VLSI.

9.2.2 Direct System Modeling

The other application of adaptive estimation techniques is in direct system modeling. Two areas where this technique has been widely applied are in noise

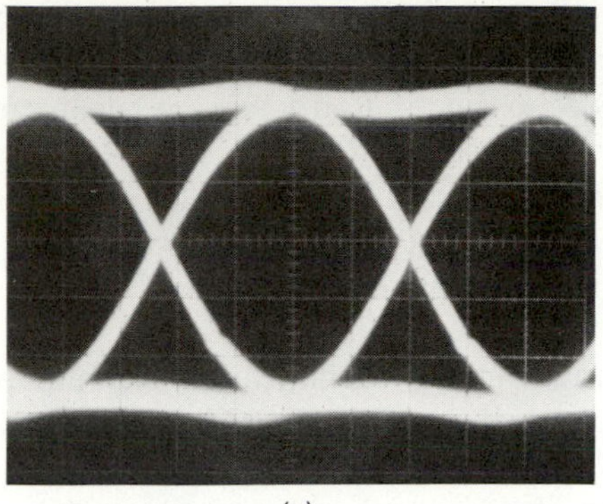
(a)

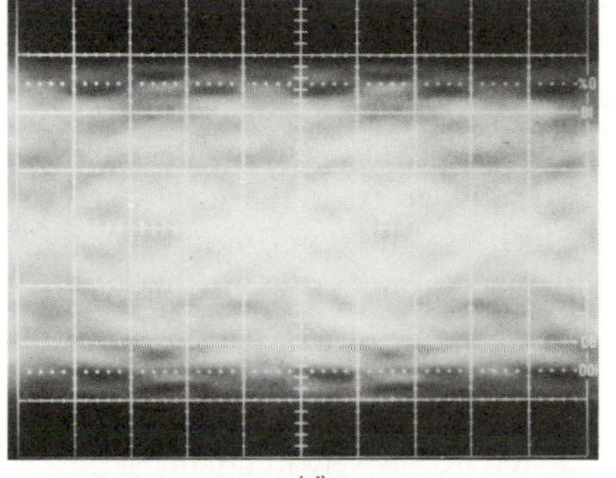
(d)

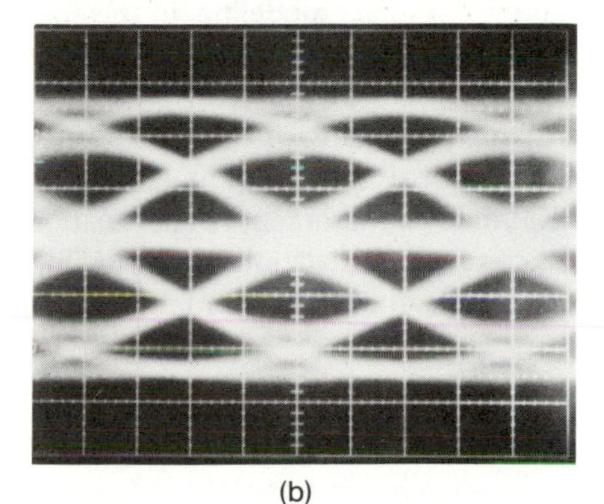
(b)

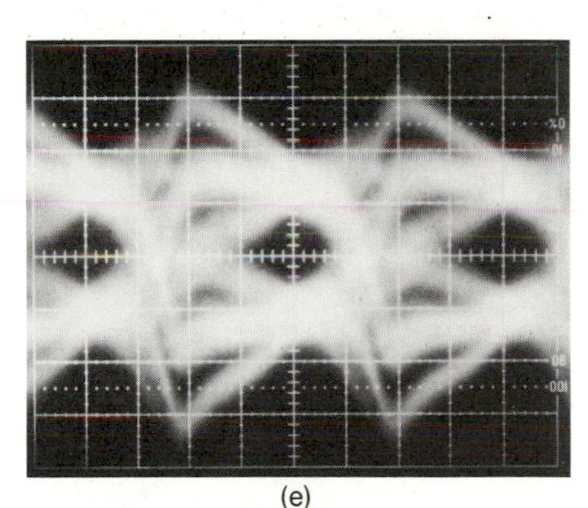
(e)

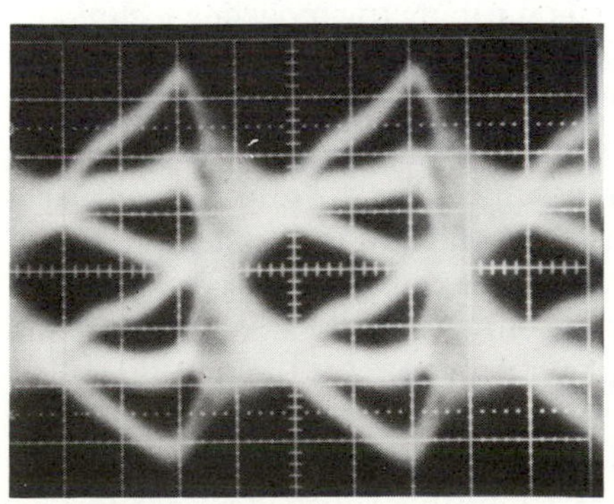
(c)

Figure 9.4 Eye diagrams show operation of practical decision feedback equalizer on microwave digital communications link. (a) Without intersymbol interference; (b) intersymbol interference introduced by the reduced bandwidth phase-shift keying; (c) as in (b) but with adaptive equalization in the receiver; (d) system-performance with multipath interference as well as reduced bandwidth transmission, without equalization; (e) as in (d) but with adaptive equalizer. (Courtesy of General Electric Company, Wembley, U.K.; after Dudek and Robinson 1981, copyright © 1981 IEE.)

cancellation [Widrow et al. 1975(2)] and in echo cancellation across the telephone line hybrid transformer, which have been discussed in Chapters 1 and 8. Other applications have been reported in a number of diverse uses, such as for echo cancellation in Teletext receivers [Voorman et al.] where integrated adaptive filters now exist, in electrocardiography, and in canceling mains interference [Widrow et al. 1975(2)].

A further application for system modeling or identification techniques occurs in adaptive control systems [Dorf, Goodwin and Sin 1984]. These have very close similarities to the adaptive filters described in this text. Adaptive control systems normally comprise an adjustable feedback loop which is employed to stabilize an overall system response in the presence of external disturbances and changes in the system parameters. Adaptive control systems [Dorf] are usually based on one of several algorithms, which include the model reference approach [Landau 1974 and 1979] and nonlinear and impulsive response techniques. Adaptive control is applied extensively in chemical process and power systems modeling, in the control of industrial machines and hydraulic drives, and in the prediction of flight paths of spacecraft and in their altitude control. Widrow and Stearns (1984) provide a brief review of the application of adaptive signal processing in control systems. Prediction filtering in navigation systems makes extensive use of the Kalman estimators described in Chapter 2. The use of Kalman estimators in this application area arises from the conflicting characteristics of very slow data rates coupled with fast tracking requirements and rapid data fluctuations due to violent maneuvering (e.g., in combat aircraft).

9.3 SPECTRAL ESTIMATION

9.3.1 Introduction

Spectral analysis or estimation [Haykin and Cadzow 1982] is another potential application area for adaptive filters. Spectral estimation can be divided into two categories: those approaches which are based on parametric modeling techniques [Haykin 1979, Durrani, Papoulis, Friedlander 1982(4)] and nonparametric approaches. Fourier analysis [Bracewell], which is widely applied in the characterization of broadband signals, is an example of the nonparametric approach.

In parametric spectral estimation it is assumed that the response to be analyzed comprised wideband noise which has been filtered in a system or passed through a transmission path. An adaptive filter is then used in the spectral estimation processor to model the inverse of the system or transmission response, and after convergence, further processing of the poles or zeros of the modeling filter provides information on the spectral properties of the input signal via the system model. These parametric modeling techniques, which are similar to the Kalman filter modeling techniques covered in Figure 2.3, further subdivide dependent on the different types of filter that are used to perform the spectral estimation. FIR all-zero filters provide inverses for autoregressive (AR) models, while fully recursive (all-pole) filters give inverses for moving-average (MA) models. Autoregressive moving-average (ARMA) models are obtained from pole–zero IIR filters. To minimize the computation and ensure stable convergence properties, it is important to know in advance the type of system model so that the optimum filter model can be employed. However, as a FIR filter can produce a similar response to a recursive filter, provided that a sufficient number of stages are employed, AR models may generally be used throughout, although this may result in less efficient use of hardware or computer processing time.

Figure 9.5 shows the effect of applying parametric spectral estimation to a signal comprising five separate sinusoids. The estimation is performed with a set of cascaded FIR prediction error filters. These progressively whiten the input signal and if, after convergence, the filter coefficients are Fourier transformed, they yield a power spectral density analyzer response. Figure 9.5 shows the transformed output from prediction error filters whose order increased progressively from 1 to 10. These simulations show that each second-order stage can control the positions of a zero pair to model the generating pole pair corresponding to one sinusoid. Thus a filter of order 10 is required for this five-sinusoid-input signal. Extending the order beyond 10 does not provide significant improvements. If the filter order is insufficiently large (e.g., a sixth-order design), it groups the sinusoids as pairs and converges to pole values corresponding to the three discrete frequencies that it can identify. As this test signal is a set of noise-free sinusoids, their respective frequencies are accurately identified in Figure 9.5, but accurate amplitude information is not provided.

The operation of the prediction error filter cascade is illustrated further in Figure 9.6, which shows how the output spectrum is whitened as one moves progressively down the cascade. In this example the filter is input with white noise convolved with the synthetic channel impulse response $0.28z^{-1} + z^{-2} + 0.28z^{-3}$ to provide the filtered input power spectral density shown at zero order in Figure 9.6.

The prediction error filter now fits appropriate zeros to model the input. This results in a progressive whitening of the spectrum as it emerges from each filter stage. Compared to Fourier analysis techniques, the adaptive parametric spectral estimation approach optimally identifies the locations of the input sinusoids, but it provides no detail in the regions in between. For a low signal-to-noise ratio at the input, the noise components in between the sinusoidal tones drive the filter from convergence, making this approach not as useful as Fourier techniques for determining the input spectral response. Thus it is generally accepted that nonparametric Fourier analysis techniques are superior for broadband analysis with low input signal-to-noise ratios (SNR). However, the resolution of this technique, which is proportional to the length of the observation window, is inferior. For time-varying spectra or short time series, such as pulsed radar returns, the observation window has to be restricted, reducing the resolution of the Fourier approach. In addition, leakage in the discrete Fourier transform (DFT) results in smearing of

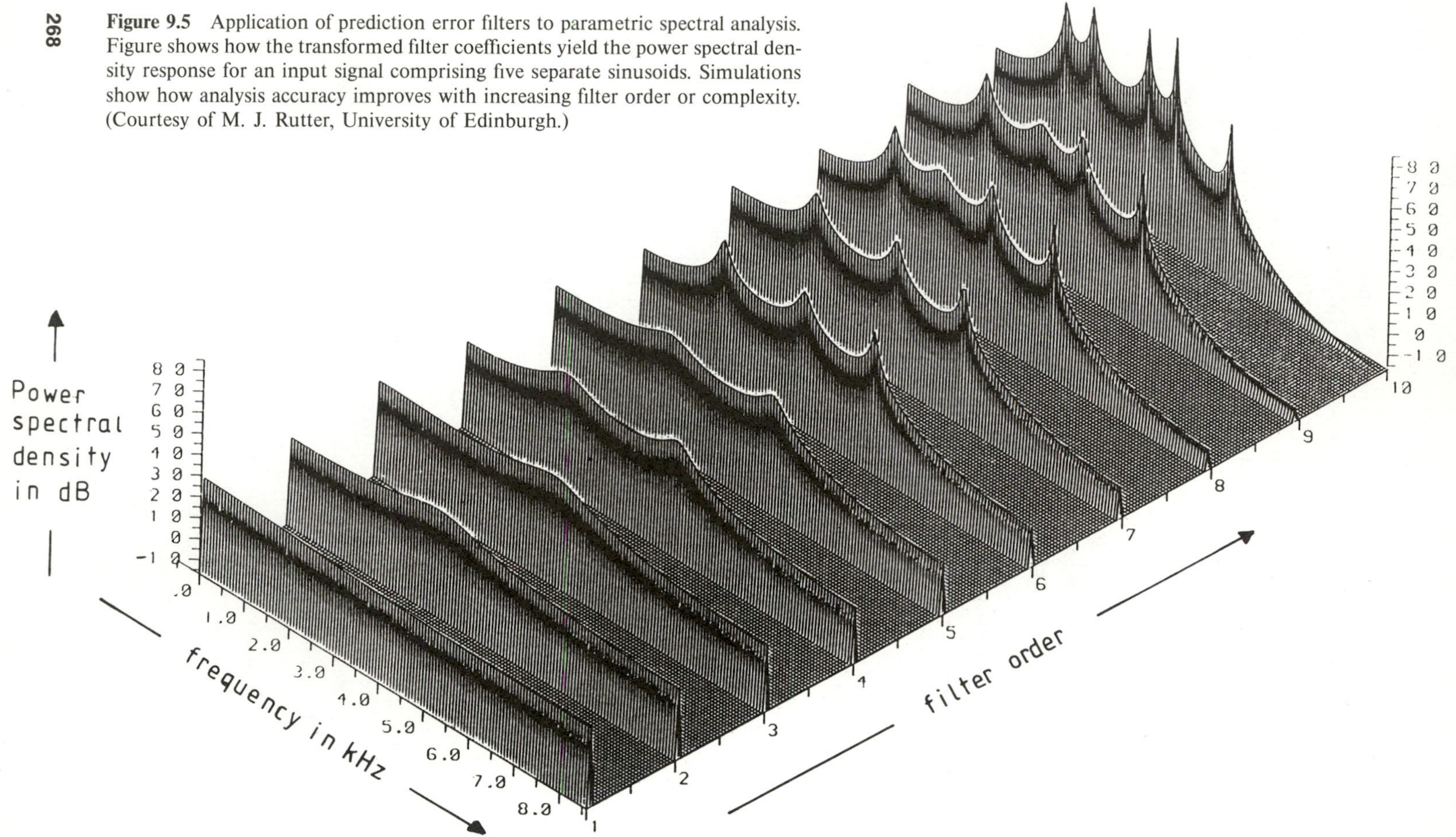

Figure 9.5 Application of prediction error filters to parametric spectral analysis. Figure shows how the transformed filter coefficients yield the power spectral density response for an input signal comprising five separate sinusoids. Simulations show how analysis accuracy improves with increasing filter order or complexity. (Courtesy of M. J. Rutter, University of Edinburgh.)

Figure 9.6 Shows how the output spectrum from a prediction error filter cascade becomes progressively whiter as the filter order increases. Input test signal is white noise convolved with the impulse response $0.28z^{-1} + z^{-2} + 0.28z^{-3}$. (Courtesy of M. J. Rutter, University of Edinburgh.)

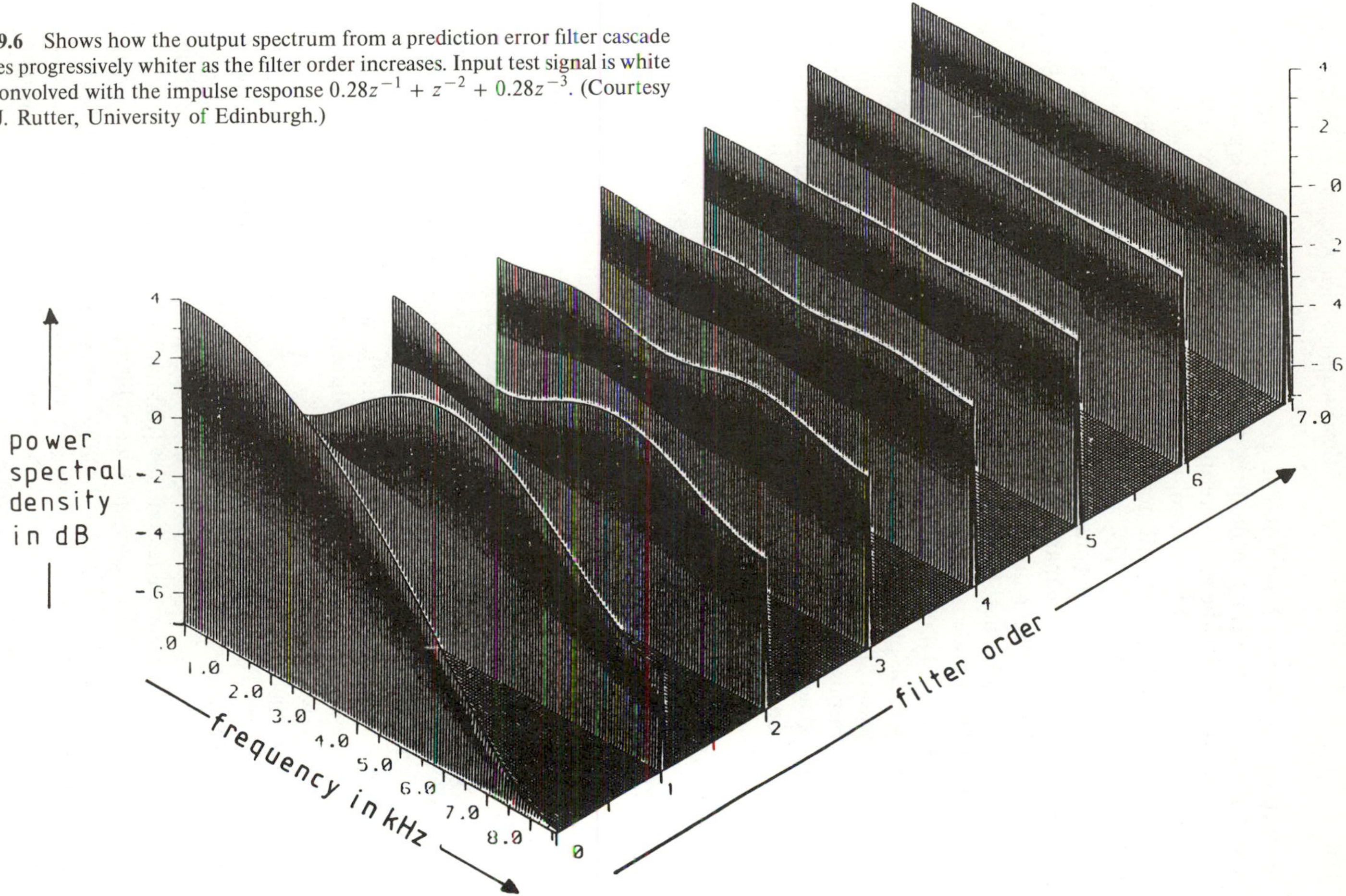

the spectral components [Brigham]. These deficiencies have spurred interest in the newer parametric modeling techniques, which obtain increased resolution by extrapolating values for the autocorrelation beyond known lags. In the DFT processor these are assumed to be zero, introducing the spectral leakage.

The maximum-entropy method (MEM) of spectrum analysis [Burg 1967 and 1975, Robinson, Haykin 1979], which is based on a FIR autoregressive model, has been shown to be a more general approach rather than a subset of AR spectral estimation techniques. The maximum-likelihood method (MLM) [Capon 1969], which is another AR technique, measures the power out of a set of narrowband filters. Unlike the DFT these filters can each have a different band shape and center frequency and they are adaptively set onto the frequencies of the signals that are present at the input. It thus produces a resolution that is superior to the DFT, but it is not quite as good as other AR methods (see Figure 9.12, which shows the results of similar processing for bearing measurement). MLM is used extensively in frequency wave-number analysis in seismic arrays. The key feature of these linear algorithm-based autoregressive techniques is that they perform well when only a few sinusoids are present, as they provide a data reduction capability, which is put to great use in applications such as speech analysis and synthesis as well as seismic processing.

9.3.2 Spectral Line Enhancement

One application of spectral estimation is in the identification of the presence of signals using enhancement techniques. An example of this is the recovery of a narrowband signal from wideband noise by adaptive line enhancement (ALE) [Zeidler et al.]. The ALE is implemented by connecting the received signal to a delay module before it enters the signal input of the adaptive filter, while the desired or training input is connected directly to the received signal. The delay is selected such that the narrowband signals, which we wish to enhance, are correlated between the signal and desired inputs while the wideband noise components are not correlated. This is similar to the results shown in Figure 7.6, which featured an adaptive self-tuning (notch) filter.

For ALE applications the adaptive filter can be either a FIR or a IIR design [Friedlander 1982(1)], depending on whether an autoregressive or ARMA processor is preferred. The ALE is a simplification of the MEM spectral analysis discussed previously, as MEM typically uses sophisticated processing algorithms [Burg 1975] or fast Kalman techniques [Kalman] which are computationally demanding. However, they rapidly provide values for the prediction error filter weights. In the simple ALE approach the prediction error filter is a simplification of this where the prediction error filter weights are derived from slower "adaptive" algorithms, such as the stochastic gradient search technique, which provide computational efficiency at the expense of a slower response.

Figure 9.7 shows simulated input and output waveforms for an ALE which is based on the frequency-domain adaptive processing techniques [Morgul et al.] dis-

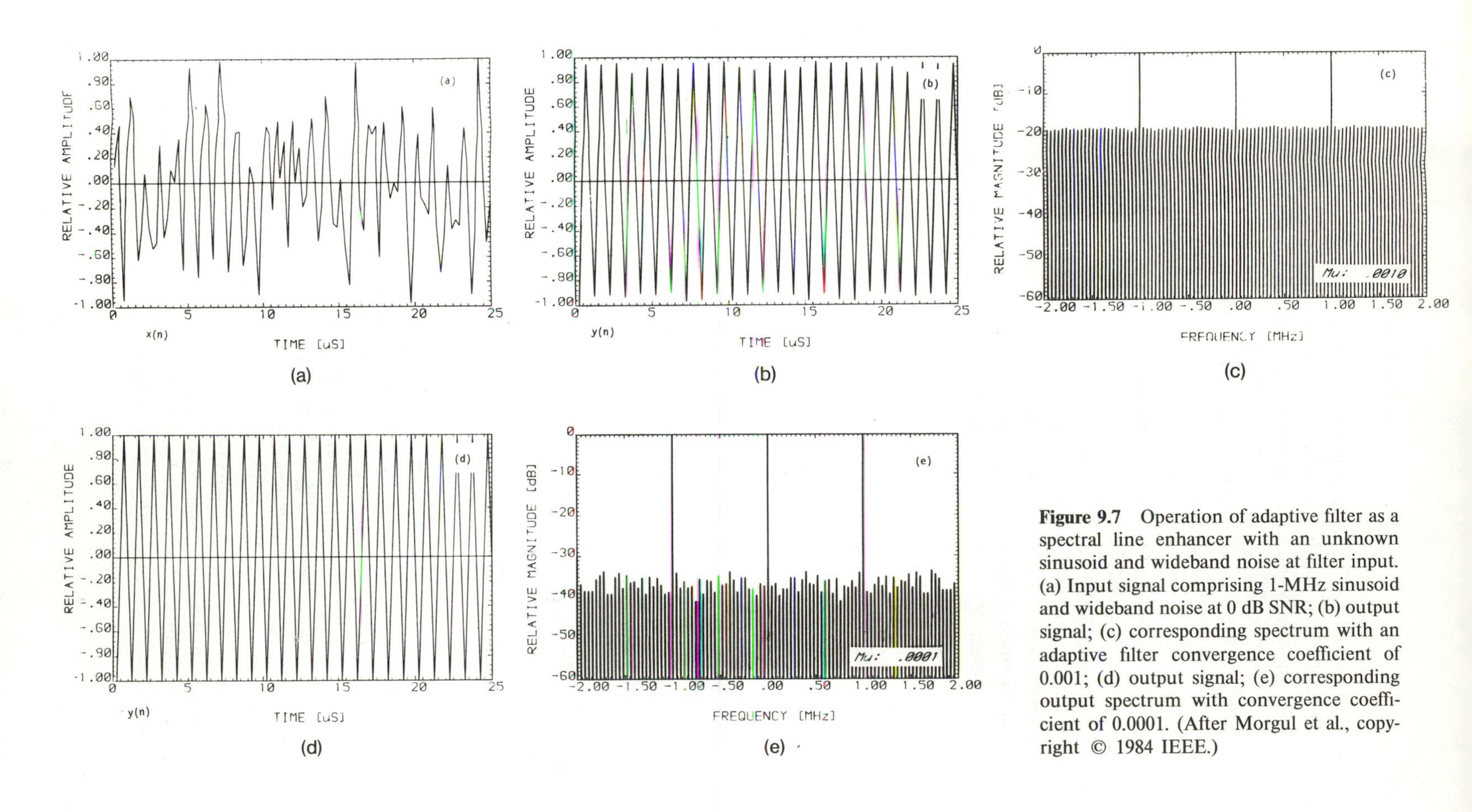

(a) (b) (c) (d) (e)

Figure 9.7 Operation of adaptive filter as a spectral line enhancer with an unknown sinusoid and wideband noise at filter input. (a) Input signal comprising 1-MHz sinusoid and wideband noise at 0 dB SNR; (b) output signal; (c) corresponding spectrum with an adaptive filter convergence coefficient of 0.001; (d) output signal; (e) corresponding output spectrum with convergence coefficient of 0.0001. (After Morgul et al., copyright © 1984 IEEE.)

cussed in Chapter 6. This uses an input signal comprising a 1-MHz sinusoid plus wideband noise at 0 dB SNR [Figure 9.7(a)]. The outputs [Figure 9.7(b) and (d)] clearly illustrate how the sinusoid is enhanced and show that the level of noise suppression [Figure 9.7(c) and (e)] is also dependent on the selected convergence coefficient μ. Examples of the application of IIR line enhancers based on optimal least-squares estimation techniques have also been reported [Friedlander 1982(1)].

For signals contaminated by white noise the FIR-based ALE performance is broadly equivalent in terms of resolution and SNR improvement to a DFT processor if the number of transform points equals the order of the FIR filter. However, as the ALE processes signals continuously whereas the DFT is a block processor, the DFT output must be averaged over several frames to integrate and obtain the same SNR improvement. The performance of the ALE is superior to the DFT approach when the input comprises either colored noise or a mix of strong and weak sinusoids. Under these conditions the weak signals are enhanced and the colored noise is suppressed by the ALE [Widrow and Stearns 1984]. A further possible advantage is the reduced computational load of the ALE.

9.3.3 Speech Processing

Spectral estimation techniques are also used in speech processing, particularly in vocoders [Blankenship], which exploit the redundancy in the speech waveform to achieve low-bit-rate ($<$2.4 kilobaud) transmission rates. Two major designs exist at present, the channel vocoder and the linear predictive coder (LPC). The channel vocoder transmits coarse spectral plus pitch information which is normally obtained with conventional analog or digital bandpass filtering or DFT techniques. Although the LPC design has been implemented by the autocorrelation method [Makhoul 1975], which uses adaptive transversal filters, a cascade of linear prediction error filters is the preferred approach, as it is less sensitive to coefficient inaccuracies. The prediction error filters remove from the signal the components that can be predicted [Markel and Gray 1973] from the previous history by modeling the vocal tract as an all-pole (AR) filter (Chapter 5).

In the LPC vocoder the analyzer and encoder normally process the signal in 30-ms frames and subsequently transmit the coarse spectral information via the filter coefficients. The residual error (noise output from the prediction error whitening filter) is not transmitted; instead, it is used to provide an estimate of the input power level which is sent along with the pitch information, and an indication as to whether the input is voiced or unvoiced (Figure 9.8). The latter can be ascertained by examining whether the first lag of the autocorrelation function of the signal lies above or below a certain threshold. If above the threshold, the value of the autocorrelation term provides the pitch information. The decoder and synthesizer apply the received filter coefficients to an AR synthesizing filter which is excited with impulses at the pitch frequency if voiced, or white noise if unvoiced. The excitation amplitude is controlled by the input power estimate information.

The adaptive lattice filter (Chapter 5) can be effectively applied to implement

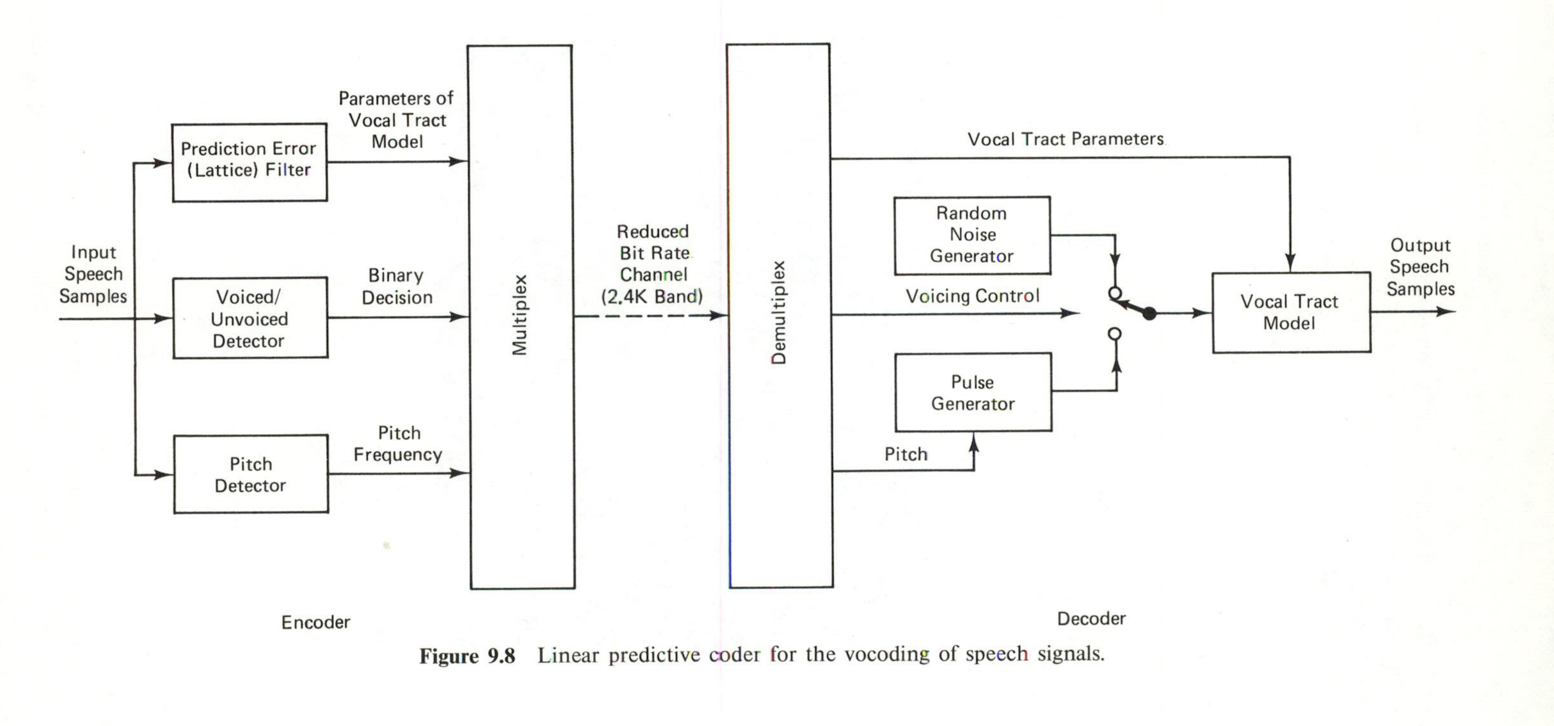

Figure 9.8 Linear predictive coder for the vocoding of speech signals.

an all-pole filter representation of the physical lossless acoustic tube model of the vocal tract (see Section 5.3.2 and Figure 5.6). This approach [Itakura and Saito 1970], which outputs a continuous representation of the filter prediction error coefficients, has received most emphasis to date because it is a regular structure that is amenable to implementation with digital LSI circuits [Ahmed et al. 1981(2)].

Further to the results shown in Figures 5.14 and 5.15, Figures 9.9 and 9.10 compare the performance of a conventional swept-frequency (nonparametric) spectrum analyzer with the autoregressive (parametric) estimator on a sample of male human speech. The vowel "ee" as in the word "feed" was sung into a tape recorder to maintain as far as possible a constant fundamental frequency. This was then played back into a commercially available spectrum analyzer. The parametric estimator simulation used to obtain the results of Figure 9.5 was applied separately for comparison purposes. The outputs are shown in Figures 9.9 and 9.10, respectively, where the individual parametric estimates, for up to 16 orders, have been interpolated to provide a fully filled display.

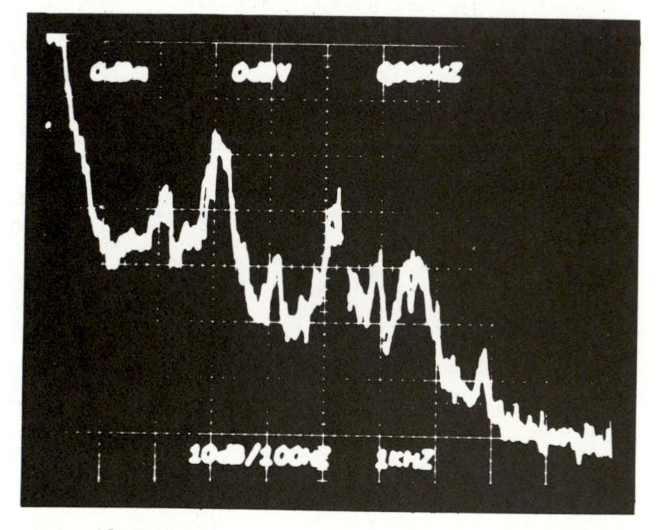

vert: 10 dB/div
horiz: 1 kHz/div

Figure 9.9 Conventional swept-frequency spectrum analyzer display of male voice saying "ee" as in the word "feed." (Courtesy of M. J. Rutter, University of Edinburgh.)

On a visual comparison one is impressed by the similarity of the results between the 12- to 16-order parametric estimator and the conventional analyzer display. Both show a similar overall shape exhibiting the 6-dB/octave reduction with frequency which is caused by the vocal tract response. However, the parametric estimator does not possess the fine detail of the swept-frequency response, due to the much shorter analysis time. Further investigation [Rutter 1983(2)] shows close agreement between approaches when estimating the overall spectral density, but

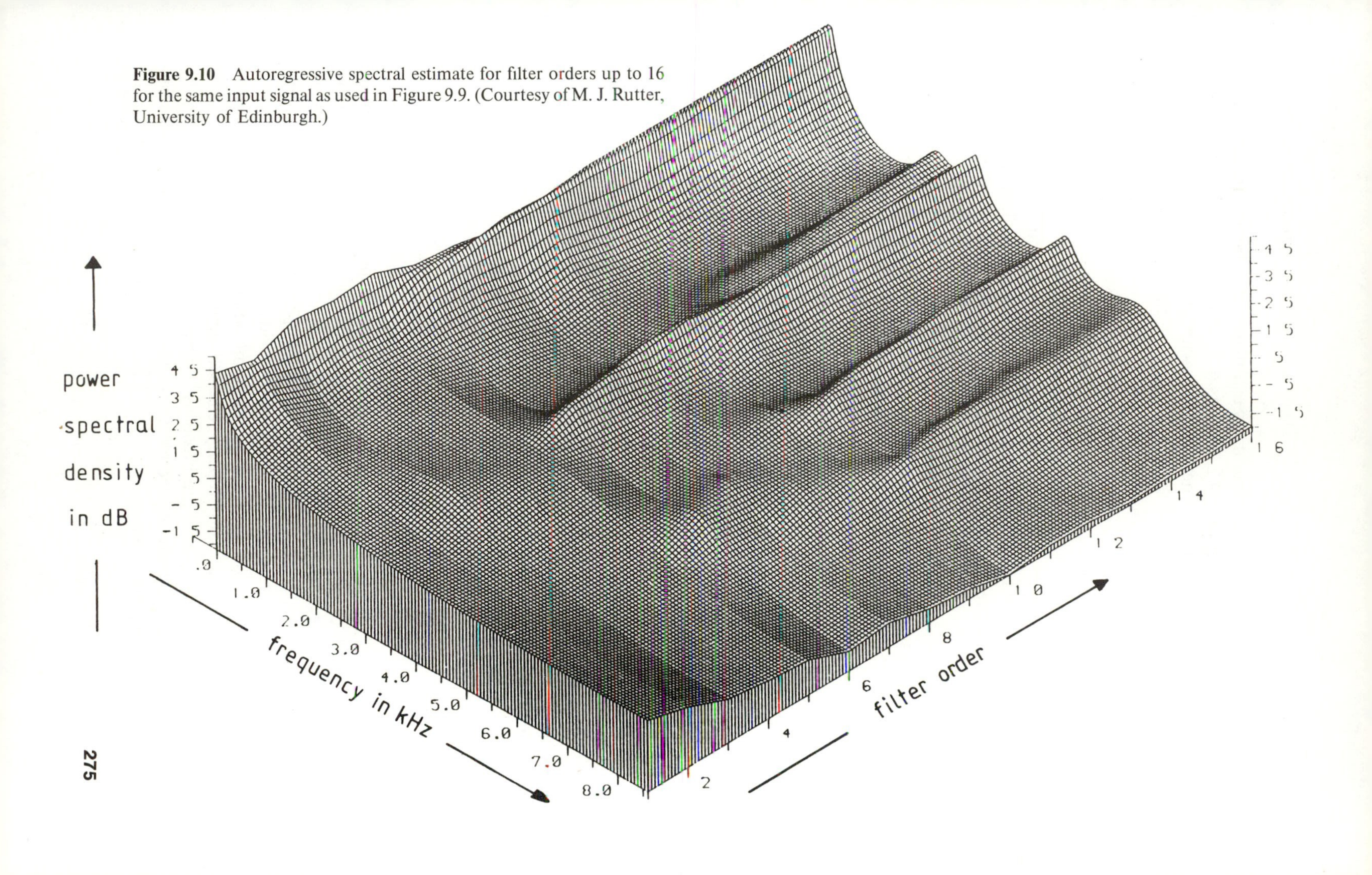

Figure 9.10 Autoregressive spectral estimate for filter orders up to 16 for the same input signal as used in Figure 9.9. (Courtesy of M. J. Rutter, University of Edinburgh.)

the parametric analyzer is less accurate in estimating the absolute frequencies of the peaks, and the minor peak at 4.2 kHz has not been detected. The test signal of Figure 9.5, which had a well-defined spectrum with sharp peaks, was modeled by accurately placed zeros close to the unit circle. In contrast, the speech waveform used here had a less sharply defined spectrum which resulted in less accurately placed zeros. Figures 9.9 and 9.10 give a broad comparison of the two approaches and show how the adaptive zero fitting in the simpler autoregressive estimator provides a sufficiently accurate spectral representation for synthetic and possibly also communications-quality speech transmission. Toll-quality transmission requires more accurate sampling techniques, such as pulse-code modulation.

Lattice adaptive prediction error filters offer a performance in terms of complexity and speed of operation which lies in between the sophisticated MEM and relatively simple ALE spectral estimation techniques. Key attractions of the lattice approach are again based on independent optimization of successive components plus the fact that there is a trade-off between convergence factor and residual error with filter length. With delays in the vocal tract of about 1 ms and typical speech sample rates of 8 to 10 kHz the number of lattice stages is normally in the range 8 to 12, with 10 being the number adopted in the integrated LPC vocoder standard, which transmits at a 2.4-kilobaud rate. Multichip microprocessor-based vocoders [Wasser and Peterson] are also available. In addition to these vocoder applications, digital lattice filters are used in several commercial speech synthesis systems [Franz and Wiggins].

9.4 ADAPTIVE ARRAY PROCESSING

Adaptive antennas [Monzingo and Miller, Hudson, Gabriel 1976(1) and (2), Taylor], which are applied to receiver designs to maximize a desired signal in the presence of interference, use processing techniques that are very similar to those of adaptive filters. They use the spatial separation between the antenna elements to provide a parallel set of signal samples rather than using the time-delayed or partly processed versions of a one-dimensional input signal. Early work on adaptive antennas [Van Atta] was concerned primarily with self-phasing systems which reradiated energy in the same direction as a received signal. These systems are less widely applied at present and the term "adaptive antenna" now refers almost exclusively to spatial nulling techniques which reduce the effect of unintentional cochannel interference or deliberate jamming.

These processors, which can automatically respond to an unknown interference environment in real time, have thus found widespread use in military radar, sonar, navigation, and communication equipment, where a high speed of adaption for fast nulling is an important property. Civilian applications such as VHF/UHF and satellite communications and broadcast systems are less demanding, as the convergence rate is usually of secondary consideration and hence less sophisticated adaptive algorithms can be employed. The interference bearing is normally fixed and the achievement of satisfactory cancellation at low cost is more important. Bearing estimation [Johnston and De Graff 1982(1), Johnston 1982(2)] is another function that can be performed with antennas, either by directly analyzing the received signals or by transforming the weight values from a converged adaptive array to determine the bearing of the source.

Adaptive antenna nulling uses a very similar processor configuration to the adaptive filter. In its simplest form it has two inputs from the main and auxiliary antennas. If these are microwave signals, they are usually down-converted to IF and the auxiliary channel is multiplied in a complex weighting network before summation with main channel (Figure 9.11). For interference cancellation the combined output is fed back and cross-correlated with the signal in the auxiliary channel to derive the adaptive weights to minimize the signal which is present in both channels.

This approach is commonly used with high-gain S- and X-band microwave reflector-based antennas to cancel out jamming which enters through the main antenna sidelobes. In this coherent sidelobe canceller (CSLC) [Howells] (Figure 9.8) the difference in gain between the main beam of the directional antenna and the wider coverage or omnidirectional auxiliary antenna result in desired signals (such as low-level radar return echos) being received only in the main channel. In a pulsed radar, where these echo returns are present only for short periods in the overall scan interval, their average energy is low and hence they do not need to be subtracted out, as they introduce only minor degradations in the processor. Jamming through the main antenna sidelobes enters both channels at approximately the same signal level and is automatically nulled out in this self-steering processor. Additional auxiliary channels must be added to the scheme of Figure 9.11 to handle multiple jammer scenarios. Lubell and Rebhun provide an example of the application of CSLC techniques in a satellite communications receiver.

Alternative adaptive antennas, based on fully phased arrays with an adaptive control loop on each of the individual elements, are employed in communications, radar, and navigation applications, but these normally require a separate sample of the desired signal to be subtracted from the combined output before it is fed back as the error signal. In theory such an N-element adaptive array can handle up to $N - 1$ interfering sources.

9.4.1 Bearing Estimation

Antenna arrays, often with only two elements, can be used for bearing discrimination [Barton]. Bearing information is translated in the antenna into a difference in timing of the received signals due to the path differences to the different elements. Thus processing of the received signals to extract the time difference of arrival (TDOA) provides information relating to the target bearing. This technique has been used for flow measurement of dangerous liquids in pipes by sensing the disturbances due to flow with a pair of transducers attached to the pipes. Cross-correlation of the two signals [Beck] yields the delay information, but this requires

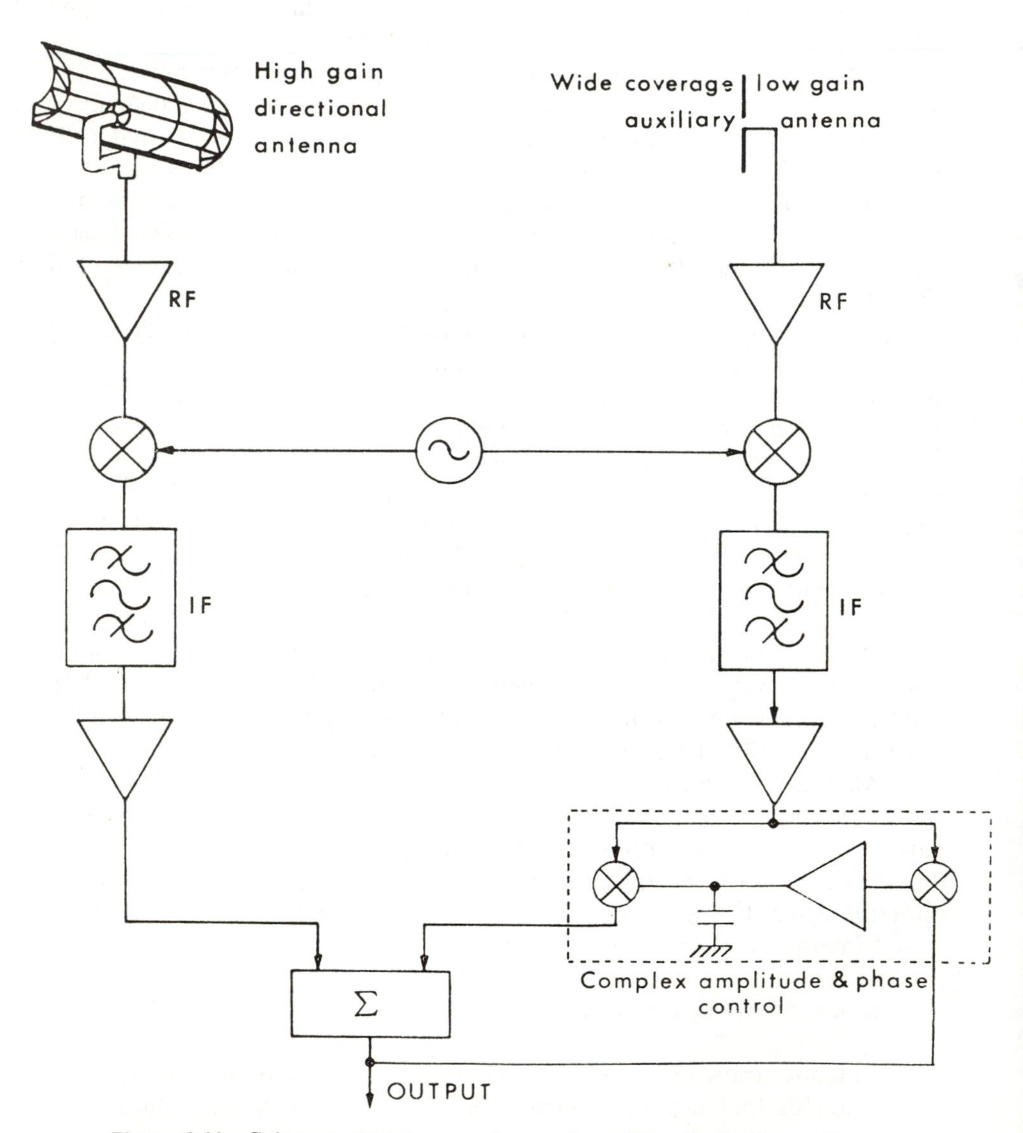

Figure 9.11 Coherent sidelobe canceler configuration of adaptive antenna. (Copyright © 1981 EW Communications Inc.)

some knowledge of the statistics of the signals and the transducer separation must be such that there is not a significant loss in correlation between the two received signals. Other applications of this correlation-based transit-time measurement system are in gas chromatography [Godfrey and Devenish], biomedical engineering [Tompkins et al.], and in sonar [Adams et al. 1980(2)].

Adaptive time-difference-of-arrival estimation techniques, which can be applied to enhance the output of these passive sensing systems, have the advantage that no prior signal information is required. One approach is based on an extension of the earlier correlation method where adaptive spectral whitening techniques are added prior to the cross-correlation processor. Such preprocessing of narrowband input signals prior to cross-correlation of the residuals reduces the incidence of multiple peaks at the output [Friedlander 1982(1)]. Most of the adaptive spectral whitening techniques reported in this text are applicable to this processor.

These concepts have also been extended to multielement arrays for accurate bearing estimation of received signals. Simple spatial Fourier analysis of the radiation field by sampling the received signals from an array gives a bearing resolution capability which is inversely proportional to the aperture of the array. Johnston and De Graaf have shown that superior resolution can be obtained (Figure 9.12) by applying the spectral estimation techniques reported earlier to find a solution to this constrained optimization problem. Adaptive spectral estimation techniques employing MLM, MEM, AR, and ARMA parameter modeling as well as the eigenvector decomposition approach [Schmidt] have all been applied to the passive sonar problem. The outputs from the filter model provide the time-delay information, which is converted into bearing estimates.

The inclusion of an adaptive processor into the array permits the beamformer to utilize the null rather than the mainlobe information for bearing measurement. As the nulls are generally much sharper, this provides a consequent resolution improvement or super-resolution capability from the deployment of adaptive techniques. Widrow and Stearns (1984) provide an analysis of the improvements gained from the use of maximum-likelihood adaptive processors [Capon et al. 1967], and Figure 9.12 [Johnston 1982(2)] shows how the resolution of a nonadaptive Fourier beamformer in (a) and (d) compares with the maximum-likelihood processor (b) and (e), for both a single source on boresight and a pair of sources at $\pm 5°$ angular separation from boresight. For this 10-element array at a 10-dB signal-to-noise ratio, the resolution improvement due to MLM processing is calculated as 18 times [Widrow and Stearns 1984]. Figure 9.12 also shows that if a linear predictive autoregressive adaptive processor is deployed, there is further resolution improvement, as shown in (c) and (f).

The resolution improvement from these adaptive spectral estimation techniques is dependent on a satisfactory received signal-to-noise ratio, and it typically requires positive values to achieve the required improvement. The future extension of this work to other application areas, such as radar and communications, can now be anticipated to produce further potential applications for adaptive spectral estimation techniques.

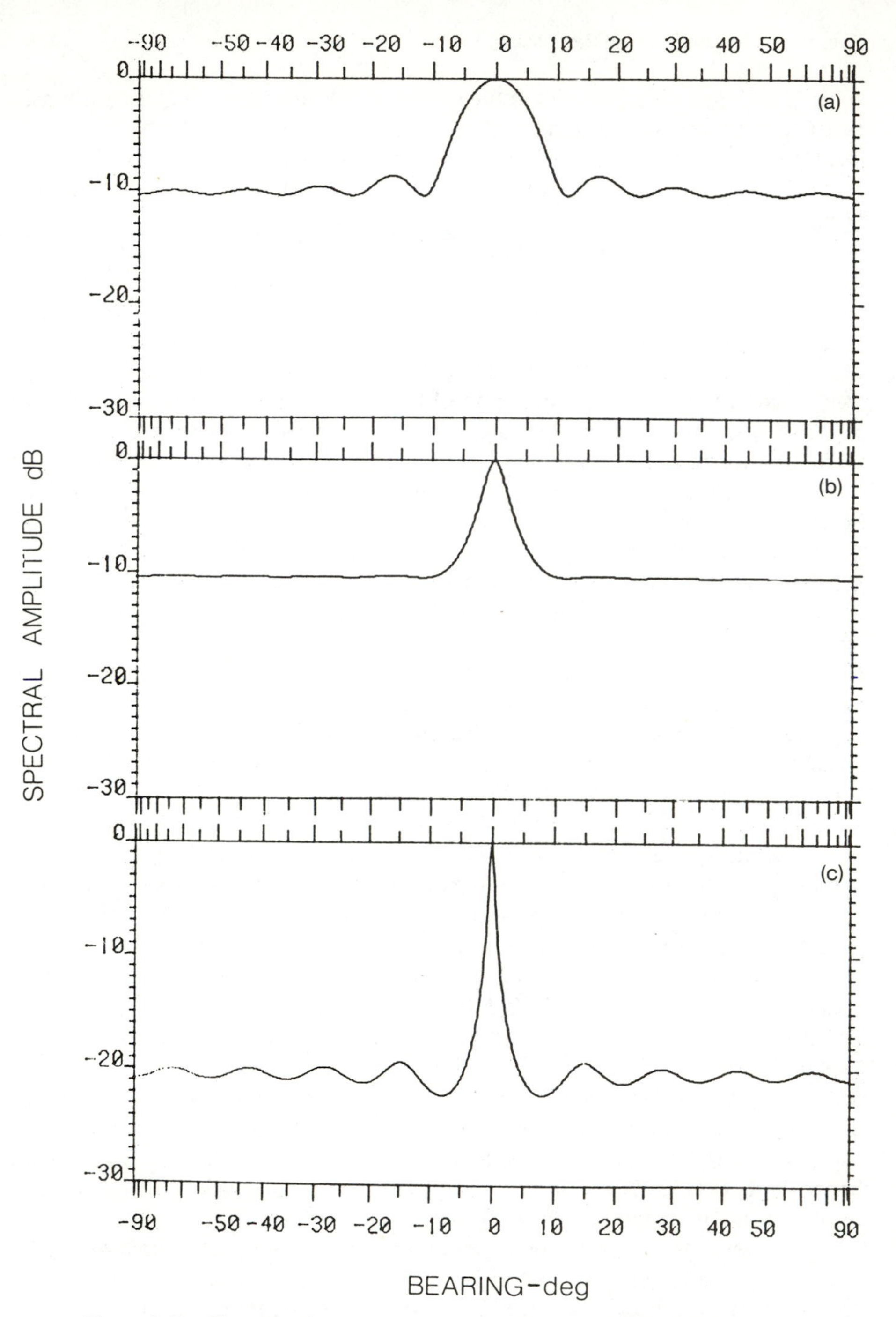

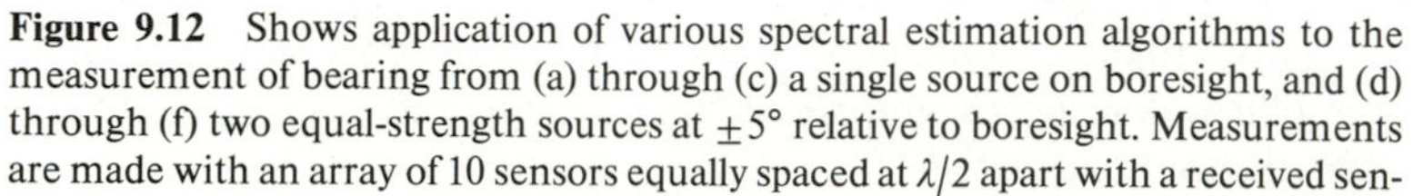
Figure 9.12 Shows application of various spectral estimation algorithms to the measurement of bearing from (a) through (c) a single source on boresight, and (d) through (f) two equal-strength sources at $\pm 5°$ relative to boresight. Measurements are made with an array of 10 sensors equally spaced at $\lambda/2$ apart with a received sen-

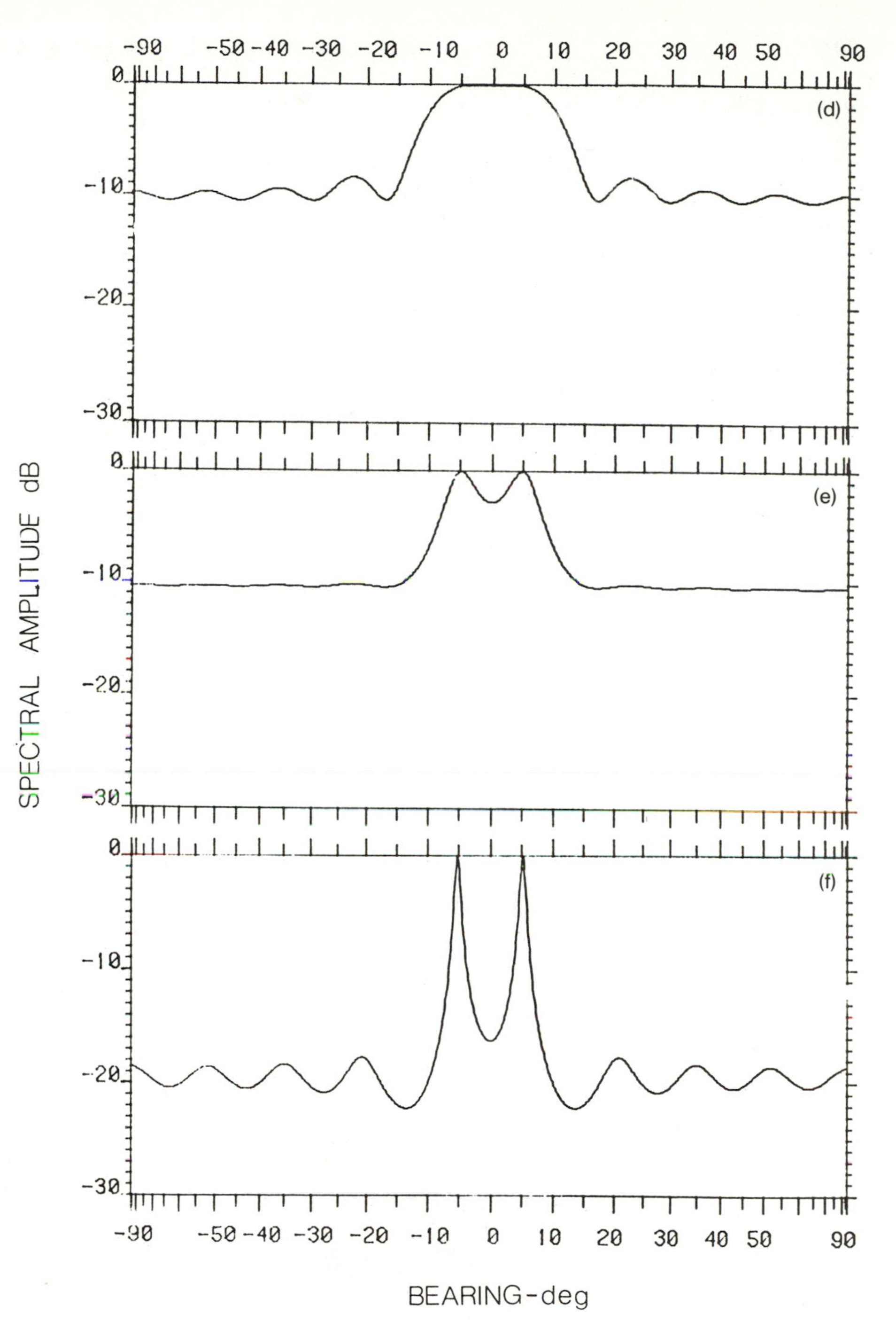

Figure 9.12 (cont.)
sor signal-to-noise ratio of 10 dB. (a) and (d) Fourier transform techniques; (b) and (e) maximum-likelihood method; (c) and (f) linear predictive estimates. (Courtesy of Rice University, Houston, TX; after Johnston 1982(2), copyright © 1982 IEEE.)

9.5 SUMMARY

This final chapter has summarized some of the many other applications in which adaptive filters may be usefully employed. At this time it can be confidently stated that these filters have matured to the point where their signal processing capabilities are well understood and documented in the technical literature. It now seems likely that their application will increase progressively as new integrated custom processors and faster general-purpose microprocessors are developed with increasing speed and accuracy which make system designers more aware of the potential performance improvements that are offered by adaptive techniques.

REFERENCES

ADAMS, P. F., "Speech-Band Data Modems," *Electronics and Power*, Vol. 26, No. 9, pp. 733–736, September 1980(1).

ADAMS, W. B., KUHN, J. P., and WHYLAND, W. P., "Correlator Compensation Requirements for Passive Time Delay Estimation with Moving Source or Receivers," *IEEE Trans.*, Vol. ASSP-28, No. 2, pp. 158–168, April 1980(2).

ADAMS, P. F., GLEN, P. J., and WOOLHOUSE, S., "Echo Cancellation Applied to Wal2 Digital Transmission in the Local Network," 2nd International Conference on Telecommunication Transmission–Into the Digital Era, *IEE Conference Publication 193*, pp. 201–204, 1981(1).

ADAMS, P. F., HARBRIDGE, J. R., and MACMILLAN, R. H., "A MOS Integrated Circuit for Digital Filtering and Level Detection," *IEEE Journal of Solid State Circuits*, Vol. SC-16, No. 3, pp. 183–190, June 1981(2).

ADAMS, P. F., and ELLIOTT, S. R., "Field Test Result of a 9600 Bit/s Echo Canceling Duplex Modem," *British Telecom Technology Executive Memorandum R9/012/83*, 1983.

ADAMS, P. F., COX, S. A., and GLEN, P. J., "Long Reach Duplex Transmission Systems for IDSN Access," *British Telecom Technology Journal*, Vol. 2, No. 2, pp. 35–42, April 1984.

AGAZZI, O., HODGES, D. A., and MESSERSCHMITT, D. G., "Large-Scale Integration of Hybrid Method Digital Subscriber Loops," *IEEE Trans.*, Vol. COM-30, No. 9, pp. 2095–2108, September 1982(1).

AGAZZI, O., MESSERSCHMITT, D. G., and HODGES, D. A., "Non-linear Echo Cancellation of Data Signals," *IEEE Trans.*, Vol. COM-30, No. 1, pp. 2421–2433, November 1982(2).

AHMED, H. M., "VLSI Architectures for Real-Time Signal Processing," Ph.D. dissertation, Department of Electrical Engineering, Stanford University, Stanford, CA, 1981(1).

AHMED, H. M., MORF, M., LEE, D. T., and ANG, P. H., "A Ladder Filter Speech Analysis Chip Set Utilizing Co-ordinate Rotation Arithmetic," *Proceedings IEEE International Symposium on Circuits and Systems (ISCAS)*, Vol. 3, pp. 737–741, April 1981(2).

AHMED, H. M., MORF, M., LEE, D. T., and ANG, P. H., "A VLSI Speech Analysis Chip Set Based on Square Root Normalized Ladder Forms," *Proceedings IEEE International Conference Acoustics, Speech, and Signal Processing (ICASSP)*, pp. 648–653, 1981(3).

AHMED, H. M., DELOSME, J. M., and MORF, M., "Highly Concurrent Computing Structures for Digital Signal Processing and Matrix Arithmetic," *IEEE Computer Magazine*, pp. 65–82, February 1982.

AMELIO, G. F., TOMPSETT, M. F., and SMITH, G. E., "Experimental Verification of the Charge-Coupled Device Concept," *Bell System Technical Journal*, Vol. 49, No. 4, pp. 593–600, April 1970.

AMITAY, N., and GREENSTEIN, L. J., "Multipath Outage Performance of Digital Radio Receivers Using Finite-Tap Adaptive Equalizers," *IEEE Trans.*, Vol. COM-32, No. 5, pp. 597–608, May 1984.

ANDERSON, B. D. O., "A Simplified Viewpoint of Hyperstability," *IEEE Trans.*, Vol. AC-13, No. 3, pp. 292–294, June 1968.

ANDERSON, C. W., BARBER, S. G., and PATEL, R. N., "The Effect of Selective Fading on Digital Radio," *IEEE Trans.*, Vol. COM-27, No. 12, pp. 1870–1875, December 1979.

APPLEBAUM, S. P., "Adaptive Arrays," *IEEE Trans.* Vol. AP-24, No. 5, pp. 585–598, September 1976.

ARNDT, H., CUDDY, D., and EGAN, E., "Canceling Noise in Aviator's Ears," *Telesis*, p. 265, June 1978.

ASSEFI, T., *Stochastic Processes and Estimation Theory with Applications*, Wiley, New York, 1979.

BAILEY, W. H., BUSS, D. D., HITE, L. R., and WHATLEY, M. W., "Radar Video Processing Using the Chirp-Z Transform," *Proceedings 2nd International Conference on CCDs, CCD-75*, San Diego, CA, pp. 283–290, 1975.

BARNDORFF-NIELSEN, O., and SCHOU, G., "On the Parametization of Autoregressive Models by Partial Autocorrelation," *Journal of Multivariate Analysis*, Vol. 3, pp. 408–419, 1973.

BARTON, P., "Direction Finding Using an Adaptive Null Tracker," *Proceedings IEE*, Vol. 130, Pts. F and H, No. 1, pp. 77–83, February 1983.

BEAUCHAMP, K. G., *Walsh Functions and Their Applications*, Academic Press, London, 1975.

BECK, M. S., "Correlators in Instrumentation: Crosscorrelation Flow Measurement," *Journal of Physics*, Pt. E, Vol. 14, pp. 7–19, January 1981.

BELFIORE, C. A., and PARK, J. H., "Decision Feedback Equalization," *Proceedings IEEE*, Vol. 67, No. 8, pp. 1143–1156, August 1979.

BELLANGER, M. G., and DAGUET, J. L., "TDM-FDM Transmultiplexer: Digital Polyphase and FFT," *IEEE Trans.*, Vol. COM-22, No. 9, pp. 1199–1204, September 1974.

BERSHAD, N. J., and FEINTUCH, P. L., "Analysis of the Frequency Domain Adaptive Filter," *Proceedings IEEE*, Vol. 67, No. 12, pp. 1658–1659, December 1979.

BEYNON, J. D. E., and LAMB, D. R., *Charge-Coupled Devices and Their Applications*, McGraw-Hill, London, 1980.

BIERMAN, G. J., *Factorization Methods for Discrete Sequential Estimation*, Academic Press, New York, 1977.

BITMEAD, R. P., and ANDERSON, B. D. O., "Adaptive Frequency Sampling Filters," *IEEE Trans.*, Vol. CAS-28, No. 6, pp. 524–534, June 1981.

BITZER, D. R., CHESLER, D. A., IVERS, R., and STEIN, S., "A RAKE System for Tropospheric Scatter," *IEEE Trans.*, Vol. COM-14, No. 6, pp. 499–506, August 1966.

BLANKENSHIP, P. E., "A Review of Narrowband Speech Processing Techniques," International Specialist Seminar on Case Studies in Advanced Signal Processing, *IEE Conference Publication 180*, pp. 108–118, September 1979.

BLUM, J., "Multidimensional Stochastic Approximation Methods," *Annals of Mathematical Statistics*, Vol. 25, pp. 737–766, 1954.

BODE, H. W., and SHANNON, C. E., "A Simplified Derivation of Linear Least Square Smoothing and Prediction Theory," *Proceedings IRE*, Vol. 38, No. 4, pp. 417–425, April 1950.

BOGUSCH, R. L., GUIGLIANO, F. W., and KNEPP, D. L., "Frequency-Selective Scintillation Effects and Decision Feedback Equalization in High Data-Rate Satellite Links," *Proceedings IEEE*, Vol. 71, No. 6, pp. 754–767, June 1983.

BOULTER, R. A., "A 60 kbit/s Data Modem for Use over Physical Pairs," *Proceedings IEEE International Seminar on Digital Communications*, Zurich, pp. H3(1)–H3(6), 1974.

BOWERS, J. E., KINO, G. S., BEHAR, D., and OLAISEN, H., "Adaptive Deconvolution Using a SAW Storage Correlator," *IEEE Trans.*, Vol. MTT-29, No. 5, pp. 491–498, May 1981.

BOYLE, W. S., and SMITH, G. E., "Charge-Coupled Semiconductor Devices," *Bell System Technical Journal*, Vol. 49, No. 4, pp. 587–593, April 1970.

BOZIC, S. M., *Digital and Kalman Filtering*, Edward Arnold, London, 1979.

BRACEWELL, R. N., *The Fourier Transform and Its Application*, McGraw-Hill, New York, 1965.

BRAYER, K., *Data Communications via Fading Channels*, IEEE Press Reprint Volume, New York, 1975.

BRIGHAM, E. O., *The Fast Fourier Transform*, Prentice-Hall, Englewood Cliffs, NJ, 1974.

BROWNLIE, J. D., JACKETS, A. E., GUNBY, D. M., and TROUSE, D. R., "Custom Designed Integrated Circuits for Data Modems," IEE Conference on the Impact of High Speed and VLSI Technology on Communications Systems, *IEE Conference Publication 230*, pp. 31–35, 1983.

BURG, J. P., "Maximum Entropy Spectral Analysis," 37th Annual International Meeting, Society of Exploratory Geophysicists, Oklahoma City, OK, October 31, 1967.

BURG, J. P., "Maximum Entropy Spectrum Analysis," Ph.D. dissertation, Department of Geophysics, Stanford University, Stanford, CA, 1975.

BUTLER, P., and CANTONI, A., "Non-interactive Automatic Equalization," *IEEE Trans.*, Vol. COM-23, No. 6, pp. 621–633, June 1975.

CAPON, J., GREENFIELD, R. J., AND KOLKER, R. J., "Multidimensional Maximum-Likelihood Processing of a Large Aperture Seismic Array," *Proceedings IEEE*, Vol. 55, No. 2, pp. 192–211, February 1967.

CAPON, J., "High-Resolution Frequency-Wave Number Spectrum Analysis," *Proceedings IEEE*, Vol. 57, No. 8, pp. 1408–1418, August 1969.

CERUTI, R., AND PIRA, F., "Application of Echo Canceling Techniques to Audioconferencing," *CSELT Rapporti Tecnici*, Vol. 10, No. 3, pp. 167–172, June 1982.

CHAMBERLAIN, J. K., CLAYTON, F. M., and COLLINS, P. V., "Reduced Bandwidth Quaternary Phase-Shift-Keyed (RBQPSK)–An Evolutionary Approach to Bandwidth Efficiency Digital Microwave Transmission," *Proceedings International Seminar on Digital Communications*, Zurich, Paper A3, 1980.

CHANG, R. W., "A New Equalizer Structure for Fast Start-Up Digital Communications," *Bell System Technical Journal*, Vol. 50, No. 6, pp. 1969–2014, July/August 1971.

CHAPMAN, R. C. (ed.), Digital Signal Processor, Special Issue, *Bell System Technical Journal,* Vol. 60, No. 7, Part 2, September 1981.

CHILDERS, D. G., *Modern Spectrum Analysis*, IEEE Press, New York, 1978.

CHU, P. L., and MESSERSCHMITT, D. G., "Zero Sensitivity Analysis of the Digital Lattice Filter," *Proceedings IEEE International Conference Acoustics, Speech, and Signal Processing (ICASSP)*, pp. 89–93, April 1980.

CHU, P. L., and MESSERSCHMITT, D. G., "Zero Sensitivity Properties of the Digital Lattice Filter," *IEEE Trans.*, Vol. ASSP-31, No. 3, pp. 685–706, June 1983.

CLAASEN, T. A. C. M., and MECKLENBRAUKER, W. F. G., "Comparison of the Convergence of Two Algorithms for Adaptive FIR Digital Filters," *IEEE Trans.*, Vol. CAS-28, No. 6, pp. 510–518, June 1981.

CLARK, G. A., MITRA, S. K., and PARKER, S. R., "Block Implementation of Adaptive Digital Filters," *IEEE Trans.*, Vol. CAS-28, No. 6, pp. 584–592, June 1981.

CLARK, G. A., PARKER, S. R., and MITRA, S. K., "A Unified Approach to Time- and Frequency-Domain Realization of FIR Adaptive Digital Filters," *IEEE Trans.*, Vol. ASSP-31, No. 5, pp. 1073–1083, October 1983.

COKER, M., private communication with E. R. Ferrara.

COOLEY, J. W., et al., "The Fast Fourier Transform Algorithm: Programming Considerations in the Calculation of Sine, Cosine and Laplace Transforms," *Journal of Sound and Vibration*, Vol. 12, pp. 315–337, July 1970. Also pp. 271–293 in L. R. Rabiner and C. M. Rader (eds.), *Digital Signal Processing*, IEEE Press, New York, 1972.

COPELAND, G. C., "Transmultiplexers Used as Adaptive Frequency Sampling Filters," *Proceedings IEEE International Conference Acoustics, Speech, and Signal Processing (ICASSP)*, pp. 319–322, May 1982.

CORL, D., "A CTD Adaptive Inverse Filter," *Electronics Letters*, Vol. 14, No. 3, pp. 60–62, February 2, 1978.

Cowan, C. F. N., Mavor, J., and Arthur, J. W., "Implementation of a 64-Point Adaptive Filter Using an Analogue CCD Programmable Filter," *Electronics Letters*, Vol. 14, No. 17, pp. 568–569, August 17, 1978.

Cowan, C. F. N., Mavor, J., Arthur, J. W., and Denyer, P. B., "An Evaluation of Analogue and Digital Adaptive Filter Realizations," International Specialist Seminar on Case Studies in Advanced Signal Processing, *IEE Conference Proceedings 180*, pp. 178–183, September 1979.

Cowan, C. F. N., and Mavor, J., "Miniature CCD-Based Analog Adaptive Filters," *Proceedings IEEE International Conference Acoustics, Speech, and Signal Processing (ICASSP)*, Denver, CO, pp. 474–478, 1980.

Cowan, C. F. N., Arthur, J. W., Mavor, J., and Denyer, P. B., "CCD-Based Adaptive Filters: Realization and Analysis," *IEEE Trans.*, Vol. ASSP-29, No. 2, pp. 220–229, April 1981(1).

Cowan, C. F. N., and Mavor, J., "New Digital Adaptive Filter Implementation Using Distributed-Arithmetic Techniques," *Proceedings IEE*, Vol. 128, Pt. F, No. 4, pp. 225–230, August 1981(2).

Cowan, C. F. N., Smith, S. G., and Elliott, J. H., "A Digital Adaptive Filter Using a Memory-Accumulator Architecture: Theory and Realization," *IEEE Trans.*, Vol. ASSP-31, No. 3, pp. 541–549, June 1983.

Cowan, C. F. N., and Adams, P. F., "Nonlinear System Modeling: Concept and Application," *Proceedings IEEE International Conference Acoustics, Speech, and Signal Processing (ICASSP)*, Paper 45.8, 1984.

David, R. A., "IIR Adaptive Algorithms Based on Gradient Search Techniques," Ph.D. dissertation, Department of Electrical Engineering, Stanford University, Stanford, CA, August 1981.

Demytko, N., and Mackechnie, L. K., "A High Speed Digital Adaptive Echo Canceler," *Australian Telecommunication Review*, Vol. 7, No. 1, pp. 20–28, 1973.

Dentino, M., McCool, J., and Widrow, B., "Adaptive Filtering in the Frequency Domain," *Proceedings IEEE*, Vol. 66, No. 12, pp. 1658–1659, December 1978.

Denyer, P. B., and Mavor, J., "Design of CCD Delay Lines with Floating Gate Taps," *IEE Journal of Solid-State and Electron Devices*, Vol. 1, No. 3, pp. 121–129, July 1977.

Denyer, P. B., Mavor, J., and Arthur, J. W., "Miniature Programmable Transversal Filter Using CCD/MOS Technology," *Proceedings IEEE*, Vol. 67, No. 1, pp. 42–50, January 1979(1).

Denyer, P. B., and Mavor, J., "256-Point Programmable Transversal Filter," *Proceedings 5th International Conference on CCDs, CCD-79*, Centre for Industrial Consultancy and Liaison, University of Edinburgh, pp. 253–254, 1979(2).

Denyer, P. B., Renshaw, D., and Bergmann, N., "A Silicon Compiler for VLSI Signal Processors," *Proceedings European Solid State Circuits Conference (ESSCIRC)*, pp. 215–218, 1982.

Denyer, P. B., Cowan, C. F. N., Mavor, J., Clayton, C. B., and Pennock, J. L., "Monolithic Adaptive Filter," *IEEE Journal of Solid State Circuits*, Vol. SC-18, No. 3, pp. 291–296, June 1983.

Derusso, P. M., Ray, R. J., and Close, C. M., *State Variables for Engineers*, Wiley, New York, 1965.

DICKINSON, B. W., "Estimations of Partial Correlation Matrices Using Cholesky Decomposition," *IEEE Trans.*, Vol. AC-24, No. 2, pp. 302–305. April 1979(1).

DICKINSON, B., and TURNER, J., "Reflection Coefficient Estimation Using Cholesky Decomposition," *IEEE Trans.*, Vol. ASSP-27, No. 2, pp. 146–149, April 1979(2).

DINN, N. F., "Digital Radio: Its Time Has Come," *IEEE Communications Society Magazine*, Vol. 18, No. 6, pp. 6–12, November 1980.

DORF, R. C., *Modern Control Systems*, Addison-Wesley, Reading, MA, 1980.

DUDEK, M. T., and ROBINSON, J. M., "A Decision Feedback Equalizer and Novel Carrier Recovery Circuit for Digital Radio Relay System Operating at up to 5 bit/Hz," *Proceedings IEEE International Communications Conference (ICC)*, Paper 41.5, 1980.

DUDEK, M. T., and ROBINSON, J. M., "A New Adaptive Circuit for Spectrally Efficient Digital Microwave-Radio-Relay Systems," *Electronics and Power*, Vol. 27, No. 5, pp. 397–400, May 1981.

DUERDOTH, W. T., "Development of Integrated Digital Telecommunications Networks," *Proceedings IEE*, Vol. 121, No. 6, pp. 450–456, June 1974.

DUFFY, F. P., and THATCHER, T. W., "Analog Transmission Performance on the Switched Telecommunications Network," *Bell System Technical Journal*, Vol. 50, No. 4, pp. 1311–1347, April 1971.

DURRANI, T. S. (ed.), Spectrum Analysis, Special Issue, *Proceedings IEE*, Vol. 130, Pt. F, No. 3, pp. 193–287, April 1983.

DUTTWEILER, D. L., "A Twelve Channel Digital Echo Canceler," *IEEE Trans.*, Vol. COM-26, No. 5, pp. 647–653, May 1978.

DUTTWEILER, D. L., and CHEN, Y. S., "A Single-Chip VLSI Echo Canceler," *Bell System Technical Journal*, Vol. 59, No. 2, pp. 149–160, February 1980.

DUTTWEILER, D. L., "Adaptive Filter Performance with Nonlinearities in the Correlation Multiplier," *IEEE Trans.*, Vol. ASSP-30, No. 4, pp. 578–586, August 1982.

EHRENBARD, C. A., and TOMPSETT, M. F., "A Baud-Rate Line Interface for Two-Wire High-Speed Digital Subscriber Loops," *Proceedings IEEE Globecom Conference* (formerly National Telecommunications Conference), pp. 931–935, 1982.

FALCONER, D. D., "Jointly Adaptive Equalization and Carrier Recovery in Two-Dimensional Digital Communication Systems," *Bell System Technical Journal*, Vol. 55, No. 3, pp. 317–334, March 1976(1).

FALCONER, D. D., "Application of Passband Decision Feedback Equalization in Two-Dimensional Data Communication Systems," *IEEE Trans.*, Vol. COM-24, No. 10, pp. 1159–1166, October 1976(2).

FALCONER, D. D., and LJUNG, L., "Application of Fast Kalman Estimation to Adaptive Equalization," *IEEE Trans.*, Vol. COM-26, No. 10, pp. 1439–1446, October 1978.

FALCONER, D. D., and MUELLER, K. H., "Adaptive Echo Cancellation/AGC Structures for Two-Wire, Full-Duplex Data Transmission," *Bell System Technical Journal,* Vol. 58, No. 7, pp. 1593–1616, September 1979.

FEINTUCH, P. L., "An Adaptive Recursive LMS Filter," *Proceedings IEEE*, Vol. 64, No. 11, pp. 1622–1624, November 1976.

FERRARA, E. R., "Fast Implementation of LMS Adaptive Filters," *IEEE Trans.*, Vol. ASSP-28, No. 4, pp. 474–475, August 1980.

FERRARA, E. R., and WIDROW, B., "Multichannel Adaptive Filtering for Signal Enhancement," *IEEE Trans.*, Vol. CAS-28, No. 6, pp. 606–610, June 1981.

FLANAGAN, J. L., *Speech Analysis Synthesis and Perception*, Springer-Verlag, New York, 1972.

FLETCHER, R., *Practical Methods of Optimization*, Vol. 1: *Unconstrained Optimization*, Wiley, New York, 1980.

FOLTS, H. C. (ed.), *McGraw-Hill's Compilation of Data Communications Standards*, 2nd ed., McGraw-Hill, New York, 1982.

FORNEY, G. D., JR., "Maximum-Likelihood Sequence Estimation of Digital Sequences in the Presence of Intersymbol Interference," *IEEE Trans.*, Vol. IT-18, No. 3, pp. 363–377, May 1972.

FOSCHINI, G. J., and SALZ, J., "Digital Communications over Fading Radio Channels," *Bell System Technical Journal*, Vol. 62, No. 2, Pt. 1, pp. 429–456, February 1983.

FRANZ, G. A., and WIGGINS, R. H., "Design Case History: Speak and Spell Learns to Talk," *IEEE Spectrum*, Vol. 19, No. 2, pp. 45–49, February 1982.

FRIEDLANDER, B., "System Identification Techniques for Adaptive Signal Processing," *Circuits Systems Signal Processing*, Vol. 1, No. 1, pp. 3–41, 1982(1).

FRIEDLANDER, B., and MORF, M., "Least-Squares Algorithms for Adaptive Linear-Phase Filtering," *IEEE Trans.*, Vol. ASSP-30, No. 3, pp. 381–390, June 1982(2).

FRIEDLANDER, B., "Lattice Filters for Adaptive Processing," *Proceedings IEEE*, Vol. 70, No. 8, pp, 829–867, August 1982(3).

FRIEDLANDER, B., "Lattice Methods for Spectral Estimation," *Proceedings IEEE*, Vol. 70, No. 9, pp. 990–1017, September 1982(4).

FROST, O. L., III, "An Algorithm for Linearly Constrained Adaptive Array Processing," *Proceedings IEEE*, Vol. 60, No. 8, pp. 926–935, August 1972.

GABOR, D., WILBY, W. P. L., and WOODCOCK, R., "A Universal Non-linear Filter, Predictor and Simulator Which Optimizes Itself by a Learning Process," *Proceedings IEE*, Vol. 108, Pt. B, pp. 422–438, 1961.

GABRIEL, W. F., "Adaptive Arrays—An Introduction," *Proceedings IEEE*, Vol. 64, No. 2, pp. 239–272, February 1976(1).

GABRIEL, W. F. (ed.), Adaptive Antennas, Special Issue, *IEEE Trans.*, Vol. AP-24, No. 5, September 1976(2).

GERSHO, A., and LIM, T. L., "Adaptive Cancellation of Intersymbol Interference for Data Transmission," *Bell System Technical Journal*, Vol. 60, No. 11, pp. 1997–2021, November 1981.

GIBSON, C. J., and HAYKIN, S., "Learning Characteristics of Adaptive Lattice Filtering Algorithms," *IEEE Trans.*, Vol. ASSP-28, No. 6, pp. 681–691, December 1980.

GITLIN, R. D., MAZO, J. E., and TAYLOR, M. G., "On the Design of Gradient Algorithms for Digitally Implemented Adjustment Filters," *IEEE Trans.*, Vol. CT-20, No. 2, pp. 125–136, March 1973.

GITLIN, R. D., and MAGEE, F. R., "Self-Orthogonalizing Adaptive Equalizer Algorithms," *IEEE Trans.*, Vol. COM-25, No. 7, pp. 666–672, July 1977.

GITLIN, R. D., and WEINSTEIN, S. B., "The Effects of Large Interference on the Tracking Capability of Digitally Implemented Echo Cancelers," *IEEE Trans.*, Vol. COM-26, No. 6, pp. 833–839, June 1978.

GITLIN, R. D., and WEINSTEIN, S. B., "On the Required Tap Weight Precision for Digitally Implemented Adaptive Equalizers," *Bell System Technical Journal*, Vol. 58, No. 2, pp. 301–321, February 1979.

GITLIN, R. D., and WEINSTEIN, S. B., "Fractionally-Spaced Equalization: An Improved Digital Transversal Equalizer," *Bell System Technical Journal,* Vol. 60, No. 2, pp. 275–296, February 1981.

GLASER, E. M., "Signal Detection by Adaptive Filters," *IEEE Trans.*, Vol. IT-7, No. 2, pp. 87–98, April 1961.

GLEN, P. J., "Simple Low-Cost Correlator and Multiplier Circuits," *Electronics Letters*, Vol. 18, No. 21, pp. 914–915, October 14, 1982.

GODARD, D., "Channel Equalization Using a Kalman Filter for Fast Data Transmission," *IBM Journal of Research and Development*, Vol. 18, No. 3, pp. 267–273, May 1974.

GODFREY, K. R., and DEVENISH, M., "An Experimental Investigation of Continuous Gas Chromatography Using Pseudo Random Binary Sequences," *Measurement and Control*, Vol. 2, p. 228, 1968.

GOLD, B., and RAEDER, C. M., *Digital Processing of Signals*, McGraw-Hill, New York, 1969.

GOODWIN, G. C., and SIN, K. S., *Adaptive Filtering Prediction and Control*, Prentice-Hall, Englewood Cliffs, NJ, 1984.

GOPINATH, B., and SONDHI, M. M., "Determination of the Shape of the Human Vocal Tract from Acoustical Measurements," *Bell System Technical Journal*, Vol. 49, No. 6, pp. 1195–1214, July–August 1970.

GOPINATH, B., and SONDHI, M. M., "Inversion of the Telegraph Equation and the Synthesis of Non-uniform Lines," *Proceedings IEEE*, Vol. 59, No. 3, pp. 383–392, March 1971.

GRANT, P. M., and KINO, G. S., "Adaptive Filter Based on SAW Monolithic Storage Correlators," *Electronics Letters*, Vol. 14, No. 7, pp. 562–564, August 17, 1978.

GRANT, P. M., and MORGUL, A., "Frequency Domain Adaptive Filter Based on SAW Chirp Transform Processors," *Proceedings IEEE Ultrasonics Symposium*, pp. 186–189, 1982.

GRAY, R. M., "On the Asymptotic Eigenvalue Distribution of Toeplitz Matrices," *IEEE Trans.*, Vol. IT-18, No. 6, pp. 725–730, November 1972.

GRAY, A. H., JR., and MARKEL, J. D., "Digital Lattice and Ladder Filter Synthesis," *IEEE Trans.*, Vol. AU-21, No. 6, pp. 491–500, 1973.

GRAY, A. H., JR., and MARKEL, J. D., "A Normalized Digital Filter Structure," *IEEE Trans.*, Vol. ASSP-23, No. 3, pp. 268–277, 1975.

GRENANDER, U., and SZEGO, G., *Toeplitz Forms and Their Applications*, University of California Press, Berkeley, CA, 1958.

GRIFFITHS, L. J., "A Continuously Adaptive Filter Implemented as a Lattice Structure," *Proceedings IEEE International Conference Acoustics, Speech, and Signal Processing (ICASSP)*, pp. 683–686, 1977.

GRIFFITHS, L. J., "An Adaptive Lattice Structure for Noise-Canceling Applications," *Proceedings IEEE International Conference Acoustics, Speech, and Signal Processing, (ICASSP)*, pp. 87–90, April 1978.

GRIFFITHS, L. J., and MEDAUGH, R. S., "Convergence Properties of an Adaptive Noise Canceling Lattice Structure," *Proceedings 1978 IEEE Conference on Decision and Control*, pp. 1357–1361, January 12, 1979(1).

GRIFFITHS, L. J., "Adaptive Structures for Multiple-Input Noise Canceling Applications," *Proceedings IEEE International Conference Acoustics, Speech, and Signal Processing, (ICASSP)*, pp. 925-928, April 1979(2).

GUIDOUX, L., and LE RICHE, O., "Digital Signal Processing in an LSI 4.8 Kbit/s Modem," *Proceedings IEEE International Conference Acoustics, Speech, and Signal Processing, (ICASSP)*, pp. 1777-1780, 1982.

HAMMING, R. W., *Digital Filters*, 2nd ed., Prentice-Hall, Englewood Cliffs, NJ, 1983.

HARTMANN, P., and BYNAM, B., "Adaptive Equalization for Digital Microwave Radio Systems," *Proceedings IEEE International Conference on Communications (ICC)*, Paper 8.5, 1980.

HAYKIN, S., *Nonlinear Methods in Spectrum Analysis*, Springer-Verlag, West Berlin, 1979.

HAYKIN, S., and CADZOW, J. A. (ed.), Spectral Estimation, Special Issue, *Proceedings IEEE*, Vol. 70, No. 9, September 1982.

HITZ, L., and ANDERSON, B. D. O., "Discrete Positive Real Functions and Their Application to System Stability," *Proceedings IEE*, Vol. 116, pp. 153-155, January 1969.

HODGKISS, W. S., and NOLTE, L. W., "Covariance between Fourier Coefficients Representing the Time Waveforms Observed from an Array of Sensors," *Journal of the Acoustical Society of America*, Vol. 59, pp. 582-590, March 1976.

HOLT, N., and STUEFLOTTEN, S., "A New Digital Echo Canceler for Two-Wire Subscriber Lines," *IEEE Trans.*, Vol. COM-29, No. 11, pp. 1573-1581, November 1981.

HONIG, M. L., and MESSERSCHMITT, D. G., "Convergence Properties of an Adaptive Digital Lattice Filter," *IEEE Trans.*, Vol. ASSP-29, No. 3, pp. 642-653, June 1981.

HONIG, M. L., "Convergence Models for Lattice Joint Process Estimators and Least Squares Algorithms," *IEEE Trans.*, Vol. ASSP-31, No. 2, pp. 415-425, April 1983.

HOPPITT, C. E., "Measurement of Echo Path Impulse Response at Wood Street Exchange Using a Pseudo-Noise Test Signal," *British Post Office Research Department Report 715*, 1978.

HOPPITT, C. E., "A Prototype Echo Cancellation for Use in a Loudspeaking Telephone," *British Telecom Technology Executive Memorandum R13/007/82*, 1982.

HORNA, O. A., "Echo Canceler with Adaptive Transversal Filter Utilizing Pseudo-Logarithmic Coding," *COMSAT Technical Review*, Vol. 7, No. 2, pp. 393-428, Fall 1977.

HORNA, O. A., " Cancellation of Acoustic Feedback," *COMSAT Technical Review*, Vol. 12, No. 2, pp. 319-333, Fall 1982.

HOROWITZ, L. L., and SENNE, K. D., "Performance Advantage of Complex LMS for Controlling Narrow-Band Adaptive Arrays," *IEEE Trans.*, Vol. ASSP-29, No. 3, pp. 722-736, June 1981.

HORVATH, S., "Lattice Form Adaptive Recursive Digital Filters: Algorithms and Applications," *Proceedings IEEE International Symposium on Circuits and Systems (ISCAS)*, pp. 128-133, 1980.

HOWELLS, P. F., "IF Sidelobe Canceler," U.S. Patent 3,202,990, August 24, 1965.

HUDSON, J. E., *Adaptive Array Principles*, Peter Perigrinus, London, 1981.

ITAKURA, F., and SAITO, S., "Analysis Synthesis Telephony Based upon the Maximum Likelihood Method," in Y. Konasi (ed.), *Report 6th International Congress Acoustics*, Tokyo, Report C-5-5, August 21-28, 1968.

ITAKURA, F., and SAITO, S., "A Statistical Method for Estimation of Speech Spectral Density and Formant Frequencies," *Electronics and Communications in Japan*, Vol. 53-A, No. 1, pp. 36–43, 1970.

JACK, M. A., GRANT, P. M., and COLLINS, J. H., "The Theory Design and Applications of Surface Acoustic Wave Fourier Transform Processors," *Proceedings IEEE*, Vol. 68, No. 4, pp. 450–468, April 1980.

JAKOWATZ, C. V., SHUEY, R. L., and WHITE, G. M., "Adaptive Waveform Recognition," *Proceedings 4th London Symposium on Information Theory*, Butterworth, London, pp. 317–326, September 1960.

JENKINS, W. K., and LEON, B. J., "The Use of Residue Number Systems in the Design of Finite Impulse Response Digital Filters," *IEEE Trans.*, Vol. CAS-24, No. 4, pp. 191–200, April 1977.

JENKINS, W. K., "Recent Advances in Residue Number Techniques for Recursive Digital Filtering," *IEEE Trans.*, Vol. ASSP-27, No. 1, pp. 19–30, February 1979.

JOHNSON, C. R., JR., and LARIMORE, M. G., "Comments on and Additions to 'An Adaptive Recursive LMS Filter'," *Proceedings IEEE*, Vol. 65, No. 9, pp. 1399–1401, September 1977.

JOHNSON, C. R., JR., TREICHLER, J. R., and LARIMORE, M. G., "Remarks on the Use of SHARF as an Output Error Identifier," *Proceedings 17th IEEE Conference on Decision and Control*, San Diego, CA, pp. 1094–1095, January 1979(1).

JOHNSON, C. R., JR., "A Convergence Proof for a Hyperstable Adaptive Recursive Filter," *IEEE Trans.*, Vol. IT-25, No. 6, pp. 745–749, November 1979(2).

JOHNSON, C. R., JR., LARIMORE, M. G., TREICHLER, J. R., and ANDERSON, B. D. O., "SHARF Convergence Properties," *IEEE Trans.*, Vol. ASSP-29, No. 3, pp. 659–670, June 1981.

JOHNSON, C. R., JR., "Adaptive IIR Filtering: Current Results and Open Issues," *IEEE Trans.*, 1984.

JOHNSTON, D. H., and DE GRAAF, S. R., "Improving the Resolution of Bearing in Passive Sonar Arrays by Eigenvalue Analysis," *IEEE Trans.*, Vol. ASSP-30, No. 4, pp. 638–647, April 1982(1).

JOHNSTON, D. H., "The Application of Spectral Estimation Methods to Bearing Estimation Problems," *Proceedings IEEE*, Vol. 70, No. 9, pp. 1018–1028, September 1982(2).

JULLIEN, G. A., "Residue Number Scaling and Other Operations Using ROM Arrays," *IEEE Trans.*, Vol. C-27, No. 4, pp. 325–335, April 1978.

KAILATH, T., *Lectures on Wiener and Kalman Filtering*, Springer-Verlag, New York, 1981.

KAILATH, T., "Time-Variant and Time-Invariant Lattice Filters for Nonstationary Processes," *Mathematical Tools and Models for Control Systems Analysis and Signal Processing*, CNRS Editions, Vol. 2, Paris, pp. 417–464, 1982.

KALLMAN, H. E., "Transversal Filters," *Proceedings IRE*, Vol. 28, No. 7, pp. 302–310, July 1940.

KALMAN, R. E., "A New Approach to Linear Filtering and Prediction Problems," *Trans. ASME, Journal of Basic Engineering*, pp. 35–45, March 1960.

KAPUR, N., MAVOR, J., and JACK, M. A., "Discrete Cosine Transform Processor Using a CCD Programmable Transversal Filter," *Electronics Letters*, Vol. 16, No. 4, pp. 139–141, February 14, 1980.

KAY, S. M., and MARPLE, S. L., "Spectrum Analysis–A Modern Perspective," *Proceedings IEEE*, Vol. 69, No. 11, pp. 1380–1419, November 1981.

KELLY, J. L., JR., and LOCHBAUM, C. C., "Speech Synthesis," *Proceedings Stockholm Speech Communications Seminar*, Paper G42, pp. 1-4, Stockholm, Sweden, September 1962.

KLEIN, J. D., and DICKINSON, B. W., "A Normalized Ladder Form of the Residual Energy Ratio Algorithm for PARCOR Estimation via Projections," *IEEE Trans.*, Vol. AC-28, No. 10, pp. 943–952, October 1983.

KRETSCHMER, F. F., JR., and LEWIS, B. L., "A Digital Open Loop Adaptive Processor," *IEEE Trans.*, Vol. AES-14, No. 1, pp. 165–171, January 1978.

KUNG, H. T., and LEISERSON, C. E., "Systolic Arrays for VLSI," in C. A. Mead and L. A. Conway, *Introduction to VLSI Systems*, Addison-Wesley, Reading, MA, 1980.

KUNG, S. Y., and HU, Y. H., "A Highly Concurrent Algorithm and Pipelined Architecture for Solving Toeplitz Systems," *IEEE Trans.*, Vol. ASSP-31, No. 1, pp. 66–75, February 1983.

LANDAU, I. D., "A Survey of Model Reference Adaptive Techniques–Theory and Applications," *Automatica*, Vol. 10, pp. 353–379, July 1974.

LANDAU, I. D., "Unbiased Recursive Identification Using Model Reference Adaptive Techniques," *IEEE Trans.*, Vol. AC-21, No. 2, pp. 194–202, April 1976.

LANDAU, I. D., "Elimination of the Real Positivity Condition in the Design of Parallel MRAS," *IEEE Trans.*, Vol. AC-23, No. 6, pp. 1015–1020, December 1978.

LANDAU, I. D., *Adaptive Control–The Model Reference Approach*, Marcel Dekker, New York, 1979.

LARIMORE, M. G., TREICHLER, J. R., and JOHNSON, C. R., JR., "SHARF: An Algorithm for Adapting IIR Digital Filters," *IEEE Trans.*, Vol. ASSP-28, No. 4, pp. 428–440, August 1980.

LAWRENCE, R. E., and KAUFMAN, H., "The Kalman Filter for the Equalization of a Digital Communications Channel," *IEEE Trans.*, Vol. COM-19, No. 12, pp. 1137–1141, December 1971.

LAWSON, C. L., and HANSON, R. J., *Solving Least-Squares Problems*, Prentice-Hall, Englewood Cliffs, NJ, 1974.

LEE, D. T. L., "Canonical Ladder Form Realizations and Fast Estimation Algorithms," Ph.D. dissertation, Department of Electrical Engineering, Stanford University, Stanford, CA, August 1980(1).

LEE, D. T. L., and MORF, M., "A Novel Innovation Based Approach to Pitch Detection," *Proceedings IEEE International Conference Acoustics, Speech, and Signal Processing (ICASSP)*, pp. 40–44, April 9–11, 1980(2).

LEE, D. T. L., MORF, M., and FRIEDLANDER, B., "Recursive Least Squares Ladder Estimation Algorithms," *IEEE Trans.*, Vol. ASSP-29, No. 3, pp. 627–641, June 1981.

LEE, D. T. L., FRIEDLANDER, B., and MORF, M., "Recursive Ladder Algorithms for ARMA Modeling," *IEEE Trans.*, Vol. AC-27, No. 8, pp. 753–764, August 1982.

LE ROUX, J., and GUEGUEN, C., "A Fixed Point Computation of Partial Correlation Coefficients," *IEEE Trans.*, Vol. ASSP-25, No. 3, pp. 257–259, June 1977.

LJUNG, L., "Analysis of Recursive Stochastic Algorithms," *IEEE Trans.*, Vol. AC-22, No. 4, pp. 551–575, August 1977.

LJUNG, L., MORF, M., and FALCONER, D., "Fast Calculation of Gain Matrices for Recursive Estimation Schemes," *International Journal of Control*, Vol. 27, No. 1, pp. 1–19, January 1978.

LJUNG, L., "Convergence of Recursive Estimators," *Proceedings 5th IFAC Symposium on Identification and System Parameter Estimation*, Darmstadt, 1979.

LJUNG, L., "The ODE Approach to the Analysis of Adaptive Control Systems—Possibilities and Limitations," *Proceedings IEEE, Joint Automatic Control Conference*, Paper WA2-C, San Francisco, 1980.

LJUNG, L., "Analysis of a General Recursive Prediction Algorithm," *Automatica*, Vol. 17, No. 1, pp. 89–99, January 1981.

LJUNG, L., and SÖDERSTRÖM, T., *Theory and Practice of Recursive Identification*, MIT Press, Cambridge, MA, 1983.

LINKENS, D. A., "Short-Time-Series Spectral Analysis of Biomedical Data," *Proceedings IEE*, Vol. 129, Pt. A, No. 9, pp. 663–672, December 1982.

LUBELL, P. D., and REBHUN, F. D., "Suppression of Co-channel Interference with Adaptive Cancellation Devices at Communications Satellite Earth Stations," *Proceedings IEEE International Conference on Communications (ICC)*, pp. 284–289, June 1977.

LUCKY, R. W., "Techniques for Adaptive Equalization of Digital Communications Systems," *Bell System Technical Journal*, Vol. 45, No. 2, pp. 255–286, February 1966.

LUCKY, R. W., SALZ, J., and WELDON E. J., JR., *Principles of Data Communications*, McGraw-Hill, New York, 1968.

LUCKY, R. W., "A Survey of the Communication Theory Literature: 1968–1973," *IEEE Trans.*, Vol. IT-19, No. 6, pp. 725–739, November 1973.

LUENBERGER, D. G., *Introduction to Linear and Nonlinear Programming*, Addison-Wesley, Reading, MA, 1973.

LYON, R. F., "A Bit-Serial VLSI Architectural Methodology for Signal Processing," in J. P. Gray (ed.), *VLSI 81*, Academic Press, London, 1981.

MCCLELLAN, J. H., and RADER, C. M., *Number Theory in Digital Signal Processing*, Prentice-Hall, Englewood Cliffs, NJ, 1979.

MCDONOUGH, K., CAUDEL, E., MAGAR, S., and LEIGH, A., "Microcomputer with 32-Bit Arithmetic Does High Precision Number Crunching," *Electronics*, pp. 105–111, February 24, 1982.

MACLENNAN, D. J., MAVOR, J., and VANSTONE, G. F., "Technique for Realizing Transversal Filters Using Charge-Coupled Devices," *Proceedings IEE*, Vol. 122, No. 6, pp. 615–619, June 1975.

MACLEOD, C. J., CIAPALA, E., and JELONEK, Z. J., "Quantization in Non-recursive Equalizers for Data Transmission," *Proceedings IEE*, Vol. 122, No. 10, pp. 1105–1110, October 1975.

MAIWALD, D., KAESER, H. P., and CLOSS, F., "An Adaptive Equalizer with Significantly Reduced Number of Operations," *Proceedings IEEE International Conference Acoustics, Speech, and Signal Processing (ICASSP)*, pp. 100–104, April 1978.

MAKHOUL, J., "Linear Prediction: A Tutorial Review," *Proc. IEEE*, Vol. 63, No. 4, pp. 561–580, April 1975.

MAKHOUL, J., "Stable and Efficient Lattice Methods for Linear Prediction," *IEEE Trans.*, Vol. ASSP-25, No. 5, pp. 423–428, May 1977.

MAKHOUL, J., and VISWANATHAN, R., "Adaptive Lattice Methods for Linear Prediction," *Proceedings IEEE International Conference Acoustics, Speech, and Signal Processing (ICASSP)*, pp. 83–86, April 1978(1).

MAKHOUL, J., "A Class of All-Zero Lattice Digital Filters: Properties and Applications," *IEEE Trans.*, Vol. ASSP-26, No. 4, pp. 304–314, August 1978(2).

MANSOUR, D., and GRAY, A. H., JR., "Unconstrained Frequency Domain Adaptive Filter," *IEEE Trans.*, Vol. ASSP-30, No. 5, pp. 726–734, October 1982.

MARKEL, J. D., and GRAY, A. H., JR., "On Autocorrelation Equations as Applied to Speech Analysis," *IEEE Trans.*, Vol. AU-21, No. 2, pp. 69–79, April 1973.

MARKEL, J. D., and GRAY, A. H., JR., "Roundoff Noise Characteristics of a Class of Orthogonal Polynomial Structures," *IEEE Trans.*, Vol. ASSP-23, No. 5, pp. 473–486, October 1975(1).

MARKEL, J. D., and GRAY, A. H., JR., "Fixed-Point Implementation Algorithms for a Class of Orthogonal Polynomial Filter Structures," *IEEE Trans.*, Vol. ASSP-23, No. 5, pp. 486–494, October 1975(2),

MARKEL, J. D., and GRAY, A. H., JR., *Linear Prediction of Speech*, Springer-Verlag, New York, 1976.

MARTINSON, L., "A Programmable Digital Processor for Airborne Radar," *Proceedings IEEE International Radar Conference*, pp. 186–191, 1975.

MASENTEN, W. K., "Adaptive Signal Processing," International Specialist Seminar on Case Studies in Advanced Signal Processing, *IEE Conference Proceedings 180*, pp. 168–177, September 1979.

MASSEY, N. R., GRANT, P. M., and MAVOR, J., "CCD Adaptive Filter Employing Parallel Coefficient Updating," *Electronics Letters,* Vol. 15, No. 18, pp. 573–4, August 30, 1979.

MAVOR, J., JACK, M. A., SAXTON, D., and GRANT, P. M., "Design and Performance of a Programmable Real-Time Charge-Coupled Device Recirculating Delay-Line Correlator," *IEE Journal on Electronic Circuits and Systems*, Vol. 1, No. 4, pp. 137–143, July 1977.

MAVOR, J. (ed.), *Proceedings 5th International Conference on Charge-Coupled Devices, CCD-79*, Centre for Industrial Consultancy and Liaison, University of Edinburgh, September 1979.

MAZO, J. E., "Analysis of Decision Directed Convergence," *Bell System Technical Journal*, Vol. 59, No. 10, pp. 1858–1876, December 1980.

MEAD, K. O., and RYDER, W. H., "A PARCOR Lattice Predictor Having No Divisions," *Government Communications Headquarters (GCHQ) Memo M/2454/1019/1/23*, Cheltenham, Gloustershire, U.K., March 1977.

MENDEL, J. M., *Discrete Techniques of Parameter Estimation*, Marcel Dekker, New York, 1973.

MITRA, D., and SONDHI, M. M., "Adaptive Filtering with Non-ideal Multipliers—Applications to Echo Cancellation," *Proceedings IEEE International Conference on Communications (ICC)*, pp. 30.11–30.15, 1975.

MITRA, S. K., KAMAT, P. S., and HUEY, D. C., "Cascaded Lattice Realization of Digital Filters," *Circuit Theory and Applications*, Vol. 5, pp. 3–11, 1977.

MONSEN, P., "Adaptive Equalization of the Slow Fading Channel," *IEEE Trans.*, Vol. COM-22, No. 8, pp. 1064–1075, August 1974.

MONSEN, P., "Fading Channel Communications," *IEEE Communications Society Magazine*, Vol. 18, No. 1, pp. 27–36, January 1980.

MONZINGO, R. A., and MILLER, T. W., *Introduction to Adaptive Arrays*, Wiley, New York, 1980.

MORF, M., "Fast Algorithms for Multivariable Systems," Ph.D. dissertation, Department of Electrical Engineering, Standford University, Stanford, CA, 1974.

MORF, M., LEE, D. T., NICKOLLS, J. R., and VIEIRA, A., "A Classification of Algorithms for ARMA Models and Ladder Realizations," *Proceedings IEEE International Conference Acoustics, Speech, and Signal Processing (ICASSP)*, pp. 13–19, April 1977.

MORGAN, D. R., and CRAIG, S. E., "Real-Time Adaptive Linear Prediction Using the Least Mean Square Gradient Algorithm," *IEEE Trans.*, Vol. ASSP-24, No. 6, pp. 494–507, December 1976.

MORGUL, A., GRANT, P. M., and COWAN, C. F. N., "Wideband Hybrid Analog/Digital Frequency Domain Adaptive Filter," *IEEE Trans.*, Vol. ASSP-32, No. 4, pp. 762–769, August 1984.

MOSCHNER, J. L., "Adaptive Filter with Clipped Input Data," Stanford University, Information Systems Laboratory, *Report 6796-1*, June 1970.

MUELLER, K. H., and SPAULDING, D. A., "Cyclic Equalization—A New Rapidly Converging Equalization Technique for Synchronous Data Communication," *Bell System Technical Journal*, Vol. 54, No. 2, pp. 369–406, February 1975.

MUELLER, K. H., "Combined Echo Cancellation and Decision Feedback Equalization," *Bell System Technical Journal*, Vol. 58, No. 2, pp. 491–500, February 1979.

MUELLER, M. S., "Least-Squares Algorithms for Adaptive Equalizers," *Bell System Technical Journal*, Vol. 60, No. 8, pp. 1905–1925, October 1981.

MURANO, K., UNAGAMI, S., and TSUDA, T., "LSI Processor for Digital Signal Processing and Its Application to 4800 Bit/s Modem," *IEEE Trans.*, Vol. COM-26, No. 5, pp. 499–506, May 1978.

MURPHY, T. P., BAKER, F. M., GARNER, C. L., and KRUZINSKI, P. J., "Practical Techniques for Improving Signal Robustness," *Proceedings IEEE National Telecommunications Conference (NTC)*, Paper C3.3, 1981.

MURTHY, V. K., and NARASIMHAM, G. V. L., "On the Asymptotic Normality and Independence of the Sample Partial Autocorrelations for an Autoregressive Process," *Applied Mathematics and Computing*, Vol. 5, pp. 281–295, 1979.

NARAYAN, S. S., PETERSON, A. M., and NARASIMHA, M. J., "Transform Domain LMS Algorithm," *IEEE Trans.*, Vol. ASSP-31, No. 3, pp. 609–615, June 1983.

NEISSEN, C. W., and WILLIM, D. K., "Adaptive Equalizer for Pulse Transmission," *IEEE Trans.*, Vol. COM-18, No. 4, pp. 377–395, August 1970.

OCHIAI, K., ARASEKI, T., and OGIHARA, T., "Echo Canceler with Two Echo Path Models," *IEEE Trans.*, Vol. COM-25, No. 6, pp. 589–595, June 1977.

OPPENHEIM, A. V., and SCHAFER, R. W., *Digital Signal Processing*, Prentice-Hall, Englewood Cliffs, NJ, 1975.

PANASIK, C. M., "SAW Programmable Transversal Filter for Adaptive Interference Suppression," *Proceedings IEEE Ultrasonics Symposium*, pp. 100–103, 1982.

PAPOULIS, A., "Maximum Entropy and Spectral Estimation: A Review," *IEEE Trans.*, Vol. ASSP-29, No. 6, pp. 1176–1186, December 1981.

PARIKH, D., and AHMED, N., "On an Adaptive Algorithm for IIR Filters," *Proceedings IEEE*, Vol. 65, No. 5, pp. 585–587, May 1978.

PARKINSON, D. W., and HARVIE, I. B., "Modem 36–A New Model for 48–72 kbit/s Data Transmission," *British Telecommunications Engineering*, Vol. 1, No. 4, pp. 234–240, January 1983.

PEARL, J., "On Coding and Filtering Stationary Signals by Discrete Fourier Transforms," *IEEE Trans.*, Vol. IT-19, No. 2, pp. 229–232, March 1973.

PELED, A., and LIU, B., "A New Hardware Realization of Digital Filters," *IEEE Trans.*, Vol. ASSP-22, No. 6, pp. 456–462, December 1974.

PELED, A., and LIU, B., *Digital Signal Processing: Theory Design and Implementation*, Wiley, New York, 1976.

PELGROM, M. J. M., WALLINGA, H., and HOLLEMAN, J., "The Electrically Programmable Split-Electrode CCD Transversal Filter," *Proceedings 5th International Conference on CCDs, CCD-79*, Centre for Industrial Consultancy and Liaison, University of Edinburgh, pp. 254–260, 1979.

PELKOWITZ, L., "Frequency Domain Analysis of Wraparound Error in Fast Convolution Algorithms," *IEEE Trans.*, Vol. ASSP-29, No. 3, pp. 413–422, June 1981.

PICCHI, G., and PRATI, G., "Self-Orthogonalizing Adaptive Algorithm for Channel Equalization in the Discrete Frequency Domain," *IEEE Trans.*, Vol. COM-32, No. 4, pp. 371–379, April 1984.

POPOV, V. M., *Hyperstability of Control Systems*, Springer-Verlag, West Berlin, 1972.

PORAT, B., FRIEDLANDER, B., and MORF, M., "Square Root Covariance Ladder Algorithms," *IEEE Trans.*, Vol. AC-27, No. 4, pp. 813–829, August 1982.

PORAT, B., and KAILATH, T., "Normalized Lattice Algorithms for Least-Squares FIR System Identification," *IEEE Trans.*, Vol. ASSP-31, No. 1, pp. 122–128, February 1983.

PRICE, R., and GREEN, P. E., JR., "A Communication Technique for Multipath Channels," *Proceedings IRE*, Vol. 46, No. 3, pp. 555–569, March 1958.

PROAKIS, J. G., "Advances in Equalization for Intersymbol Interference," in *Advances in Communication Systems*, Vol. 4, Academic Press, New York, 1975.

QURESHI, S. U. H., "Fast Start-Up Equalization with Periodic Training Sequences," *IEEE Trans.*, Vol. IT-23, No. 5, pp. 553–563, September 1977.

QURESHI, S. U. H., "Adaptive Equalization," *IEEE Communications Society Magazine*, Vol. 21, No. 2, pp. 9–16, March 1982.

RABINER, L. R., and GOLD, B., *Theory and Application of Digital Signal Processing*, Prentice-Hall, Englewood Cliffs, NJ, 1975.

RABINER, L. R., and SCHAFER, R. W., *Digital Processing of Speech Signals*, Prentice-Hall, Englewood Cliffs, NJ, 1978.

RAMSEY, F. L., "Characterization of the Partial Autocorrelation Function," *Annals of Statistics*, Vol. 2, No. 6, pp. 1296–1301, 1974.

REDINGTON, D., and TURNER, J., "Recursive Ladder Autoregressive Modeling of Electrophysiological Data," to be published in *Electroencephalography and Clinical Neurophysiology*.

REED, F. A., and FEINTUCH, P. L., "A Comparison of LMS Adaptive Cancelers Implemented in the Frequency Domain and Time Domain," *IEEE Trans.*, Vol. ASSP-29, No. 3, pp. 770–775, June 1981.

RIDOUT, P. N., and ROLFE, P., "Transmission Measurements of Connections in the Switched Telephone Network," *Post Office Electrical Engineers Journal*, Vol. 63, No. 2, pp. 97–104, July 1970.

RIDOUT, I. B., and HARVIE, I. B., "The Principles of Scramblers and Descramblers Designed for Data Transmission Systems," *British Telecom Engineering Journal*, Vol. 1, Pt. 2, pp. 111–114, July 1982.

ROBBINS, M., AND MONROE, S., "A Stochastic Approximation Method," *Annals of Mathematical Statistics*, Vol. 22, pp. 400–407, 1951.

ROBINSON, E. A., "An Historical Perspective of Spectrum Estimation," *Proceedings IEEE*, Vol. 70, No. 9, pp. 885-907, September 1982.

RUDIN, H. J., "Automatic Equalization Using Transversal Filters," *IEEE Spectrum*, Vol. 2, No. 1, pp. 53–59, January 1967.

RUMMLER, W. D., "A New Selective Fading Model: Application to Propagation Data," *Bell System Technical Journal*, Vol. 58, No. 5, pp. 1037–1071, May-June 1979.

RUTTER, M. J., GRANT, P. M., RENSHAW, D., and DENYER, P. B., "Design and Realization of Adaptive Lattice Filters," *Proceedings IEEE International Conference Acoustics, Speech, and Signal Processing (ICASSP)*, Boston, pp. 21–24, April 1983(1).

RUTTER, M. J., "Theory Design and Application of Gradient Adaptive Lattice Filters," Ph.D. thesis, Department of Electrical Engineering, University of Edinburgh, Edinburgh, September 1983(2).

SAMSON, C., and REDDY, V. U., "Fixed Point Error Analysis of the Normalized Ladder Algorithm," *IEEE Trans.*, Vol. ASSP-31, No. 5, pp. 1177–1191, October 1983.

SATORIUS, E. H., SMITH, J. D., and REEVES, P. M., "Adaptive Noise Canceling of a Sinusoidal Interference Using a Lattice Structure," *Proceedings IEEE International Conference Acoustics, Speech, and Signal Processing (ICASSP)*, pp. 929–932, April 1979(1).

SATORIUS, E. H., and ALEXANDER, S. T., "Channel Equalization Using Adaptive Lattice Algorithms," *IEEE Trans.*, Vol. COM-27, No. 6, pp. 899–905, June 1979(2).

SATORIUS, E. H., and SHENSA, M. J., "Recursive Lattice Filters—A Brief Overview," *Proceedings 19th IEEE Conference Decision and Control*, Albuquerque, NM, pp. 955–959, December 10–12, 1980.

SATORIUS, E. H., and PACK, J., "Application of Least Squares Lattice Algorithm to Adaptive Equalization," *IEEE Trans.*, Vol. COM-29, No. 2, pp. 136–142, February 1981.

SCHMIDT, R., "Multiple Emitter Location and Signal Parameter Estimation," *Proceedings RADC Spectral Estimation Workshop*, Griffiss AFB, Rome, NY, pp. 243–258, 1979.

SHENSA, M. J., "The Spectral Dynamics of Evolving LMS Adaptive Filters," *Proceedings IEEE International Conference Acoustics, Speech, and Signal Processing (ICASSP)*, pp. 950–953, 1979.

SHENSA, M. J., "Recursive Least Squares Lattice Algorithms—A Geometrical Approach," *IEEE Trans.*, Vol. AC-26, No. 3, pp. 695–702, June 1981.

SINGLETON, R. C., "An Algorithm for Computing the Mixed Radix Fourier Transform," *IEEE Trans.*, Vol. AU-17, No. 2, pp. 93–103, June 1969.

SODERSTRAND, M. A., and FIELDS, E. L., "Multipliers for Residue Number Arithmetic Digital Filters," *Electronics Letters*, Vol. 13, No. 6, pp. 164–166, March 17, 1977(1).

SODERSTRAND, M. A., "A High-Speed Low-Cost Recursive Digital Filter Using Residue Number Arithmetic," *Proceedings IEEE*, Vol. 65, No. 7, pp. 1065–1067, July 1977(2).

SODERSTRAND, M. A., VERNIA, C., PAULSON, D. W., and VIGIL, M. C., "Microprocessor Controlled Adaptive Digital Filters," *Proceedings IEEE International Symposium on Circuits and Systems (ISCAS)*, Houston, TX, pp. 142–146, April 1980(1).

SODERSTRAND, M. A., and VIGIL, M. C., "Microprocessor Controlled Totally Adaptive Digital Filter," *Proceedings IEEE International Computer Communications Conference (ICCC)*, pp. 85–89, October 1980(2).

SONDHI, M. M., and BERKLEY, D. A., "Silencing Echoes on the Telephone Network," *Proceedings IEEE*, Vol. 68, No. 8, pp. 948–963, August 1980.

SOUTH, C. R., HOPPITT, C. E., and LEWIS, A. V., "Adaptive Filters to Improve Loudspeaker Telephone," *Electronics Letters*, Vol. 15, No. 21, pp. 673–674, October 11, 1979.

STEARNS, S. D., and ELLIOTT, G. R., "On Adaptive Recursive Filtering," *Proceedings 10th Asilomar Conference on Circuits, Systems, and Computers*, pp. 5–11, November 1976.

STEARNS, S. D., "Error Surfaces of Adaptive Recursive Filters," *IEEE Trans.*, Vol. ASSP-28, No. 3, pp. 763–766, June 1981.

STEIN, M., "Sematrans Modems, A Study," *Philips Technical Review*, Vol. 40, No. 4, pp. 291–300, December 1982.

STEINBUCH, K., and WIDROW, B., "A Critical Comparison of Two Kinds of Adaptive Classification Networks," *IEEE Trans.*, Vol. EC-14, No. 5, pp. 737–740, October 1965.

SUNTER, S., CHOWANIEC, A., and LITTLE, T., "Adaptive Filtering in CCD and MOS Technologies," *Proceedings 5th International Conference on CCDs, CCD-79*, Centre for Industrial Consultancy and Liaison, University of Edinburgh, pp. 261–267, 1979.

SUSSMAN, S. M., "A Matched Filter Communications System for Multipath Channels," *IRE Trans.*, Vol. IT-6, No. 3, pp. 367–372, June 1960.

TAYLOR, N. G., (ed.), Adaptive Antennas, Special Issue, *Proceedings IEE*, Vol. 130, Pt. F, No. 1, pp. 1–151, January 1983.

TOMPKINS, W. R., MONTI, R., and INTAGLIETTA, M., "Velocity Measurement by Self-Tracking Correlator," *Review of Scientific Instrumentation*, Vol. 45, No. 5, pp. 647–649, 1974.

TREICHLER, J. R., LARIMORE, M. G., and JOHNSTON, C. R., JR., "Simple Adaptive IIR Filtering," *Proceedings IEEE International Conference Acoustics, Speech, and Signal Processing (ICASSP)*, pp. 118–122, April 1978.

TREICHLER, J. R., "Transient and Convergent Behavior of the Adaptive Line Enhancer," *IEEE Trans.*, Vol. ASSP-27, No. 1, pp. 53–62, February 1979.

TREICHLER, J. R., and AGEE, B. G., "A New Approach to Multipath Correction of Constant Modulus Signals," *IEEE Trans.*, Vol. ASSP-31, No. 2, pp. 459–472, April 1983.

TREITEL, S., and ROBINSON, E. A., "Seismic Wave Propagation in Layered Media in Terms of Communication Theory," *Geophysics*, Vol. 31, pp. 17–32, 1966.

TURIN, G. L., "Introduction to Digital Matched Filters," *Proceedings IEEE*, Vol. 64, No. 7, pp. 1092–1112, July 1976.

TURIN, G. L., "Introduction to Spread-Spectrum Anti-multipath Techniques and Their Applications to Urban Digital Radio," *Proceedings IEEE*, Vol. 68, No. 3, pp. 328–353, March 1980.

TURNER, J., DICKINSON, B., and LAI, D., "Characteristics of Reflection Coefficient Estimates Based on a Markov Chain Model," *Proceedings IEEE International Conference Acoustics, Speech, and Signal Processing, (ICASSP)*, pp. 131–134, April 1980.

TURNER, J., "Application of Recursive Exact Least Square Ladder Estimation Algorithm for Speech Recognition," *Proceedings IEEE International Conference Acoustics, Speech, and Signal Processing (ICASSP)*, pp. 543–555, May 1982.

UNGERBOECK, G., "Theory on the Speed of Convergence in Adaptive Equalizers for Digital Communication," *IBM Journal of Research and Development*, Vol. 16, No. 6, pp. 546–555, November 1972.

UNGERBOECK, G., "Adaptive Maximum-Likelihood Receiver for Carrier-Modulated Data Transmission Systems," *IEEE Trans.*, Vol. COM-22, No. 5, pp. 624–636, May 1974.

UNGERBOECK, G., "Fractional Tap-Spacing Equalizer and Consequences for Clock Recovery in Data Modems," *IEEE Trans.*, Vol. COM-24, No. 8, pp. 856–864, August 1976.

UNKAUF, M. G., "An Acoustic Surface Wave Modem for Time-Varying Dispersive Channels," pp. 485–493 in J. Fox (ed.) *Optical and Acoustical Micro-electronics*, MRI Symposium Series, Vol. 23, Polytechnic Press, Brooklyn NY, 1975.

VAN ATTA, L. C., "Electromagnetic Reflection," U.S. Patent 2,908,002, October 6, 1959.

VAN GERWEN, P. J., and VERHOECKX, N. A. M., "A Digital Transmission Unit for the Local Network," *Proceedings IEE Communications '82 Conference*, pp. 65–69, 1982.

VERHOECKX, N. A. M., VAN DEN ELZEN, H. C., SNIJDERS, F. A. M., and VAN GERWEN, P. J., "Digital Echo Cancellation for Baseband Data Transmission," *IEEE Trans.*, Vol. ASSP-27, No. 6, pp. 768–781, 1979.

VOLDER, J. E., "The CORDIC Trigonometric Computing Technique," *IRE Trans.*, Vol. EC-8, No. 3, pp. 330–334, September 1959.

VOORMAN, J. O., SNIJER, P. J., BARTH, P. J., and VROMANS, J. S., "One-Chip Automatic-Equalizer for Echo Reduction in Teletext," *IEEE Trans.*, Vol. CE-27, No. 3, pp. 512–529, August 1981.

VRY, M. G., and VAN GERWEN, P. J., "Digital Signal Transmission to the Subscriber Using a 1 + 1 System," 2nd International Conference on Telecommunications Transmission–Into the Digital Era, *IEE Conference Publication 193*, pp. 197–199, March 1981.

VRY, M. G., "Digital Local Network Systems: The Impact of Signal Processing," *Proceedings IEEE International Conference Acoustics, Speech, and Signal Processing (ICASSP)*, pp. 1785–1788, 1982.

WAIT, J. R. (ed.), Applications of Electromagnetic Theory to Geophysical Exploration, Special Issue, *Proceedings IEEE*, Vol. 67, No. 7, July 1979.

WALKER, G., "On Periodicity in Series and Related Terms," *Proceedings Royal Society*, Vol. A131, p. 518, 1931.

WALMSLEY, C. F., and GOODING, J. N., "50 MHz Time Delay and Integration CCD," *Proceedings 5th International Conference on CCDs, CCD-79*, Centre for Industrial Consultancy and Liaison, University of Edinburgh, pp. 341–346, September 1979.

WALTHER, J. S., "A Unified Algorithm for Elementary Functions," *Proceedings IEEE Joint Spring Computer Conference*, pp. 379–385, July 1971.

WALZMAN, T., and SCHWARTZ, M., "Automatic Equalization Using the Discrete Frequency Domain," *IEEE Trans.*, Vol. IT-19, No. 1, pp. 59–68, January 1973.

WASSER, S., and PETERSON, A. M., "Medium-Speed Multipliers Trim Cost, Shrink Bandwidth in Speech Transmission," *Electronic Design*, February 1, 1979.

WATANABE, K., INOUE, K., and SATO, Y., "A 4800 BPS Microprocessor Data Modem," *Proceedings IEEE International Conference on Communications (ICC)*, pp. 47.6.252–47.6.256, 1977.

WECKLER, G. P., and WALBY, M. D., "Programmable Transversal Filters: Design Tradeoffs," *Proceedings 5th International Conference on CCDs, CCD-79*, Centre for Industrial Consultancy and Liaison, University of Edinburgh, pp. 211–221, 1979.

WEINSTEIN, S. B., "Echo Cancellation in the Telephone Network," *IEEE Communications Society Magazine*, Vol. 15, No. 1, January 1977(1).

WEINSTEIN, S. B., "A Passband Data-Driven Echo Canceler for Full-Duplex Transmission on Two-Wire Circuits," *IEEE Trans.*, Vol. COM-25, No. 7, pp. 654–666, July 1977(2).

WEINSTEIN, E., "Stability Analysis of LMS Adaptive Filters," submitted for publication, 1983.

WEISS, A., and MITRA, D., "Digital Adaptive Filters: Conditions of Convergence, Rates of Convergence, Effects of Noise and Errors Arising from the Implementation," *IEEE Trans.*, Vol. IT-25, No. 6, pp. 637–652, November 1969.

WEN, D. D., "A CCD Video Delay Line," *Proceedings IEEE International Solid-State Circuits Conference (ISSCC)*, Philadelphia, pp. 204–205, February 1976.

WHITE, S. A., "An Adaptive Recursive Digital Filter," *Proceedings 9th Annual Asilomar Conference on Circuits, Systems, and Computers*, pp. 21–25, November 1975.

WHITE, M. H., MACK, I. A., BORSUK, G. M., LAMPE, D. R., and KUB, F. J., "Charge-Coupled Device Adaptive Discrete Analog Signal Processing," *IEEE Journal of Solid-State Circuits*, Vol. SC-14, No. 2, pp. 132–147, February 1979.

WIDROW, B., and HOFF, M., JR., "Adaptive Switching Circuits," *IRE WESCON Convention Record*, Pt. 4, pp. 96–104, 1960.

WIDROW, B., MANTEY, P. E., GRIFFITHS, L. J., and GOODE, B. B., "Adaptive Antenna Systems," *Proceedings IEEE*, Vol. 55, No. 12, pp. 2143–2159, December 1967.

WIDROW, B., "Adaptive Filters," pp. 563–587 in R. Kalman and N. DeClaris (eds.), *Aspects of Network and System Theory*, Holt, Rinehart and Winston, New York, 1971.

WIDROW, B., MCCOOL, J., and BALL, M., "The Complex LMS Algorithm," *Proceedings IEEE*, Vol. 63, No. 4, pp. 719–720, April 1975(1).

WIDROW, B., GLOVER, J. R., MCCOOL, J. M., KAUNITZ, J., WILLIAMS, C. S., HEARN, R. H., ZEIDLER, J. R., DONG, E., and GOODLIN, R. C., "Adaptive Noise Canceling: Principles and Applications," *Proceedings IEEE*, Vol. 63, No. 12, pp. 1692–1716, December 1975(2).

WIDROW, B., and MCCOOL, J. M., "A Comparison of Adaptive Algorithms Based on the Methods of Steepest Descent and Random Search," *IEEE Trans.*, Vol. AP-24, No. 5, pp. 615–637, September 1976(1).

WIDROW, B., MCCOOL, J. M., LARIMORE, M. G., and JOHNSON, C. R., "Stationary and Non-Stationary Learning Characteristics of the LMS Adaptive Filter," *Proceedings IEEE*, Vol. 64, No. 8, pp. 1151–1161, August 1976(2).

WIDROW, B., and STEARNS, S. D., *Adaptive Signal Processing*, Prentice-Hall, Englewood Cliffs, NJ, 1984.

WIENER, N., *Extrapolation, Interpolation and Smoothing of Stationary Time Series*, Wiley, New York, 1949.

WONG, K. M., and JAN, Y. G., "Adaptive Walsh Equalizer for Data Transmission," *Proceedings IEE*, Vol. 130, Pt. F, No. 2, pp. 153–160, March 1983.

YULE, G. U., "On a Method of Investigating Periodicities in Distributed Series, with Special Reference to Wolfer's Sunspot Numbers," *Philosophical Transactions*, Vol. A226, p. 267, 1927.

ZEIDLER, J. R., SATORIUS, E. H., CHABRIES, D. M., and WEXLER, H. T., "Adaptive Enhancement of Multiple Sinusoids in Uncorrelated Noise," *IEEE Trans.*, Vol. ASSP-26, No. 3, pp. 240–254, June 1978.

INDEX